FROST & FIRE

NATURAL ENGINES, TOOL-MARKS & CHIPS

WITH

SKETCHES TAKEN AT HOME AND ABROAD

BY A TRAVELLER

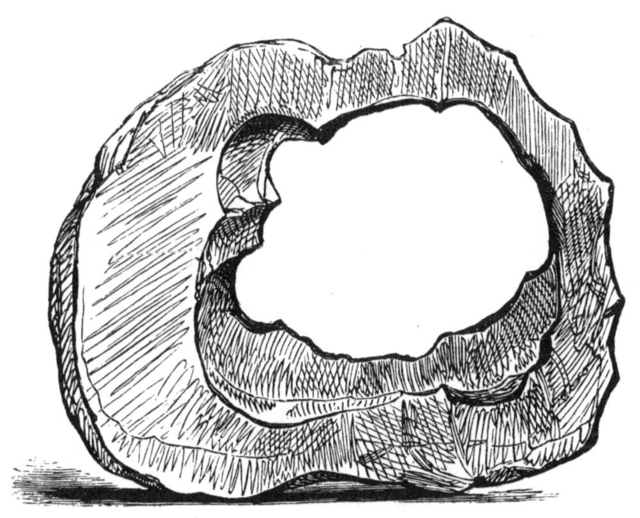

THE AINSA METEORITE.

VOL. II.

PHILADELPHIA: J. B. LIPPINCOTT & CO.

MDCCCLXV.

CONTENTS OF VOL. II.

CHAPTER XLII.

CHAPTER XLIII.

CHAPTER XLIV.

CHAPTER XLV.

CHAPTER XLVI.

CHAPTER XLVII.

CHAPTER XLVIII.

CHAPTER XLIX.

CHAPTER L.

CHAPTER LI.

CHAPTER LII.

CHAPTER LIII.

CHAPTER LIV.

CHAPTER LV.

CHAPTER LVI.

CHAPTER LVII.

CHAPTER LVIII.

ILLUSTRATIONS TO VOL. II.

Chiefly selected to illustrate forms which result from the action of certain forces, and from movements caused by them. Marks of Denudation, Deposition, and Upheaval: Gravitation, Radiation, and Rotation: Forces: and Will.

Waves and Beaches. Denudation and Deposition.

Upheaval.

CHAPTER XXVIII.

BALTIC CURRENT—BRITISH ISLES.

When facts have been gathered, sorted, and piled, the mound is an observatory. When a train of machinery has been explored, from the dial-plate even to the axis of one small wheel, the dial may be read though the entire engine may still be incomprehensible. When an engine has been seen to work, the tool-marks may be used as records of work done. When a creature has been seen to make tracks the old spoor may be followed. In the preceding pages an arctic current has been followed; a pile of facts gathered; part of an engine explored; tool-marks studied; a spoor learned; a theory has been built on a pile of ice; it will fall to the ground if ill founded. The way to test it is to work up stream, from delta to source, from circumference to centre, from the spoor to the deer, from old ice-marks to melted ice, from tool-marks back to the wheels which carved out hills and hollows. Old marks in the British Isles will serve to test the theory of an old Baltic Current; and the following pages give the result of an attempt to read and translate the record.

It has been shown that a current probably flowed from the polar basin through the Gulf of Bothnia, over Southern Scandinavia and Denmark, and parts of England, if ever central Europe was under water; and if so its tracks should remain in the British Isles.

If men wish to know from what quarter the wind is

blowing they look up to the nearest chimney for a stream of smoke; to a steeple for a weathercock; to mist on a hill; or to clouds moving freely in air. They do not watch eddies near the ground which whirl round corners and posts in streets, or past rocks and glens in hilly countries; and which pack sand and whirling autumn leaves in curved ridges and furrows in every sheltered nook.

The weather-wise look up to some high point in the general air-current, where the wind is not altered by impediments. If we wish to know the direction in which the wind commonly blows, we look for a tree growing in some exposed place, and note the bend in the trunk and branches (vol. i. pp. 31, 59). It is vain to look at sheltered trees, or at trees in glens where the wind eddies and whirls in all directions, while the main stream blows steadily on above. If we want to find out the course of an old arctic current which brought glacial drift to grind British rocks, we must in like manner look up. It is vain to search sheltered glens for marks of a general system of glacial denudation, and for tracks of polar ice moved by ocean-currents. If such marks exist they can only be found at exposed places; on wide plains; on hill-tops; on high ridges, where trees and plants are bent by the wind.

To find out whence British glacial drift came, British hill-tops near the coast, and far inland, must be searched for marks, and the marks followed from hill to hill. Marks of old local glaciers, and old local glacial systems, must be sought in hollows, for glaciers like rivers flow in hollows down-hill. But marks of ocean-currents and ice-floats must be sought along some ancient sea-level, for ocean-currents move on the curves of the globe.

Hunting is healthy pastime, and hunting for ice-marks upon hill-tops may be combined with other sport. The spoor leads to the haunts of grouse, deer, and ptarmigan ; to grand scenery and to regions of fresh air.

In the following pages an attempt is made to show the result of a search for high ice-marks along some of the curves on the maps at pages 232 and 496, vol. i.

The spoor.—Before starting on any pursuit, be it the spoor of an animal or an arctic current, the marks must be learned. A Highland deer-stalker, an Indian tracker, a Bushman, or any practised hunter, will follow a deer where a stranger sees no track ; and so it is with ice-marks, they must be studied before they can be followed. An attempt has been made to show how some ice-marks are now made ; the old marks relied on are shortly these—

1. *Polishing.*—Upon certain hard rocks which will take a fine surface, and over which ice is passing, or has lately passed ; beneath glaciers, or near them, or near moving sea-ice ; the stone surface shines when wet, feels perfectly smooth, and is neither "joint" nor "cleavage plane," nor "bedding." It is worn, ground, and polished by the continual passage of hard heavy ice, clay, and fine sand. As no other natural engine now produces like work, and ice always does, a polished surface "*in situ*" proves the passage of ice, even over a hill-top.

2. *Striæ.*—According to the direction in which ice moves, so is the direction of the mark made. The polished surface is usually varied by grooves. On the surface of the rock parallel straight lines of various dimensions are often ruled, and these lines point out the direction in which the polishing engine moves. It may not be easy to recognise these marks at first, and there seems always a lurking wish to show that they were made by something familiar. It is told that a number of geologists once met at a quarry, to hold solemn

conclave over certain marks on the stone. Much breath and some brain-work were expended, and no solution of the mystery found. At the end of the meeting a workman, who was going home, appeared above, and slid down the rock with hob-nailed boots. The denuding engine was seen to make tracks, and there was an end of this question. When glaciers have been seen at work, their tracks are as easily known as the print of a shoe. Striæ are only skin deep; they do not, in any way, correspond to the structure of the rock, or if they do at one place, they do not elsewhere. They sometimes cross each other at small angles; but so far as each line extends, it follows a straight course, up one end of a rising ground, over it, and down the other, or along the sides of a mound or hollow. These grooves are part of the polished surface, and follow the track of ice. Where they are found they mark out the path like a spoor, and they are of many kinds.

3. "*Sand-lines.*"—These are fine as a hair, and are like the marks of the finest sandpaper; they extend a few inches only, and are very easily overlooked.

4. "*Scores.*"—These are deeper, and are sometimes made by hard gravel, or by points in larger blocks, fixed in moving ice. Stones have been found under glaciers, fixed in ice, and placed in the end of a new groove. Scores are like a firm line, cut with a small gouge, or a grooving plane with a round iron. They often contain sand-lines, and a pencil will rest in them. They fade gradually away, but many are two or three feet long. They are often attributed to ploughs and harrows.

5. *Grooves.*—These are deeper, a walking-stick will rest in them, and some are eight or ten feet long; some are dinted, as if a stone had started and rolled while making the groove. Cart-wheels get the credit of these sometimes; they often contain scores and sand-lines.

6. *Deep grooves.*—These are long rounded hollows ⌣ which would fit a man's body. When freshly made or well preserved, they are fluted, and often contain grooves, scores, and sand-lines. They generally occur where great pressure has been exerted; on the weather-side of a point; in the bed of a river-glacier; on the weather-side of an island, which has become a hill; at a sharp turn in a glen at the dot **S.** when moving ice has been forced to curve, and has run full tilt against the bank, as in Justedal (vol. i. p. 197) and Romsdal. Ice can be squeezed into a mould; so ice under pressure is forced into hollows; and stones, sand, and clay, frozen in and fixed in ice, deepen the groove, and flute the hollow sides.

7. *Hollows* ⌣.—These are but larger grooves, and often contain all the others, though the smaller marks may be buried in bogs, or drowned in lakes.

8. *Glens* ⌣.—These are marked on good maps, and many of them seem to be large ice-grooves worn in rock by glaciers, local systems, and ocean-currents, as shown above. Many glens may have been hollows produced by contortions and disturbances of the earth's crust at first; but many are hollows worn by some engine, and these generally retain all the marks above described, though they may also contain beds of drift, alluvial plains and rivers, lakes and arms of the sea. If glens are ruts in which ice moved, for the reasons above given, their direction in a wide tract of country must be considered in spooring.

Hollows in Southern Scandinavia (chap. xviii.) and in Iceland (chap. xxv.) have been attributed above to the passage of arctic currents, like the stream which has been followed from Spitzbergen to Newfoundland. All these are but grooves of various sizes ⌣, which large engines might cut.

9. *Roches Moutonnées.*—When any ground surface covers

a large area, it is pretty sure to take in rocks of various hardness, which wear unequally. If a bit of wood is rubbed with fine sandpaper and a soft pad the grain rises. If a bit of slate is rubbed, the beds wear unequally. An ice-ground rock-surface wears unequally, and the rock takes the "mammillated" form which suggested the Swiss name of "muttoned rocks." They look like bosses, domes, waves, rounded tables, saddle-backs, hog-backs. In Devonshire, rocks of this shape go by the name of "tors." The word is good ancient British for "mound;" so it is used as shorter than the usual glacial slang terms, "roches moutonneés," and "mammillated surfaces." An example on the large scale is drawn on the margin of the map; the Λ shape of hills in Gairloch, 4000 feet high, is there contrasted with the curved shape ⌒, which only reaches to about 2000 feet. Examples symbolized by a convex curve are given in woodcuts in the preceding pages. This mark may be used to determine the point on the horizon from which the grinding force moved. As a rule, the longest slope is up-stream or up-hill, and the steepest end down-hill or

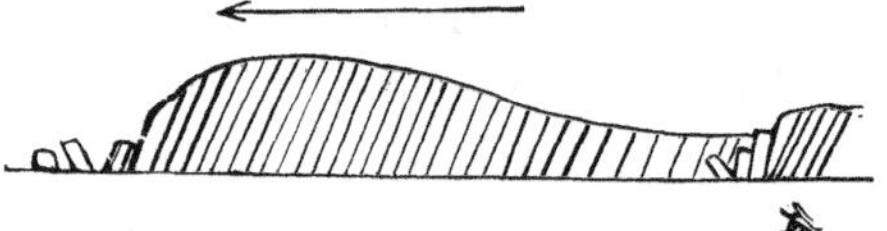

FIG. 65. A SMALL EXAMPLE OF "ROCHE MOUTONNEE," WALES.

down-stream. The woodcut was made as an illustration of this fact. It shows the form of a small slate "tor" in Wales. The arrow shows the direction in which ice slid down-hill, the lines show cleavage, the direction in which the rock breaks; the case was selected because the ice-plane had worked against the grain of the stone, and had made fine work nevertheless.

10. *Broken tors.*—If the smooth surface ends abruptly, the

broken end generally faces the shelter. Joints and bedding generally weaken the stone vertically, and a force acting horizontally tends to push, drag, or tear away the end of a worn ridge, where the resistance is least. After a time the upper edge of the fracture is worn and rounded off by a force which works both vertically and horizontally, as heavy sliding ice does. Another shove breaks off another slice ; and so a rock is worn and broken, and the fragments pushed and rolled downhill or down-stream.

11. *Jointed tors.*—The weather-end of a ridge is some-

FIG. 66. JOINTED TORS, CONNEMARA.

times displaced as if the rock had been broken and shaken loose by a thrust or heavy blow.

The woodcut is from a sketch made near Inver in Connemara.

The rest of the marks in the neighbourhood seem to prove that ice generally moved from A towards B, and so wore the granite into long ridges, all pointing one way. In this case the ends next A have been carried off; several ridges are jointed and shaken loose ready to be moved, but the sheltered end of the ridge next B is still solid.

If such a fracture came to be worn, the steep end would be on the weather-side at first.

So far these marks are all fixtures; they are *in situ*:—in the place where the form was hewn out of the solid rock. They are tool-marks of glacial denudation, and show the direction in which the graving-tools worked. Even large hills and whole countries seem to be hewn into these two forms—

◡ ◠.

Besides these fixed marks others are used.

12. *Quarried blocks.*—Large stones are sometimes partly hewn and ground, and partly broken out of the solid rock, and pushed a few inches or yards from their beds, so that each block might again be fitted into its place.

The direction in which the stone has been moved is that in which some force pushed or dragged it, and many of these blocks are so large that no common stream of water could well move them.

13. *Wandering blocks.*—These are similar stones of all sorts and sizes, more or less worn or fractured, of the pattern above described, but moved further from the quarry. As an example, granite blocks have been moved some hundred yards from the granite hills of Arran, and are left upon slate hills 1200 feet high. They are so placed that they could not possibly roll to the spots where they are poised; but they have been moved so far, that the hole from which the stone was taken can no longer be identified. Kane gives examples of similar transport and deposition by arctic ice in Greenland, and numerous examples of transport by ice are mentioned above. The highest wandering boulders yet found at home, by the writer, are above Loch Ericht, as shown on the margin of the map (vol. i. p. 496), and on the shoulder of Ben Wyvis. The last is a large mass of mica-schist dropped nearly 3000

feet above the sea, and wholly cut off from any hill of the same material. Antrim flints have been somehow carried to the south of Ireland; zircon syenite, which is found in Norway, has been carried to Galloway; and rocks supposed to be of Scandinavian origin have been carried to Poland and London. If the kind of stone thus transported can be identified with the parent rock, the direction of movement is thereby shown. But the mark taken alone is uncertain.

Granite may have come from the polar basin, or from lands which have disappeared. The test is good for land-glaciers which must flow one way, but bad for ice-floats.

If a similar test were used to discover the prevailing direction of the wind, it would fail, even though the wind may have a prevailing direction. Winds in the British Isles drive thistle-down, and thistles grow where the seed lights. Some thistles are cultivated, so the direction in which a new variety spreads from field or garden marks the spoor of the wind. If there were a constant wind, thistles would spread from the garden down-stream, but thistle-down, which moves every way, like a British weathercock, would never mark out the prevailing south-west wind which bends British trees. Marks in the solid rock are fixed, and, like the trees, show the prevailing current; wandering blocks, like flying seeds, may show eddies and occasional currents, and stray ones may drift wherever a gale can blow an ice-float.

14. *Perched blocks* are wandering blocks, placed upon hill-tops or hill-shoulders, or balanced one upon the other, or on "tors" and ridges, on points where they must have been gently placed by something strong enough to lift them, and carry and lay them down. Ice floating over a hill might drop a stone on the top, or land-ice, grounding at high-water, might place a stone, and break away when the tide

ebbed. The woodcut was drawn on the block, and represents a stone perched near Inver in Connemara. There are many

Fig. 67. Perched Block, Connemara.

other examples in the neighbourhood, but this one is remarkable, for it looks like a work of art.

Fig. 68. Dropped Block, Connemara.

15. *Dropped blocks.*—These seem to have fallen so far as to break where they fell. The cut was drawn on the wood, and represents a large mass of granite near the police station at Inver. It is mentioned again below.

15. *Trains.*—These are rows of large stones, some perched, some dropped and broken, which probably fell from drifting ice. If so, the lines point out the course of the moving rafts, and the run of the stream which moved them, but this test is uncertain. If a bit of a glacier, with a medial moraine, were launched, and then stranded and melted, the row of big stones might cross the stream. A slice of ice-foot might swing any way, and drop its wandering beach so as to leave a ridge with any bearings (vol. i. p. 404).

16. *Drift.*—This word applies to confused heaps of stones, of many kinds, shapes, and sizes ; some larger than haycocks, others as big as casks, kegs, turnips, apples, nuts, and peas, generally imbedded in sand or clay.

17. *Old moraines* are land-ice chips, piled in conical mounds at the mouths of glens, and composed of stones which are found *in situ* in higher grounds.

18. *A terminal moraine* marks the end of an old glacier (vol. i. p. 181.)

19. *A medial moraine* is similar stuff in the middle of a rock-groove, generally near the rivulet.

20. *A lateral moraine* is similar stuff on one or both sides of a glen. Stones on the right come from hills on the right, stones on the left from the left.

21. *A moraine* formed in water must differ in shape from all these, and samples of all kinds abound in the Alps, Scandinavia, Iceland, and the British Isles. True moraines indicate land-glaciers, and are sure marks, which can easily be compared with moraines on existing glaciers. Sea-moraines, formed under water, cannot be compared with existing sea-glaciers, but their shape may be inferred from models, and from the movements of land-ice in Spitzbergen, Greenland, etc. (chaps. xxiii. to xxvi.)

These are all specimens of "drift," but the term is generally used to express piles of loose rubbish, widely spread over a whole country or continent, in glens and on plains and hill-sides. The formation has lately been divided into stratified and unstratified, and in America it has been subdivided largely. The lowest beds are "unstratified," contain scratched boulders, and rest upon grooved rocks. The upper series are stratified, that is to say, packed in layers. The deposition of these geological formations has still to be explained. According to one theory, the unstratified drift is the debris of land-ice, and the stratified glacial drift was dropped by floating ice, and packed by streams of water in a deep sea. It has been argued above that the drift is the moraine-work of large floating glaciers like the Arctic Current, with its icebergs and sea-ice.

22. *Boulders* which belong to these formations are known by their forms. Those which belong to the lower boulder clay, which rests upon grooved rocks, are often washed out by the sea, or by rivers, or picked out by men. They are found on beaches, in walls, in houses, in fields newly reclaimed. One side is generally flatter than the rest ; and, when freshly moved, the polish on the surface is nearly as fine as the material is capable of taking. Striæ of all sizes run every way, but most commonly along the longest axis of the flattest surface. It seems as if the drift were the polishing powder with which the rocks were ground, left in the tool-marks of the polishing engine. The drift seems to consist of stones of all sizes, partially rubbed and ground to clay, frozen into a con glomerate and pushed onwards, till climate changed and the ice melted. The worn stones bear marks of each other and of the rock ; the rock bears marks of the drift, and these mark the direction in which the drift was last moved. If most of the stones in any patch of drift belong to any known forma-

tion, the line of movement is shown by the nature and position of the stones moved. For example, the majority of the stones in a hill of drift near the sea, at Galway, are bits of scratched mountain limestone, and that kind of stone is found *in situ* to the north-east. The direction in which this hill of drift moved was from N.E. to S.W., because striæ and loose stones point to the same conclusion. But the hill also contains specimens of many other rocks ; so it may have belonged to ice which had sailed far, like that which is drifting along the coast of Labrador, loaded both with foreign and native drift.

23. *Weathering.*—As all kinds of rock wear when exposed to the atmosphere, ice-marks on rocks and boulders wear out when the dressed surface is bare.

First, the fine polished skin gets rough and pitted, as rain and air and lichens decompose parts of the stone. Then "striæ" wear out in the order of their depth. Then deep grooves become shallow, from the weathering of their sides and edges. Then larger grooves, and hollows, and tors, and ridges between them, assume new shapes. Beds and joints weather and widen, till an old tor looks like a pile of stones. Then valleys and hills change their form. Rivers dig smooth pits and jagged angular ruts in hill-sides, and these split, and crumble, and fall, and join, leaving weathered glens, peaks, and needles at last. This spoiling process may be watched, and the work may be seen in all stages, in the mountains of Northern Europe. But still the last bit of an ice-ground surface may sometimes be found left at the very top of a hill, whose sides have crumbled and fallen away to make heaps of talus, cliffs, and cairns of stone.

The ridge ⌒ or the peak Λ is least worn by falling water, so it lasts longest.

24. *Shape.*—Because of weathering, old ice-marks are not to

be found without search. But so long as any part of the outline of an ice-ground hill retains its shape, a practised eye can detect ice-work; and a careful search at likely spots will generally unearth some one or all of the marks above described. Two or three will suffice to determine the direction in which ice moved, and a few well-chosen spots will serve to map out a large district.

25. *Rocks.*—Different rocks weather in different ways and at different rates.

It is hopeless to search for any but large marks upon coarse materials like sandstone. Limestones, unless protected from rain-water by clay, lose the marks readily. Granites protected from the air retain even sand-lines, and the finest polish; when exposed they become rough, and some kinds crumble. On some granite-hills in Arran even deep grooves are obliterated, though slate-hills close to them retain a fine polish and the whole series of ice-marks.

Where quartz rock has not split up, it retains the finest marks; but quartz rock is very liable to break and fall away. So marks on quartz are rare.

Trap, whin, and greenstone, etc., last well, retain striæ, and lose the polish, but some kinds of trap weather easily and crumble to dust.

Hard blue clay-slate appears to resist the weather best of all. Ice-marks still exist on bare slate-rocks in Wales, Scotland, and Ireland, which could hardly be distinguished from marks on rocks beneath existing glaciers.

It follows that the best material for inscribed monuments is the slate which still retains fine sand-lines, made when British hills were 2000 feet deeper in the sea, or up to their shoulders in land-ice.

26. *Searching.*—In searching a country for old ice-marks,

it is best to look out for a hill of slate, quartz, or trap, which has a rounded outline ⌒.

Try the hill-top first for old marks, then beat the sides about burns, new-made turf-dykes, quarries, and other such places where the rock has been laid bare. If no marks of a general movement can be found at the upper levels, try the glens for the spoor of glaciers, and such small game.

There are few parts of Northern Europe where an old scratch may not be found by careful searching.

27. *Copying.*—Rock-surfaces and ice-grooves cannot be carried away, and specimens are bulky, heavy, and hard to carry when quarried. Drawings take a long time to make, photographic apparatus are grievous impediments, but rock-surfaces may be quickly and accurately copied thus :—

Lay a sheet of foolscap on the rock with the longest edges in the meridian, as nearly as a compass or the sun will show. Hold the paper fast and rub it with a pencil, a bullet, a coin, a burnt stick, a bit of black coal, or a bit of heel-ball. The pattern below will be copied :—raised points dark ; hollows light. The experiment may be tried on the cover of this book, which is copied from a rubbing made from a striated rock beside the "Queen's Drive," on Arthur's Seat, at Edinburgh. The copy and the original may be compared, so as to test the method ; and then other copies, and descriptions of marks, will have more value if the paper, the book, and the rock, are found to correspond when compared.

When the copy is made mark the north, and from the centre of a circle draw arrows pointing at any hill or hollow which might influence the movements of glaciers ; or currents of water moving from the horizon to the spot, at the level. Small outline sketches may be drawn at the ends of the arrows if there is time.

Note the name of the place ; the names of conspicuous

points on the horizon ; their bearings are given by the arrows. Note the height of the spot by aneroid barometer ; the distance by pedometer from the last place of observation in a day's walk ; the kind of rock ; the dip and strike by clinometer and compass ; the slope of trees, and anything else worthy of note ; and do all this as much as possible without moving the paper from the rock.

The finished sheet is a portable, accurate, pictorial record of a set of observations at one spot, which may be transferred to a map, or otherwise combined at leisure. Ranged in order with dates, each record becomes a page in a journal. The woodcut below is a reduced copy of a sheet which was thus prepared, on a rock-surface, on the hill-shoulder which is represented on the margin of the map at the end of vol. i.

The dark marks within the circle are ridges ⌒ between striæ ⌣ on a very smooth surface of fine-grained hard quartz rock. The direction in which the engine moved is shown by the arrow. The loch is Loch Maree in Scotland, and the sea horizon is open to the W. of N., and to the E. of S. ; to Greenland, and to Scandinavia. To the west are tall hills of the Λ pattern, and higher ice-ground rocks of the ⌒ pattern ; to the east is a deep ice-ground glen ⌣ running parallel to the striæ, and beyond it are high hills of the ⌒ pattern, and higher hills of the Λ shape, and numerous ice-marks, none of which point at the peaks.

The dip of the rock is towards D, the white marks in the rubbing are chinks and fractures.

At this spot on the backbone of Scotland, at 1800 feet above the present sea-level, ice moved past peaks of the Λ pattern over hills of the ⌒ pattern, from the direction of the Baltic towards the Atlantic, horizontally. The spoor is so fresh that sand-lines need a fine lens to make them out, while other grooves would hold the mast of a ship ; and the hill-side is thus worn, for a height of nearly 2000 feet, throughout an area of many square miles.

If this plan of copying had been devised twenty years sooner, observations made would have had more value. With such a plain spoor as this ice-tracking is easy work.

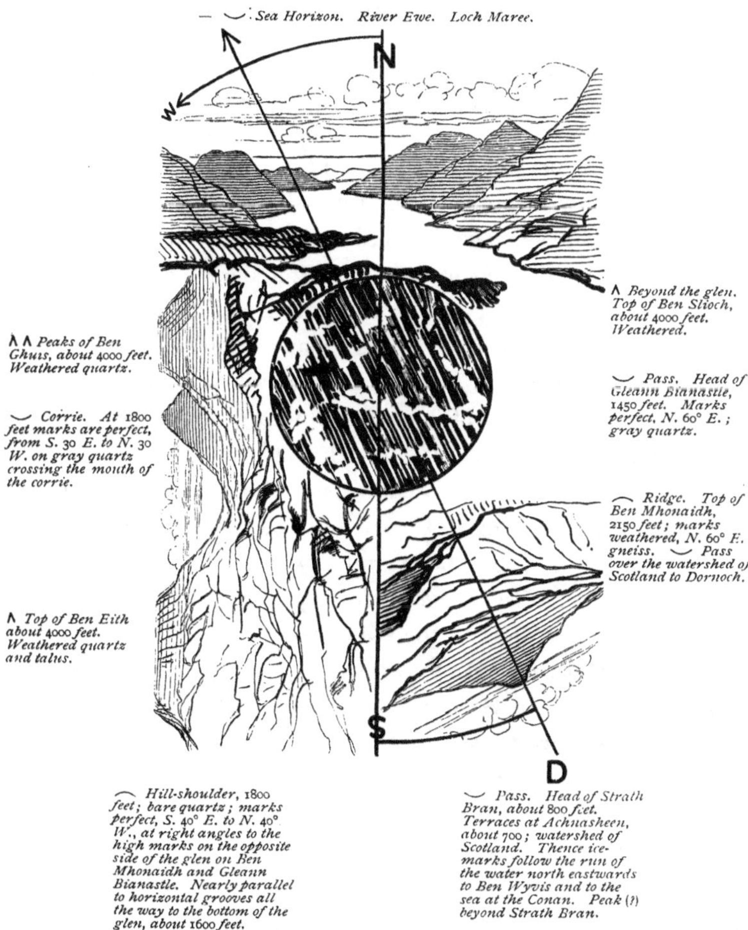

FIG. 69. FOREST OF GAIRLOCH. Ice-marks on a hill-shoulder ⌒ of gray quartz, at about 1350 feet above the sea; level with the opposite edge of the glen ⌣.

CHAPTER XXIX.

BALTIC CURRENT 2—BRITISH ISLES 2—IRELAND 1—CONNEMARA GALWAY AND WESTPORT CURVES.

In the map of the northern hemisphere (end of vol. i.), a series of curves are drawn from the Pole towards the Equator.

The space between two of these corresponds roughly to the existing Arctic Current between Spitzbergen and Newfoundland; and to low grounds in North America which are strewed with glacial drift, and where many large hollows and small ice-marks on shore point south-westwards. The space between another couple of curves includes Novaya Zemlya, part of Russia, Scandinavia, Denmark, and the British Isles. It corresponds to the supposed course of an arctic Baltic Current, which, according to theory, only ceased to flow south-west in this tract when the Scandinavian isthmus rose and turned the stream. In the map (vol. i. p. 232), similar curves are drawn, and one ends in the sea at Galway.

In a systematic attempt to test the soundness of this theory founded on marks in Scandinavia, a search should begin as far to the south-west as possible. A stick laid in an ice-groove on a hill-top points out the way, and it should be honestly followed. If it leads to the marks already mentioned, and the whole series point one way, the Baltic Current theory may be launched like a big boulder to find its own resting-place amongst other rough blocks.

The west coast of Ireland is at the tail of the fossil stream ; so the west of Ireland is the place to search for marks of ice-floats like those which now cumber the Straits of Belleisle.

London can be got at from any part of the world, and the western coast of Ireland is very easily reached from London, between morning and midnight.

FIG. 70. CLOCH CORRIL AND THE TWELVE PINS OF CONNEMARA.
Drawn from nature on the wood, 1863. (Reversed).

Forms characteristic of the action of ice are well seen by the way. Running into Chester by railway, the N.E. corner of Wales appears in profile, and on leaving the station the hills are conspicuous. They rise gradually from a plain strewed with glacial drift and water-worn boulders, and from the sea. They are green and cultivated ; their bones are hid beneath a skin of clay and soil, and covered by a rich mantle of green and yellow ; but rounded rocks appear, as the skeleton does in a living creature. Where a quarry or railway cutting has torn a rent, or cut a gash, the sandstone frame appears broken and angular ; but the hills are all rounded and smooth.

This is denudation, but not the work of water. There is not one ravine **V** between Chester and Rhyll, nor is there a cliff **L**, though the line runs over a raised beach between the sea and an old margin all the way.

At Conway the hills are steeper and higher, but the glens still are rounded, and in them fresh ice-marks abound, as will be shown below.

Near the Menai Bridge glens have the peculiar forms of glaciation. Many quarries and cuttings, faults and fractures in the slate, show that the rounded outlines of these hills and glens are not due to fracture and disturbance, but to some wearing action; and boulders and beds of clay all tell of ice.

The N.E. end of the Snowdon range is seen in profile from Anglesea. It has a sloping outline ⌒ like the north-eastern corner of Wales ; but the rocks are harder, the slope is steeper, and some hill-tops are broken and weathered.

Anglesea is all ice-ground. Near Holyhead, amongst some drifting sand-hills, glaciated rocks rear their heads amongst the bent. They are smooth and round like the sand-dunes, and their longest slope, like that of the hills, is still towards the N.E. The waves which roll in from the S.W., driven by the wind, have their longest slope towards the S.W. If Wales were a new country, the shape of it would suggest the glaciation which is proved by a closer search.

From Dublin to Galway the country is boggy, low, and flat. A depression of 500 feet would sink it beneath the Atlantic.

The first glance at the country about *Galway* shows the action of ice. Large boulders piled and scattered broadcast everywhere, low rounded hills, beds of clay stuck full of rounded stones, walls built of boulders—all suggest glacial denudation on the large scale; but no high mountains are to be seen to account for land-glaciers. Close to the town, on the beach, but above high-water mark, numerous ground rocks show smaller ice-marks distinctly. The scores and grooves point from N.E. to S.W., or thereby. At *Blackrock*, the favourite bathing-place, these guides point out into Galway Bay, where the track is lost in the Atlantic.

About three miles to the west of the town the sea has undermined a long round-backed hill. It is broken short off at the end, leaving a perpendicular cliff about 50 feet high, with a beach of boulders under it. The hill is called *Cnoc-a-Bhlaka* or Blake's Hill, and the point Cnoc-na-Carrig or the Hill of the Stones.

The sea-cliff is a section of the boulder-clay, and ice-work of the most striking character. A matrix of hard, compact, bluish-yellow gray clay is stuck full of rounded "subangular" blocks; some are three or four feet long, others as big as a man's head, others small, like apples, nuts, and peas; and the beach is made of them. They stand out from the clay where the rain has washed it down, like plums in an iced pudding. Every stone is scratched, grooved, and scored; and the marks are as plain as if they had just been made with rasps, files, and sandpaper. Many surfaces are polished so brightly that they shine in the sunlight. New-fallen stones, stones *in situ*, and stones picked out of this cliff, all are polished, ground, scored, and scratched in many directions, and on all sides. There are specimens of red and yellow, coarse and fine granite, fossiliferous dark blue limestone, and other rocks. The hill is a museum of transported stones, gathered long ago by wandering ice, and pushed into Galway Bay.

Near the place, specimens of the same stones, weathered

and water-worn, may be compared with these boulders. In the dykes, where mountain limestone has been weathered, fossils stand out in relief, showing the minutest detail. In the cliff where the ground surface has been preserved from weather by hard clay, fossils can only be distinguished by their colour. On the beach away from the cliff, rolled pebbles are rounder and dinted; the scratches have disappeared. Where these sea-rolled stones have been weathered, they retain the finished oval shape which sea-waves gave them, after ice had blocked them out. The waterworn and the weathered surfaces are wholly different from the old ice-mark. Here then, at the most western coast-line of northern Europe, are the works of ice; and here too the prevailing S.W. direction of the wind is pointed out by growing trees.

If the direction of the wind is pointed out by a weathercock, and its prevailing direction by a bent tree on a hill, it is equally well shown on a plain by sand-drifts or grass tufts. If the direction in which a large ice-system moved is well shown by grooves upon hill-tops, it is equally well shown by grooves on a wide plain, where there are no high hills to interfere with the general movement. So at Galway the striæ tell of a general system of glacial action, not of local glaciers. On the tops of low hills, by road-sides, in fields, and generally in the neighbourhood, whatever the kind of rock laid bare may be, grooves have a general N.E. and S.W. direction.

One end of a long stick laid in a groove points N.N.E. or N.E., and the other end aims a little to the outside of Black Head, past the end of the Clare mountains.

This direction agrees neither with the slope of the country nor with the flow of rivers, nor with the present run of the tides; it only agrees with a system of large hollows which cross Ireland, and are marked as valleys and sea-lochs on the best Irish map.

The movement was not a result of sliding, for there are no hills to the N.E. of Galway from which ice could slide. This is no part of a local glacier system, but there are clear traces of the general movement, which also left its marks on Scotland, Yorkshire, Wales, and Devonshire, as will be shown below.

A good map of Ireland shows the large grooves which correspond to the curves on the map. The northern and southern end of the country is crossed by diagonal valleys, whose general direction agrees with that of the Menai Strait, the Caledonian Canal, the Forth and Clyde Canal, and other Scotch and English hollows. The ice-stream certainly floated over the low grounds of Ireland, and part of it poured out between the mountains of Clare and Connemara, through Galway Bay.

Curves drawn from Galway in the direction pointed out by ice-grooves upon hill-tops near the town, cross Ireland by way of Carrick-on-Shannon, the end of Lough Conn, and north of Belfast Lough. They pass between the Mull of Ceantire and Portpatrick, into the Firth of Clyde. In Ireland they pass over a low flat country, in the neighbourhood of lakes, canals, and lines of railway. In Scotland they join a system of large wide glens, which traverse that country. Let this be called the Galway curve, and traced back as far as it will lead.

Travelling northwards, other curves should be crossed if this were a general movement. From *Galway to Oughterard*, the road skirts the north-eastern side of a low range of hills in Moyculleen, and coasts Lough Corrib. The hills on this side are all rounded and strewed with large wrecked boulders, but on the other side they are steeper, and the rock is bare.

The low country beyond the lake, the shores of the lake, and the lake itself, all are strewed with enormous stones and patches of clay. Low down, boulders and gravel are everywhere, but the hill-sides are generally rock with a thin covering of soil or peat, or bare.

Where limestone is the foundation of the country, the general outline of glacial denudation alone remains. The rock is furrowed and drilled into the most fantastic shapes, apparently by water and weather.

When granite is the rock, the general form is nearly the same, and the surface is still weathered. Crystals stand up separately, veins stand out and run over the backs of rounded tors and ridges. The veins are sharp and angular, but the rocks are all round like Devonshire tors, and the hills to the very top retain shapes into which ice ground them ⌒.

Beyond *Oughterard* a road leads over a low col down into a wild tract of country where the rocks are bare or smothered in bogs.

The surface is generally weathered, so that striæ and grooves are hard to find, but when the morning sun is shining across the grooves, the marks come out clearly, as blue lines of shadow on long ridges of warm gray granite, which raise their backs in the dark moor.

Low down, at the sea-level, and on hills about 400 feet high, the direction is from N.N.E. or N.E. to the opposite points.

At *Furness Lake,* which lies close beneath the Moyculleen Hills, grooves, ridges of granite, and trains of large stones, point the same way.

The cut was sketched from nature. It shows part of the Moyculleen Hills, on which ice-marks are plain, and part of the low country, which is strewed with drift and trains of blocks. The district is one of the best samples of an ice-ground country that is to be found in Western Europe.

These grooves do not aim at the hills ; they run along the hill-foot, and aim at a large groove ⌣. A pass about 500 feet high.

At *Sgriob Bridge* the direction is still the same ; at *Inver Lodge*, at *Luggeen Lough*, at *Lough Corrib*, the low grooves

Moycullecn, a 1100 *feet, marks N.E. by N.*

FIG. 71. TRAIN OF BLOCKS NEAR FURNESS LAKE AND MOYCULLEEN HILLS.
Drawn from nature on the wood, 1863. (Reversed.)

all point nearly one way. They do not aim at mountains which surround the low bogs of Connemara and the sea-lochs, but point at glens which lead to the low country beyond the hills, and to great lakes. One of these mountains stands alone. It goes by the name of *Cnoc Ourid*, and is about 1300 feet high. It is about two miles from *Shan Folagh*, which is N.N.E. of it, 2000 feet high, and the end of the Mam Turk range. A valley more than 1000 feet deep separates Cnoc Ourid from the higher range, and Shan Folagh is joined to

Mam Turk by a col. To the S.S.W. is a third isolated hill called *Cnoc Mordan*, and about 1100 feet high. It is separated from Cnoc Ourid by a boggy plain more than four miles wide, and but little above the sea-level. To the S.E. is a range of low hills in *Moyculleen*, which makes one side of a block of high land, and is separated from Shan Folagh by the glen of Oughterard.

These four high points are well situated for ascertaining the direction of the general movement, which has so ground and altered the whole face of this country.

Cnoc Ourid.—In mounting Cnoc Ourid from the north side from Rusheen Lake, the rock is seen to be upheaved and strangely contorted. It contains fragments of other rocks, broken and rounded, and is folded about the fragments in waving lines. Ice polished the rock across the edge of the beds, and the surface has been weathered so as to leave the structure of the rock in low relief. Upon ridges and domes of this gray moss-grown gneiss large boulders are perched.

At the foot of the hill deep grooves are well preserved, and they point at Mam Turk and Shan Folagh, past the shoulder of Cnoc Ourid. Here then are the works of cold and heat—contorted gneiss, upheaved and altered by fluid granite, ground down by ice, and weathered afterwards. Five hundred feet up the hill the rocks are all of the same pattern as those in the plain below, and on them rest large angular blocks of gneiss, and smaller boulders of various hard rocks—quartz, greenstone, etc. These last must have travelled far. Eight hundred feet up is a large block of gray trap freshly broken, and near it is a block unbroken, and perched upon a rounded saddle of gneiss. Eleven hundred and sixty feet up, on the top of the northern shoulder, striæ and grooves are well preserved on gneiss. They point N.N.E. at the end of the higher range beyond the valley, and S.S.W. out of Camus Bay at the Atlantic. These marks are unlike those which are made by river-glaciers; they are like writing made by a shaking hand, for they waver and vary slightly in direction, so as to cross each other at a small angle.

Thirteen hundred feet up, by aneroid barometer, on the top, the view is wild and desolate. Lakes appear to lie in every possible direction, in a wilderness of water, stone, and bog, which fades away into a shallow sea, full of low islands, stones, and rocks, scattered broadcast in bays and sea-lochs. Galway Bay is seen over Moyculleen; Lough Corrib and Lough Mask, and a wide stretch of low land, are seen past the shoulder of Shan Folagh. There is no hill far or near to account for glaciation by land-ice at this spot and in this direction, and yet ice-marks are there, and well preserved. A stick laid in a groove points S.W. by S. at the shoulder of Cnoc Mordan, out of Camus Bay, at the sea-horizon, and N.E. by N. through a notch in the hills, at a sea of lakes and bogs bounded by a land-horizon as flat as the sea. The notch is the col which joins Shan Folagh to the Mam Turk range, and the nearest hill-top of equal height is beneath the horizon, if not beyond the sea. Descending the hill on its eastern side, a block is perched at 1200 feet; and near it, where the woodcut was sketched, a solitary goat had perched himself upon a saddleback of gneiss. His family and friends were scattered about picking up a scanty supper amongst the bare rocks. They kept peering at the stranger, bleating, stretching their long necks, wagging their gray beards, and flourishing their horns over the sky line. The click-click of a sparring-match between two old bucks was the only sound besides the sough of the evening wind, and the red light of sunset made the old gray rocks and their gray inhabitants glow like fire.

It was a different scene when the block was dropped by ice 1200 feet above the present sea-level, and when ice floated over the top of Cnoc Ourid. This hill is joined by a low col about 500 feet high to a range of low granite and gneiss hills, on the S.E. At the top of this col the grooves point N.E. by N. over a wide flat moor, which leads to Lough Corrib and Lough Mask. There is no high hill in that direction for many miles. A line drawn on the map passes north of Belfast. Patches of hard yellow clay are deposited in sheltered hollows on this col, and these contain small boulders of black

Fig. 72. Perched Block on Rounded Tor, Cnoc Ourid, 1200 feet.

limestone, mica schist, very hard trap, quartz rock, gray porphyry, and other rocks which are foreign to this hill, but which may be found in the direction of the grooves. The limestone in particular is like rocks near Oughterard on the low shores of Lough Corrib, and the trap is like Antrim trap. The north-eastern slope of the hill and of the col is less steep than the south-western.

Cnoc Mordan, the second hill, is even more isolated. It makes the north-western horn of Camus Bay, and no hill of the same height is near it.

At the sea-level the striæ are well seen; they point N.E. by N., S.W. by S. Large granite boulders are scattered about in the moor. One shaped like a chipped pebble, near Invermore Lake, measures 18 × 12 × 9 feet, and many are still larger. Ascending the north-eastern slope, the angle is less steep than the south-western side of Cnoc Ourid. There are rounded surfaces and perched blocks to the very top. At 600 feet the grooves are N.W. by N.; at 700 a groove points N. and S.

FIG. 73. PERCHED BLOCK, CNOC MORDAN, 1100 feet. (Reversed.)

At 1100 feet above the sea a great angular mass of granite is stranded upon a shelf, like a boat ready for launching. It goes by the name of Cloch mor Binnen na gawr—the big stone of the goat's peak. A lot of bare-footed Celts, two pretty girls, two men and a small boy, were clustered about when the sketch was made; while a party of fishermen out for a walk took shelter from a S.W. breeze, and smoked under the lee of a rock. Behind the stone, Cnoc Ourid and Shan Folagh rose up to the N.E. beyond the lakes of Inver and the endless bogs of Connemara.

The top of the hill is flat, boggy, and strewed with small boulders, and every rock-surface is ground. Grooves are well marked everywhere, though weathered, and their general direction is N.N.E., S.S.W. The hill is very like a small Dartmoor. Granite tops, which rise out of the moss, are miniature tors, with joints beginning to open and weather. The work is the same though it is further advanced in Devonshire.

A great change has come over Great Britain since these rocks were thus ground at a height of 1300 feet, and yet the marks are so fresh that the change must have happened recently. Granite weathers and crumbles, but these mountain-tops upon which tempests beat, and where rain falls in torrents; mountain-sides, where torrents gather and pour down after every shower; river-beds, lake-basins, and sea-margins—all retain the marks of ice moving diagonally on meridians in a general south-western direction over this corner of Ireland.

Shan Folagh (the Hill of Flesh) is the third hill in this row. It is 2000 feet high by the Ordnance map, and by aneroid barometer. The top is about ten miles from Inver Lodge by pedometer. It is the eastern end of Mam Turk (the Range of the Boar), and the top is isolated.

At 800 feet on the south-western side the rock is stratified gneiss, dipping at a high angle, and the whole outline of the hill is rounded; but the surface on this side is much split and weathered. The hill is very steep. At the head of the glen, near the col, the angle is 45°. Few boulders are to be seen, and few grooves; but those which do remain at this height point N.N.E. over the shoulder of the hill at the col which joins it to the range, and S.S.W. out to sea past Cnoc Ourid and Cnoc Mordan.

They are parallel to the deep glen below them, and to

several chains of lakes which are seen in the plain, and they correspond to marks on the hill-tops at which they point.

From this height it is easy to understand how brittle plates of ice of great thickness, like those which drift about off Labrador, might float and slide over low hills of granite in the hollow between Mam Turk and Moyculleen; for the wide valley—six or seven miles across—seems almost a plain. In particular, it is easy to see how ice-floes might split and ground upon the tops of Cnoc Ourid and Cnoc Mordan; score them, break them, stick to them, pick up fragments, and drop them in the lee.

Supposing these hill-tops to be awash in a frozen sea moving south-westward, the stream and the ice which it carried would curl round the hill-tops, as a stream curls round a big stone, and it would spread out when it had passed the Straits of Oughterard.

At 1450 feet the tops of Cnoc Ourid and Cnoc Mordan sink below the sea-horizon of Shan Folagh, and at that level a groove upon a rounded table of gneiss points S.S.W. over the top of Cnoc Ourid down Camus Bay at the sea-horizon.

At 2000 feet, on the very top of Shan Folagh, the rock is gray quartz traversed by white veins. The beds are nearly vertical; the surface rounded and polished wherever it has not broken and split from weathering.

On the north-eastern side of the top, the rocks are polished and scored in the most remarkable manner, and from their hardness the surface is exceedingly well preserved. Great flat tables, sloping towards the N.N.E. at an angle of 54° or thereby, are ground perfectly smooth, and rounded off at the upper edge. Grooves run upwards in various directions, from N., N.N.E., and N.E. by N., and they are peculiar. Some marks are rounded dints, as if the polished rock had been struck and ground at one spot by something which was afterwards pushed over the hill-top. Bits of this polished surface are easily picked out, for joints in the stone make it a sort of smooth mosaic work.

Looking towards places at which these grooves point, there is no higher land to account for this manifest glaciation. The grooves point 2000 feet over Lough Mask, or 800 feet over Slieve Patry, or level at hills twenty miles off, over glens, and through deep glens, and over the end of Killary Harbour, which shines like a glass amongst the dark hills.

These certainly are grooves made by floating ice, which grounded upon this hill-top, 2000 feet above the present sea-level, when the whole land was under water.

The whole aspect of the hills seen from this high station is that of something ground at about this level. Moyculleen seems to be a rolling plateau of rounded tops, like those which exist in the valley. Slieve Patry is a block of high land deeply furrowed by glens, but the top is a smooth even rounded slope. Beyond it lie Castlebar, Lough Conn, Ballina, and Sligo. In one direction only, to the northward, higher mountains seem peaked; but the northern line, when drawn on a map from the top of Shan Folagh, passes through a deep glen forty miles off, beyond Clew Bay. Standing upon glaciated rocks 2000 feet above the sea, and looking at a horizon 54 miles away, it seems almost certain that these ice-ground Irish hills rose in the midst of an arctic current which flowed amongst them and altered their forms. So here the first impression suggested by the shape of the country is amply confirmed by closer examination of details.

Glaciers.—A marine glacial period ending in a rise of land should have produced land-glaciers, and local systems of marks; and these marks do in fact remain.

The col and corrie between Shan Folagh and Mam Turk certainly contained a small glacier, for the marks are there. The top of the col is bare ice-ground rock, and the glen has the rounded shape of a glacier valley. There is hardly any talus, though the rocks split easily. Looking downwards from the steep slope at the head, the glen seems to fade away into the boggy plain. There are few large stones in it, and these seem to have rolled down from broken rocks above them. Cnoc Ourid seems nearly to fill the mouth of the glen, and Cnoc Mordan is seen to the right, over the shoulder of Mam Turk. Between them are Camus Bay and the sea-horizon nearly level with the distant hill-tops.

The col was a sea-strait when Cnoc Ourid was awash, and the glen ought to be full of wrecked drift dropped in the shelter. It seems to have been swept clean. The hill-sides are ground from top to bottom, for the glen is a trench dug transversely through nearly vertical strata.

But when the mouth of the glen is reached, the small river is found to have cut through a bed of boulders and clay nearly fifty feet thick. A green hillock is found to be part of a moraine, and most of the stones contained in the clay seem to be derived from hills which make the sides of the glen. Lower down, ice-ground rocks peer up through the brown moss, and the river washes a grooved rock-surface, which it has failed to spoil. But this moraine has been washed out of shape.

Shan Folagh was a sunken rock; then awash; then a low island at the end of a point; then a peninsula with small glaciers at the isthmus; then a hill in a plain: and then the glacier seems to have come to a sudden end, for the moraine stops short in the jaws of the glen. The glacial period probably ended when the land had risen to a certain point.

At the moraine-level, about 200 feet above the sea, the

low hills between Mam Turk and Moyculleen, and those upon the borders of Lough Corrib, and near Galway, Ballina, and Sligo, would be like rocks which now fill the sea-loughs; and ice might still drift and carry boulders through straits which are now county Galway, and the glen in which the road has been made to Inver Lodge.

At the present level of sea and land, the Arctic Current is shut out by Ireland, Great Britain, Denmark, Scandinavia, and Lapland, and the Gulf Stream flows up in the lee. If the sea were 2000 feet higher on this region of the earth's northern surface generally, the Arctic Current would overflow the dam which separates the Gulf of Bothnia from the White Sea. Then the Equatorial Current might be driven elsewhere, and then the climate would be changed.

When Celts named the "hill of flesh," and the "range of boars," the "lake of stags," and similar places, they found other creatures in Connemara than snipes and hares. When they composed the long poems which Connemara peasants still repeat, the pastime of their lives and the burden of their songs were love, war, and hunting; but before there were elephants, elks, and men, to be hunted and smothered in Irish bogs; the wide Atlantic covered the whole land; and marks an eighth of an inch deep, made by floating ice on the highest top of Shan Folagh, have not been worn out by all the rain which has fallen there since the day of Finn MacCool, MacArt, MacTreunmor, and since Shan Folagh peered above the waves.

Leaca Donna.—Shan Folagh, Cnoc Ourid, and Cnoc Mordan, being on one side of a strait, the other side is a gneiss hill, called Leaca Donna, or brown slabs. It makes the western corner of the block of high land in Moyculleen, the highest point of which is about 1200 feet above the sea. The western face of this block is rounded, and almost bare of soil and vegetation. From the road at Sgriob Lake to the top is about three and a half miles.

At the head of Sgriob, Shan Folagh is seen to the north-east as a rounded, conical, isolated hill. Slieve Patry is seen past the eastern shoulder as a block of hills with a smooth sloping top ; and to the westward, in the Moyculleen range, a wide rounded valley runs half a mile eastwards into the hills.

About the lake in the low grounds loose blocks of granite are scattered in every direction, and the rocks are all ground and scored. The grooves at high-water mark at this spot run north and south.

At the same level, a mile and a half eastwards, grooves are well seen ; they point N.E., S.W., and *cross* the mouth of the small glen, which seems made to be the habitation of a glacier. If these grooves were made by land-ice they would point due west *out* of the glen.

Half a mile nearer to the hills the ground is strewed with the débris of a small moraine, which makes a curved sweep across the mouth of the glen. It marks the spot where a small glacier ended, at about the same level as the Shan Folagh glacier. This moraine is washed out of shape.

In this sheltered nook a village built of boulders, fields fenced with rounded stones, green corn, blighted potatoes, and worm-eaten cabbages, show a better soil than bare granite and wet peat, which make the plain.

The base of the hill on the right of this glen, up to 350 feet, is thickly strewed with large loose blocks. Above that level—which would join Lough Corrib to the sea, make Moyculleen an island, and Ireland an archipelago—the ice-ground hill is swept bare ; but every here and there perched blocks riding on granite saddles hang on the steep hill-side, where a good push would send them rolling to the bottom.

The rock generally is rough and weathered, but every here and there a vein of hard quartz stands up half an inch from the gneiss. The quartz surface is smooth, polished, shining, and marked by sand-lines and scores. The edges of the ribs are still angular. Elsewhere hard patches preserve their smooth surface for a couple of square yards. At 700 feet the grooves and finer sand-marks point N.N.E. and S.S.W. along the face of the hill, past Slieve Patry, over Lough Mask, at the Firth of Clyde in one direction, and out to sea in the other.

At 1000 feet a well-marked groove on the top of a shoulder points N.E. by N., S.W. by S., near Arran in Scotland, and at the Irish Arran Islands.

At 1130 feet by barometer the hill-top is a boggy rolling plateau, with low rocky saddlebacks peering up through black moss. Sea and bog; hills, islands, lakes and mountains; Galway Bay, Lough Corrib, and the low grounds of central Ireland—are spread out like a map, and there is not a hill in sight to account for this glaciation by land-ice.

In the foreground of this wild landscape a wild group of figures completed the picture. In a dark wet hollow, where a stream oozed out of a bog, a thin blue smoke curled up into the sunlight. Two bare-footed, black-haired girls, dressed in patched red garments, shaded their eyes from the sun, and peered doubtfully at the intruder. Three men and a boy, picturesque and wild, unkempt, bare-footed, ragged, and polite, paddled about in the black peat. Barrels, casks, noggins, baskets, creels, peats, malt, a copper still, sweet worts, the worm in its tub, a pile of potatoes for supper, and the black holes from which the whole gear had been dug, showed a poteen distillery in full work. The Oughterard gauger—bad luck to him—found it out.

From the ice-period to the period of poteen in Connemara is a long time, but the weathering of gneiss during that time has been less than half an inch ; for it can be measured from the polished surface of a rib of quartz to the rough surface above which it rises. Space could be turned into time if the rate of weathering were known. Surely works of human art, obelisks, pyramids, or sculptured stones, might give the rate of weathering, and so fix the date of the glacial period in Ireland.

Thus, on four isolated hill-tops within sight of each other, but far apart, at a height of 2000 feet and at the sea-level, the Galway curve is repeated in well-marked ice-grooves upon fixed rocks in Connemara.

The boulders which ice carried are very remarkable in this district. They seem to spread like a fan from the pass. Close to the road-side, near the police barracks at Inver, lies a great block of granite (p. 10). It measures 36 × 12 × 10 feet, and it rests upon rounded granite, where it fell.

It is broken into seven pieces, which retain their positions. The upper side is ground like other neighbouring surfaces ; one end, the rest of the sides, and the fractures, are angular and unground. It is evident that this great stone was a bit of the granite surface of the country ; that it was lifted bodily, carried some distance, and dropped where it lies broken. Perhaps it broke when it fell ; perhaps it split afterwards.

It lies in the jaws of a glen, which was a strait at the foot of a rounded granite hill, Shan na Clerich (the Clerk's Hill), which is about 400 feet high. The hill is scored and ground all over. Perched blocks are scattered over it ; but all about it, and chiefly on S.W., or lee-side, enormous blocks of granite are thickly strewn. A great many of these are broken, and most of them are rounded on one side or another. Some few are rounded on all sides, and chipped at the lower edge, as if they broke them when they fell down. Sometimes they are ranged in rows, which point N.E. by N. over the shoulder of the hill towards the low pass, through which the road leads from Oughterard.

Nearly all these blocks rest upon bare rock, but here and there the rock is covered by compact hard beds of gravel and reddish clay. The gravel is chiefly granite, but the clay encloses small boulders of greenstone, and quartz rock of various kinds and colours. These are foreigners, for there are no rocks of the kind within ten miles at least. Where the clay has been moved to make roads, the granite-surface beneath is perfectly preserved in many places. Crystals of quartz and felspar no longer stand out in relief to give a firm hold to hob-nailed boots, but crystals and strings of harder rock are all smoothed to a fine polished surface ; upon this grooves which a pencil fills and finer marks remain. Hob-nails make almost as clear a mark when they slide upon the rock. The polish on the pillars of the Colosseum is not better preserved, and the marble of the Parthenon is far more weathered than this ice-ground Connemara granite where protected by the clay, which helped to smooth it. All these grooves, great and small, high and low, point nearly N.E. by N.

There can be no doubt that ice scraped along, carrying boulders and grinding rocks, and the rocks show whence some of these boulders came ; others may have come from Antrim.

Amongst the large blocks, and trains of blocks, ridges of granite of the same kind rise up in the moor. They have strange weird shapes, and suggest gray monsters crawling eastwards out of the moss. They are the sides ⌒ of rock-grooves ‿, in which peat-moss gathers and grows, and the dragons and giant caterpillars and maggots are tors and ridges, ready to be jointed, quarried, and carried away to make granite boulders, for the stone is already split.

Some, as in the woodcut (p. 7), are actually moved, and left loose in the place where they were first ground into shape, and then quarried and pushed out by ice. These are chiefly to be found at the north-eastern end of ridges, where they were struck and shaken.

At other places the angular nest, from which a stone has been pushed, lifted, or dragged, remains, but the stone has disappeared. At some places the granite has been worn so near to a joint that it can be split off in thin layers. Elsewhere it is solid, and the fracture is never round like the worn surface.

All over the moors and bogs, chiefly on the lee-side of isolated hills, these blocks are scattered and ranged in rows. Many are of enormous size. One, near Inver Lake, measures 14 × 11 × 12 feet, and must weigh about 130 tons.

Cloch Corril (p. 19) is still larger; it stands on the bank of Lough Corril, and it probably came from Shan Folagh, ten miles off. The circumference is 66 feet, and the height about 24. The upper side is rounded, the under hollowed and smoothed. The sides are angular, and coincide with the natural fracture of the stone, for it is splitting up and falling in large masses, which lie about it, and the rain drips through it into the hollow beneath. It stands upon a rounded table of granite, on which straw is laid; it is smoked, for fires are burned beneath it; and it is rumoured that malt dries there. The lake is a rock-basin full of big stones, and the striæ upon its islands point the usual way, towards Cnoc Mordan and Mam Turk. It is a beautiful spot to look at, and "a fine place for brewing poteen," as a native remarked. It has a bad name, so it is seldom visited. It is haunted by "each uisge," the water-horse, and other dangerous beings—so few people go there except to fish or brew spirits; heather, blaeberries, ivy, yew, holly, birch, and oak scrub, flourish upon the islands; white goats caper about amongst the stones, and nibble the bark of the trees; it is a green spot in the midst of a wilderness of brown boggy moor, surrounded by the distant blue hills of the "Joyces' country," and the Twelve Pins of Connemara. The chief feature in the landscape is the old gray boulder, which is very like one upon the Unteraar glacier (vol. i. p. 153). That stone has given shelter to many a tourist—to Saussure, Forbes, and to masters and students of glacial action. The Swiss stone rests on ice which is grinding rocks; the Irish stone upon rocks which are ice-ground. Ice is carrying one, and ice certainly carried the other.

Such a stone must have a legend, and thus the biggest boulder in Connemara has one of its own. It was the plaything of a Celtic hero, Corril, who crushed his finger and left the mark in the hollow stone, when he threw it from Mam Turk at Mordan, the father of Goll MacMorna, who stood on his own hill about ten miles off.

There can be no doubt that this tract was ground for a depth of 2000 feet by ice moving from N.E. or N.N.E. to the opposite points. All marks, from general forms of hill and dale, down to minute sand-lines, tell one story. If this be glacier-work, the snowshed was beyond Scotland. If it be the work of a current with floats, similar work is going on in corresponding latitudes within ten days' sail.

Surely it was sea-ice which carried Cloch Corril (p. 19), and set it gently down on its base. Surely it was a fusible raft which planted a block upon end like a pillar on a big stone pedestal at the foot of Cnoc Ourid, on a rock in the midst of a bog. When the sketch was made on the wood, two gray horses stood beside the stone, lazily switching their tails to keep away a host of flies. When it was gently placed upright on its base, sea-horses, seals, and bears, may have played about the hill-sides, where goats now browse. There are "seal-meadows" further south on the opposite coast.

These sea-monsters, and the end of the Irish glacial period, may have been seen by the ancestors of the men who are now migrating westward after the glacial period. Celts owned the land at the earliest historical date, the ice-marks are as fresh as Roman and Egyptian sculpture, and all Celtic tribes in the British Isles, from Cornwall to Sutherland, people their lakes and seas with water-horses, water-bulls, dragons, and sea-monsters. Their popular tales speak of ice-mountains, of hills of glass, of islands with fire about them, rising from the sea; of wicked cities and plains sinking beneath the waves.

According to a Connemara man, Finn and his warriors once chased a deer till they lost their way, "and all but two were frozen and starved, so that they died of cold and starvation." The survivors did many marvellous feats. If these myths be of native growth, they must surely be tracks which a recent glacial period has left on human minds. The belief in mythical sea-monsters, large deer and birds, is fresh and vivid, plain and clearly marked, amongst all ancient Britons, as are the ice-marks upon these Irish hills and plains in Connemara.

CHAPTER XXX.

BALTIC CURRENT 3—BRITISH ISLES 3—IRELAND 2—CONNEMARA 2—NORTH-WESTERN, AND NORTH-EASTERN COASTS—GALWAY, WESTPORT, AND DERRY VEAGH CURVES.

THE broad trail of the Galway curve is well marked.

The fact of glaciation in a certain south-westerly direction for a height or depth of 2000 feet, and a breadth of thirty miles, being established at one point on the western coast of Ireland, the next step is to look to the configuration of the country. Books on geology—*The Antiquity of Man* by Lyell, Jukes' *Manual of Geology*, and other works of authority—show that the sea-level has varied greatly on Irish hills. Shells are found high up, and peat, which grows on shore, is found below low-water mark; and for numerous reasons it is taken to be an established fact that most of Ireland was under water after its hills had assumed their present general form.

If the contour line of 500 feet is traced, and assumed to be an ancient sea-level, Ireland becomes an archipelago. Fifteen groups of islands are disposed about a central strait, which ends at Galway and Oughterard. If the level of 2000 feet, the top of Shan Folagh, is taken to be the sea-level, very little of Ireland remains. (See map, *Antiquity of Man*, p. 276.)

The western coast at the present sea-level is indented by a series of bays running northwards and eastwards—Donegal Bay, Clew Bay, Galway, Shannon, Dingle, Kenmare, Bantry, etc. Most of the high mountains to the west are on promontories which separate these bays. If these western mountains were groups of islands stretching along the lines of movement already indicated, it is easy to understand how a north-eastern current ran amongst them, and to know where to look for conspicuous ice-marks upon Irish plains and hill-tops.

The north-eastern corner of each block of high land ought to bear the strongest marks of ice drifting south-westwards; and curves drawn through glens which were sounds and straits ought to bear reference to main lines drawn by greater streams in the widest openings.

The course of a rivulet passing through a row of stepping-stones; the run of larger streams which split and join in passing a salmon weir; the run of the ebb in a sea-loch studded with rocks and islands; the curves in the tail of the Gulf Stream where it passes northwards and eastwards amongst islands off Hammerfest and the north of Norway; the Mediterranean Current off Gibraltar; the Baltic Current off the south of Sweden, and the windings of the Arctic Current off Greenland and North America, all are illustrations of the movements of an old Arctic Current striking upon Irish hills. The theory is simple; but a theory, however formed, is worth little till it has been well tried. If it stands examination, it rises in value by every new test.

North-western coast.—A curve drawn below the 500 level from Galway to Newport joins Clew Bay to Galway Bay, and cuts off a large block of high land which would be a group of islands if the sea were less than 500 feet above its present level. The Twelve Pins of Connemara form part of the group.

Roads wind about amongst the mountains in this district and follow the lowest levels, towns are built near the coast; so ice-marks which occur near roads and towns must either be marks of glaciers sliding from the hills, or of streams flowing in shallow sounds. If a main stream flowed in from the N.E., about Belfast and Londonderry, it must have found its way out by glens, into bays, which open to the Atlantic at Galway, Westport, and Donegal. Ice-marks do follow curves which agree with this supposed movement of an arctic current amongst islands.

In travelling *from Oughterard to Clifton*, the road leads along the foot of Mam Turk and the Twelve Pins of Connemara. If ice-grooves were made by land-glaciers, they would cross the road ; if they were made by floating ice and an arctic current this was a place for an eddy in the stream, and the grooves should run along the foot of the hills.

At the foot of Mam Turk, in the lee, there are thick beds of glacial drift; the large boulders are buried in moss, and the rocks are hidden, but the hill-sides are ground to the very top. On nearing *Ballynahinch*, after passing a deep glen, the rocks appear, and grooves point back at Shan Folagh, the promontory round which a north-eastern stream 500 feet deep must have turned to reach this spot. The marks run nearly E. and W.

At Ballynahinch Lake, near *Canal Bridge*, the rock is slate, and much contorted. The ground surface is well preserved near the road, and the grooves point E. N. E. along the foot of the Twelve Pins at the shoulder of Mam Turk. In the other direction, they point out to sea over the lake, wherein fishermen disport themselves and salmon plunge.

At Clifton, a glen, a hill-side, and well-marked grooves, point E. and W. out of a deep gorge in the mountains at the sea.

Further on, in a wide boggy plain, a rounded boss of whinstone has grooves which point N. W. and S. E. at the end of the Twelve Pins. Thus, in passing along the foot of the hills on the lee-side, the grooves turn gradually, till at the point they cross the main current at right angles, as eddy-streams do behind a stone. (See vol. i. p. 127, and map, p. 496.)

From this place the road bends back, and passes up-stream into a deep gorge at *Letterfrack.* Here large mounds of boulders are piled below steep mountains, which are swept bare higher up. A few large boulders are strewed about the foot of the hills which border Kylemore, and woods of birch and other trees fringe the lakes, and explain the name of Greatwood. At the mouth of this pass the drift is arranged in terraces, and these look like sea-work.

The valley divides the Twelve Pins from Ben Coona, and after passing a low col the road descends about 300 feet to the Killaries.

Here a very small depression would join the sea to Lough Mask, and make the hills a group of long islands separated by narrow sounds.

Up to 700 feet these hill-sides are certainly ice-ground, and they seem to be ground to the top in the direction of the valleys. Low down, the rocks are strewed with boulders; high up, they are swept clean.

At *Leenan* the road comes to the end of a long sea-loch, and runs up-stream in a deep glen in the direction of Castlebar and the Ox Mountains, N.E. by N. At the head of the sea-loch is a mass of drift packed in level terraces.

From Leenan the road follows a deep gorge, with steep hills on both sides. On the right, cross-glens run far up. A few moraines cross the mouths of these glens. The rock is silurian, a series of beds of conglomerate; mica-slate and clay-slate much upheaved. Where the road passes out of the glen, at heights of about 600 and 700 feet, ice-grooves are exceedingly well preserved on blue slate. The bottom of the glen elsewhere is full of drift. Here, near the col, the rock is bare, or covered only by peat. Torrents have cut a few shallow angular trenches in the steep hill-sides, but here, at the top of the pass, is evidence of a current 700 feet deeper than the present sea flowing in from the low centre of Ireland. The grooves are clear as well-preserved sculpture on a slate tombstone a year old, and in ascending the hill they turn gradually round till they get clear of obstructions, and point the same way as the high Shan Folagh grooves already described.

At the bridge they point E.N.E. over the shoulders of a hill at the head of the pass.

At 300 feet, a little further on, N.E. at a notch. At the head of the glen, 700 feet, they point N.E. by N. over everything at the Ox Mountains twenty-five miles away and beyond a glen.

A glance at the map shows that in this district minor valleys all agree with these marks. From large and small grooves it seems that the stream, which ran out by Galway and Oughterard, split upon the hard block of land which is now the Twelve Pins of Connemara, and glanced off northwestwards through the Killaries and Kylemore.

Looking back over Slieve Patry, which makes the northeastern corner of this block, the outline is smooth and the slope small, though the outline is along the strike of strata which dip away from the ridge on both sides. It seems clear that little weathering or river-work has been done amongst these hills since they were last ground by floating ice.

On leaving this glen the road passes across the supposed stream, and over a plateau varied by ridges of low hills, strewed with large blocks.

Near *Westport* these become very numerous. The whole country is covered with big stones, and wherever the peat has been cut away the drift appears.

Many stones are scored and grooved, walls are museums of transported stones. Red sandstone, gray and blue and black limestone, white quartz, coarse conglomerates, whinstones, grits, and granite, are piled up in houses and fences; and no ice-groove in the neighbourhood points at the holy Croagh Patrick, which towers up 2510 feet on the left. It must have been a tall island when the rest of Ireland was nearly all drowned.

At Westport the head of Clew Bay is reached. A curve drawn N.E., or thereby, 500 feet above the sea-level, passes up a valley to Castlebar, through a gap in the hills at the end of Lough Conn, past Ballina, over a flat country to Sligo, and so through Donegal Bay to Lough Foyle. (See vol. i. p. 232.)

It cuts off two blocks of high land; one which ends in Achill Head, and a second to the north of Donegal Bay, which ends about Letterkenny and Rossan Point. Let this be called the "Westport curve," and followed wherever it will lead.

Westport curve.—If a stream ran in by Lough Foyle, out by Donegal Bay, branched off through the gap at Lough Conn, between the Ox Mountains and Croagmoyle, and struck upon Croagh Patrick, the northern shore of Clew Bay would be in the lee, and the rush would be at the narrows at the end of Lough Conn; at Westport; and at the end of Donegal Bay. The western mountains—Achill, and those near that island—would all be sheltered by hills to the east. The road to Achill is in the supposed lee, and the country supports theory.

The whole of the northern shore of Clew Bay is thickly covered by drift, and the hills are clothed to the top with heather, so that the rock is hidden. The bay is a wide arm of the sea studded with islands. These seem all to be of one pattern. They have rounded slopes towards the head of the bay, and many are broken short off to seaward. The drift upon the mainland is piled up in great heaps, mounds, and beds. Many of the stones are a very coarse conglomerate of white quartz pebbles, as large as pigeons' eggs. Where these have been long exposed the cement weathers out, leaving surfaces which resemble a modern sea-beach. But many surfaces have been ground, so that one front of a bed of pebbles is flat and smooth, while the sides are round. Amongst these are specimens of gray mica schist, red sandstone, and other rocks, imbedded in hard yellowish clay.

Achill Island, the Isle of the Cell, is separated from the mainland by a narrow shallow sound. The low grounds are covered by very deep peat-mosses, in which bog-pine and bog-oak abound. Beneath the peat are thick beds of boulders and clay. Several large hills occupy the rest of the space, and these end in steep slopes or perpendicular sea-cliffs. These hills have the usual long north-eastern slope and rounded forms, and piles of drift-like moraines fill up the ends of mountain hollows. Where rocks do appear they have the shape of ice-ground rocks, and some few have grooves, but bare rocks are hard to find in Achill. *Cruachan,* 2222 feet high according to the survey, and 2200 and odd by observation, is the highest point.

On the eastern shoulder, at 600 feet, a rock-surface, very much weathered, is exposed, and a deep groove, which can still be traced there, points east and west. A few blocks are perched upon rounded rocks at this spot, and higher up at 800 feet. These are clear ice-marks. At 1000 feet the ground is covered with large loose stones, laid flat and closely packed. They are of many kinds. At 1500 feet stones still cover the ground, but they are smaller, and some patches of yellow clay peep out. At the top the ground is still thickly covered with large loose rounded stones, and the rock-surface is hidden.

To the *eastwards* a small glen has been hollowed out of the slope of the hill, and swept bare. A small lake has formed behind a mound, which seems to be the moraine of a small glacier which once nestled here and swept a trench in the drift. To the *north* the hill has been broken. It has a steep scarped face more than 2000 feet high, along which men and sheep can barely scramble, and at many places the slopes end in sheer cliffs.

The end of Achill is a ridge which projects westward into the Atlantic. Sheep and shepherds scramble along the face of the cliffs by paths on which even natives hesitate to venture. Perched on the verge of this cliff, 830 feet above the Atlantic, when the wind is high, the whole rock seems to shake and quiver. It is a grand specimen of ocean-work, and a striking contrast to the ice-marks in Connemara. There everything is round; here all is angular, the hills are ground from above, but the cliffs are undermined and broken from below by the sea. Even where black rocks peer through broken white water off the extreme point; where the run of the tide is the strongest, and Atlantic waves are of the largest size; even there rock-forms are sharp and angular. Water-work and ice-work are very different.

On a fine morning after a westerly gale has blown itself out, great rolling masses of cloud gather and ground upon these high western points. They seem to anchor themselves upon the peaks and stretch slowly away to leeward, 1000 feet above the sea, dropping showers as they drift. Their tall white heads roll upwards and shine like snow in the sun, while the ribs and keels of these air-ships, dyed blue and purple, cast deep indigo shadows on the heather. As these clouds now drift steadily and ground upon the hill-tops, so ice once drifted and grounded; and here, on the lee-side of a group of hills, boulders which ice carried and dropped are strewed, 2000 feet above the sea, at the edge of cliffs which the sea is now breaking down.

Here, too, is evidence of the persistence of ocean-movements which result from the earth's rotation, and from heat and cold. Where ice-grooves of an arctic current point seawards towards America, the Equatorial Current now brings tropical seeds to land. The people constantly pick up "nuts," and they are the "horse-eyes" and "brown purses" which are the playthings of English children in Jamaica, "fairy eggs" in the Hebrides, and "Ljusne sten" in Iceland.

In Achill, according to theory, there ought to be drift in the lee, and there is so much of it that rock-surfaces are almost wholly concealed. At Westport and Lough Conn, at the north-eastern end of this high ground, the rock ought to be swept bare.

On leaving *Westport* the road passes up-stream over a low hill about 400 feet high. It separates the bay from the inland plain, and it stands in the way of a current flowing in from the N.E. It is swept bare of drift, and the rock is much ground. Trees point from W.N.W. and show the usual run of currents of air; rock-ridges point W.S.W. out into the bay, and E.N.E. up a wide valley at the low lands of central Ireland. From this hill the road descends into a rich, well-cultivated plain, which seems to be made of drift, for rocks and large boulders are hidden.

At *Castlebar* rock-surfaces begin to appear, and they seem to be ground from the N.E.

Thence to *Cullen Lake* the road passes over a tract of low country, where numerous boulders, large blocks, beds of boulder-clay in hollows, and glaciated rocks and ridges abound. The country is flat and boggy, but the block of high land of which Achill Island forms part is close to the plain. The plain is about 300 feet above the sea-level. The hills are about 2000. Ice-furrows run along the road-side, gradually sweeping round the foot of the hill till they point at the narrows between *Lough Conn* and Lough Cullen. Here, according to theory, rocks at a north-eastern corner, on a weather-side, and in a low pass, ought to be much ground, and swept clear of drift; and here in fact rocks are as bare as hill-tops in Scandinavia, or the straits at Oughterard.

It is a beautiful spot. The road winds along the shore, and passes between the two lochs, beneath gray rocks, amongst which berries, heather, fern, and graceful birch-trees find shelter and room to grow. Distant blue hills are mirrored in the calm water, and beaches of yellow sand and mica glow and glitter in the sun like gold and diamonds. High up, on large bosses, ridges, and tors, great rounded boulders and rocking-stones hang poised where legends tell that Finn and his giants cast them, and a pretty salmon river curls under a bridge and joins the lakes. It is a bit of Sweden planted in the midst of Ireland, and the same agent has done similar work in both countries. More conspicuous ice-work could scarcely be found, and yet there is no indication of land-ice. Large ridges, and grooves upon them, all point at low lands along the course which was chosen to make a level road through the pass which was a strait at the 500 feet level.

The lines come in from N.N.E. near the river, pass S.S.W. through the strait, and turn gradually westward as they pass round the foot of the hill, past Castlebar and over the plain to the bare hill behind Westport. There the tall cone and saddle-back of Croagh Patrick blocks the way, and turns the course of currents of air ; it seems to have thrown the water-stream westwards into Clew Bay, to join another branch which came in from Lough Conn to Newport ; and these two probably dropped their burdens of drift in the lee of the hills.

From *Ballina to Sligo* the road passes up-stream over a low flat country which is generally well cultivated. Large blocks of stone and smaller boulders are scattered about, and stand up like monuments in the green fields. Wherever the soil is broken glacial drift appears, and where rivulets have cleared their beds, the rock-surface below the drift is ground. For many miles the cone of Croagh Patrick may still be seen past the shoulder of a hill of the same Λ pattern, which rises west of Lough Conn, and divides the glens which lead to Newport and Westport.

So two groups of hills in Galway and Mayo appear to record that they were groups of islands in a frozen sea which moved south-westward.

To the right is a block of high land which reaches to Enniskillen ; to the left are the mountains of Donegal beyond the bay ; and in front is the deep groove which crosses Ireland, and holds Donegal Bay and Lough Foyle.

According to theory, a N.E. current entered between Innishowen and Ballycastle, and split upon hills about Enniskillen. The Westport branch ran down past Ballyshannon and Sligo, through Donegal Bay, and branched off into Clew Bay at Lough Conn; the other joined a stream which came in by Belfast, and ran out by way of Lough Mask, Lough Corrib, Oughterard, and Galway. Both came from Scotland. The Derry and Donegal stream came along the north side of Ceantire ; the Belfast and Galway stream came from the Firth of Clyde, and they were kept separate by the mountains of Antrim and by Ceantire.

In travelling from *Ballyshannon to Enniskillen* these

two streams are crossed. The south-western bank of Lough Erne is the block of high land which stretches to Lough Conn; the north-eastern bank is low and undulating. A depression of a few hundred feet would sink the plain, and make these hills islands. They are beds of grit and limestone nearly horizontal, and from Sligo to Enniskillen the hill-faces resemble broken sea-cliffs. At Enniskillen the eastern side has the same form, but the low grounds about the foot of the hills, and the hill-tops, are rounded. The lake itself seems to be a rock-basin filled with mud, boulders, and water. If an ice-laden current beat upon the edge of a stratum of limestone it would tend to make sea-cliffs.

From Enniskillen to Lough Foyle the stream is crossed again by a railway. The country is low and flat, thickly covered with deep soil and beds of clay and boulders, and no rocks are to be seen by a passing traveller. At *Ballyshannon,* where a salmon stream worthy of Norway is cutting a drain for Lough Erne through limestone, fossils are weathered out, and the rock-surface is pitted like that of weathered limestone elsewhere. In the plain the rocks are hid, striæ cannot be seen, but the general shape of the country remains, and it tells of ice. Hollows and low ridges have one general direction, and point from or towards the bays which here approach each other and make Donegal a peninsula.

From *Strabane to Letterkenny* the sea of rolling hills and glens is crossed at the isthmus. Every here and there a great round stone in a corn-field, a dam built of boulders, a gravel-pit, or a bed of clay in a burn, appears to give evidence in favour of ice-floats. So from the end of Lough Foyle to Achill Head and Galway the evidence agrees so far.

At the highest point on the road between Letterkenny and Strabane, 400 feet or thereabouts, the boulders include granites of various sorts, gray and white quartz rock, and traps of various colours. Many of these must have travelled far :—some perhaps from the Giant's Causeway. The lines point at Aberdeen, and the granites resemble Aberdeen granites ; according to theory they may have come thence, but there is granite close at hand in Donegal.

From *Letterkenny to Gweedor* a coast-road makes nearly half a turn round the north-eastern corner of the Donegal mountains, or the weather-side of a group of islands.

On leaving *Letterkenny* glaciated rocks appear at about 400 feet above the sea-level. Ridges run N.E. and S.W., but the rock is too much weathered for small marks. Further on, at the turn, the rocks are swept bare and much ground, but it is very difficult to determine the direction. Thence all the way to Gweedor the rocks near the sea are glaciated, but broken into low cliffs. A range of lofty hills—Muckish, big and little Ach, and Aracul—stand out from the Derry Veagh range ; and on the top of the most northern mountain, about 2000 feet high, a bed of fine white sand is worked for glass-making. It is hard to understand how it got there, or why it has not been washed away. The road bends south-westwards along the base of these mountains, which are separated from each other by deep glens.

If these hills were islands in a north-eastern current, and exposed to the Atlantic, the inn at *Gweedor* would be at the end of a sea-strait, and in the lee of the stream. The weather-side has been swept clean ; in the supposed lee a large deposit of glacial drift is piled at the end of the strait. The heap crosses the glen below the lake, and rises more than 500 feet on the hill-flanks. Small rivulets have made sections, which show these low hills to consist of sand, gravel, large and small boulders, all mixed confusedly and resting upon sandstone. The river which drains the lake cuts through the mound in a wide gap which looks as if a glacier had ploughed it out after the land rose. Many of the larger stones in these mounds are scored. The sweep of the Atlantic and the prevailing wind is from the S.W. If sea-waves driven by S.W. winds piled such heaps, these would be in the lee at the north-eastern end of the range, which in fact is swept clean, so the evidence tells for movement from the N.E.

Aracul is the highest mountain in this tract. After leaving the inn, glaciated rocks begin to appear close to the foot of the hill at about 400 feet. The ascent from this side is very steep. After passing over a series of cairns of angular quartz blocks which seem to have fallen from the hill, a steep slope of talus, angle 35°, leads up to the foot of a large whin dyke. This stands out from the loose stones like a great cyclopean wall. No better specimen of the works of fire is to be found in Iceland. It runs south through the hill. In that direction a quarry has been opened which yields excellent crystalline white marble. It is fine and white as that of Pentelicus.

At about 2200 feet these cliffs are passed, and a steep slope of stones, with patches of heather, grass, and moss like green velvet, leads to the top. From this point, on a showery day, with a S.W. wind, the march of clouds over the Atlantic is seen in perfection. When a shower is coming, a low ragged fringe blots out the horizon to windward, and advances steadily upon the mountain, seeming to eat up the coast-line, the low country, and the lakes. Then a puff of mist like a wreath of gray smoke sweeps up the hill-side, and then the whole cloud sweeps round the top and a sudden darkness wraps everything as in a thick veil. The lower world disappears; the rain patters down and splashes against the stones, and the wind sweeps past with a rushing noise like the sound of the sea. There is nothing for it but to crouch under a stone, and smoke the pipe of resignation. In ten minutes the cloud passes on its way ; light dawns as suddenly as it disappeared ; coast-line, plain, corn-land, hill and moor, seem to grow out of the gray sea of mist. The sun wades out into the blue sky, the tail of the cloud creeps over the highest peak of the hill, the sough of the wind dies away, and the shower and the cloud are gone.

If the cloud were ice, the wind an arctic current, and the rain boulders, it is easy to comprehend how rocks would be marked, and drift scattered.

On the sides of this particular hill there is no vestige of ice-work, for it is a broken ruin. Looking down from the peak, loose stones, which rains have freshly washed from the crumbling sides, radiate in yellow winding streams, like the floods which carried them to lower grounds. This hill is weathered. But lower down, rocks on cols have the familiar ice-shape, and nearly all the lower hills to the south are manifestly ice-ground. On the very top of the highest peak of Aracul one only patch of the original surface seems to be preserved. It is a hard gray quartz rock about three square yards in area, and smoothed across the joints. The surface appears to be scored N.E. by N., S.W. by S. ; the height is 2450 feet above the sea.

This mark is uncertain, but about 1000 feet lower down ice-marks are plain. On a col about 1500 feet above the sea-level, on a knob of hard gray quartz, grooves cross the col from S.E. to N.W., in the direction which a stream would take if it flowed through *Glenveagh* and branched off seawards upon the cone of Aracul. In the glen at which these grooves point are heaps of broken stones piled confusedly, as if swept there by streams or a glacier. On the col are several large rounded boulders of granite, which contrast strangely with the angular gray quartz of the broken mountains. One great granite pebble is nine feet long by six broad. At a height of about 900 feet, in the pass by the road-side, the rocks are hidden beneath a mass of boulders and clay, and the great bulk of the stones are foreign to the rocks upon which they rest. At the top of the pass of Glenveagh, about 1100 feet on the side of Benduich, are many well-preserved granite surfaces, upon which grooves point E.N.E. over the shoulder of a hill, at the mouth of the Caledonian Canal, in Scotland. Many perched blocks of large size are balanced upon these bare granite rocks. Burns and gravel-pits by the road-side show the whole of the low grounds in this pass to be paved with drift beneath a carpet of peat-moss, but the col is swept bare, and high up on the sky-line, to the south, great stones are poised in ranks, as if the inhabitants had ranged them there to hurl upon offending Saxons.

The quartz hills to the north have none of these conspicuous ice-marks; they are weathered quartz peaks, but granite has withstood the weather, and the hills to the south are manifestly ice-ground. On one side are talus, soil, and vegetation; on the other, bare rock and perched boulders. Lower down on the weather-side there is little drift and much glaciation; jointed tors and long ridges abound, and the hills are rounded to the very top. At Lough Veagh another great pass runs S.W. through the hills, and here a patch of drift or a moraine makes a dam and a beautiful lake. At the weather-end of the next ridge a series of grooves point N. and S., at an elevation of about 500 feet. Soon after this the north-eastern end of the Donegal peninsula is passed, and the direction of ice-grooves changes. They pointed across the stream at the end of the ridge, where the streams split ; when the end is passed they point along the side of the ridge, and into glens which converge about the head of Donegal Bay. The spoor seems to record movements like those which are roughly shown on the margin of the map (vol. i. p. 496).

Here, too, the rock changes—granite is left, flags are reached, and heather and bog give place to grass and corn-fields. But still the old rocks, with their old-world inscriptions, peer out all the way down to the sea at Lough Swilly.

At the holy rock of *Tobar-an-doon,* where sick pilgrims resort from all parts of Ireland, from Scotland, and even from America ; where a garden of planted crutches and walking-sticks bears flowers and a foliage of bows and rags, the votive offerings of those who believe that the holy well beside the rock cured, or will cure, their ailments ; the old rock upon which Irish kings were crowned in the olden time—is an ice-ground tor ; and here in the low grounds the direction is once more N.E. and S.W.

So the trail is clearly marked for a height equal to that of the highest hills in the north and west of Ireland, all the way from Galway to Gweedor, and the lines all aim diagonally across meridians, northwards and eastwards, except at places where a current would split or eddy behind an island, as the wind now eddies behind the Irish hills.

Three curves are thus started from Galway, Westport, and Derry Veagh.

North-eastern coast.—The western coast gives a broad clear trail, and it points to the N.E. coast of Ireland. It was crossed from Galway to Gweedor northwards; the next cast, like a steady pointer's range, should be southwards, the other way.

The north-eastern corner of Ireland is about *the Giant's Causeway.* From Derry a line of rail leads over a flat, up-stream to Coleraine, and the first high hill is at Ballycastle.

Looking N.E. from the Causeway, on a fine day, the landscape fades in the Sound of Jura. A north-eastern line passes near Loch Awe in Scotland, and clears the land of Ceantire; a S.W. line passes over low lands towards Enniskillen and Galway. The rocks of the district are basalt or chalk, and the boulder clay seems chiefly to contain blocks of basalt. But on the beach and elsewhere, specimens of various kinds of granite, of a dark limestone, of sandstone, and of gray quartz, are found.

Near the top of the cliff ice-striæ are well marked upon whinstone, near a wall. They point N.E. by E. along the north shore of Ceantire, and S.W. by W. along the shore of Lough Foyle. In a field near this spot is a large wandering block of trap, and near it are several boulders of sandstone, greenstone, and granite, some of which are grooved. This direction agrees with the run of the flood-tide, which splits off the Giant's Causeway. One branch pours up Lough Foyle in the old groove, the other passes outside of Innishowen, and so north in an eddy. A depression of 500 feet would let the flood pour through Donegal Bay. Parallel to the sea-cliffs, at some distance from the shore, is a line of submarine cliffs, well known to fishermen, who get fish in the deep water.

If heavy ice were now floating in the Irish Channel, and grounding upon the top of this lower shelf, some 200 feet below the sea, ice-floats would make parallel marks similar to those which now exist on the top of the upper shelf, about 300 feet above the sea. If the upper cliff were under water half Ireland would be submerged. If it were 2000 feet under water, and the sea over Shan Folagh, large bergs, like those which now pass Cape Farewell, might ground at the Giant's Causeway. If the depression was general in Europe, the sea-way would be open to the polar basin. (See map, vol. i. p. 232.)

There can be no doubt as to these marks; they are ice-grooves crossing each other at a small angle. They are precisely the same in kind as grooves which are found on the top of basaltic cliffs, within sight of glaciers, near the edge of the Arctic Current, at the foot of Snæfell at Stapi in Iceland (chap. xxv.) There the grooves point at glaciers, basalt, and lava, and at the top of a volcano; here they point at low lands and sounds, where the tide still moves in curves parallel to the old ice-grooves. And here the works of fire are as manifest as they are at Staffa and Stapi.

From *Ballycastle to Cushendal* the road passes over a spur of the Antrim hills, and reaches as high as 800 feet. The higher it goes the more drift there is, and at the highest point the rocks are ground but weathered. To the N.E. is the Mull of Ceantire, so this part of the coast was in the lee of the Scottish Land's End, between two streams or tides which passed through Lough Foyle to Donegal Bay, and through Belfast Lough to Galway Bay.

From *Cushendal to Glenarm* the road coasts along the sea-margin beneath cliffs of chalk capped with whin. The contrast of white and brown, with all possible shades of green and blue and purple, on land and sea, and in the distance, make these cliffs very beautiful. The beach is composed of boulders, chiefly whinstone, but pink granite is to be seen here and there.

When rocks whose colours are so conspicuous are thus placed, transported fragments are like thistle-down which a deer-stalker throws up to find out the direction of a breeze

A bit of "Irish limestone" used to form part of a child's museum, on the opposite coast; a flint is a rare stone beyond the Giant's Causeway. There are none on the opposite coasts of Scotland—flints were buried with their owners in Ross-shire and in Arran. Boulders on the opposite Scotch coasts are chiefly gray quartz, like hills to the north and east of the Hebrides. But if the south-western line is followed, Irish drift is full of chalk and trap. Professor Jukes says (*Manual of Geology*, p. 675)—"Chalk flints and pieces of hard Antrim chalk are found in the drift in the counties of Dublin and Wicklow, up to heights of one or two hundred feet, and along the whole eastern and southern coast of Ireland, at least as far as Ballycotton Bay, on the coast of Cork."

The tides run both ways, but this drift went S.W., which again supports a theory of a Baltic current.

Opposite to the Antrim hills at *Clandeboye*, in County Down, an isolated hill of slaty quartz rises upon the southern point of Belfast Lough. The hill is ice-ground, and the striæ at about 600 feet point N.E. by N. at Arran, and S.W. by S. at the shoulder of the Mourne Mountains, in the direction of Galway. From "Helen's Tower," on the top of this hill, a magnificent panorama includes the Isle of Man, and the opposite coasts from the Mull of Ceantire to Cumberland.

Belfast stands at the head of a long lough, in a hollow which stretches far inland. The hollow is bounded on the N.W. by a range of hills, extending south-westward from Larne. These are of trap or chalk, and where they are not broken away in cliffs they are rounded. At 600 feet a large wandering block of whin stands in a green field, where it must have been carried. At 1450 feet, on the top of one of these hills, another large block is planted. It has been split by gunpowder, but the rounded forms of the fragments contrast with the fracture, and betray the origin of the stone. From this point the ground slopes in all directions, and long heather slopes stretch inland towards Lough Neagh. A long search on these hill-tops failed to discover a rock-surface. Some snipes, a grouse, a collie-dog, and a keeper were found, and the latter, on being questioned, exclaimed, "What, in heaven's name, do you want with rocks?" Quarries in the hill-side show that the rounded forms of these hills are due to denudation, and the glen gives the same direction as the grooves at Helen's Tower. The form remains, but the exposed surface and all small marks have crumbled away.

Another hill of about the same height gave a similar result. On the side of Cave Hill a large quarry facing Belfast gives a fine section of the chalk, with its dykes and cover of trap. A thin bed of red and yellow baked flints divides the two. The dykes appear to have cooled, and set at the sides of the fissures through which the melted stone rose, and the chalk in the walls of the vein of trap is hard and brittle as if it had been heated.

Above the trap is a layer of loose brown earth, containing numerous rounded stones, chiefly trap. The chalk from this quarry is used for ballast, and ballast when done with is thrown overboard ; ships from Belfast sail far, so a lump of Antrim chalk on a beach must not be taken as evidence of natural movement in the sea. About 1000 feet up this hill is a large rounded stone, different from the rock beneath it. At the top, 1300 feet, are more loose stones, but the rock is hidden. The sea-face is a cliff. The chalk has been undermined, and the trap has split off and sunk down like the Undercliff in the Isle of Wight. Looking towards central Ireland from this hill-top, there is no high land to stop the movement which marked the hill at Clandeboye. The Mourne

Mountains are there, but they fade away inland. At 600 feet the whole land from Belfast to the Mourne hills would be a wide strait. It is now the line of various canals and railways, works which follow level ground and avoid mountains. Far as the eye can reach is a level horizon or an undulating plain.

When all the lines thus found ruled upon a few Irish hills are laid down on a map, and carried at the proper level from hill to hill ; over plain, glen, and sea ; they are found to have a common general direction. Galway lines point towards Antrim hills. Lines at Clandeboye point along the south side of Ceantire at Arran in Scotland. Lines near Westport point at Lough Conn, and there lines point at Lough Foyle. At the Giant's Causeway, at the mouth of Lough Foyle, lines point along the north shore of Ceantire towards Inverary and Oban. At Glen Veagh lines point towards Mull and the Caledonian Canal. The lines seem to agree with hollows laid down on good maps. Either the lines of movement were governed by the form of the land, or the form of the land was altered by the movement. But it is admitted that the form of the rock-surface is a result of denudation, and where ice is working in earnest now, as it is off Labrador, rocks seem to crumble like mole-hills before the mighty force. Looking to the geology of Ireland, harder rocks are in the hills, and softer generally in hollows. Looking to the ice-marks, it is clear that ice has worked in Ireland up to a height of 2000 feet. Taking the whole evidence, it seems that denudation, and transport of a great mass of debris, have resulted in northern Ireland from a general south-westerly movement in a current laden with heavy ice, which continued to flow till land rose and stopped the movement.

The people of Antrim and the N.E. of Ireland hail from

Scotland, as they say. The lines drawn by ice on Irish rocks aim back at Scotland; so the next cast must be taken beyond the sea, and this time northwards.

FIG. 74. ACHILL HEAD.

CHAPTER XXXI.

BALTIC CURRENT 4—BRITISH ISLES 4—SCOTLAND—GALWAY CURVE—ARRAN.

THE ice-lines on the east coast of Northern Ireland seemed to converge on Arran, Ceantire, and Loch Linne; so the Irish spoor must be followed past the Mull of Ceantire by the Galway and Westport curves.

Galway curve, Firth of Clyde, Cumbrae.—Steamboats follow the Galway curve up-stream from Belfast to Ardrossan. On that coast no observations are recorded, and none were made on this journey; but ice-marks abound in Ayrshire.

On the Cumbraes, an arrow on Mr. Geikie's map* points nearly south, out of the Firth. It is a low-level mark corresponding to the run of the ebb.

Arran.—On Arran no arrows are marked by Geikie. The hills are well seen from the Ayrshire coast, and to them the high grooves in Connemara and Antrim point.

The high ground forms a block which is still surrounded with water. The granite mountains differ in shape from the granite hills of Connemara; they are higher, and down to a certain level, about 2000 feet, Goatfell and his giant brethren are broken weathered peaks Λ. They are like jagged mountains which tower above ice in Spitzbergen and in the Alps. But in Arran, and elsewhere about the Clyde, hills below 2000 feet are rounded like ice-ground hills everywhere ⌒. Above *Lamlash,* a long glen and a steep

* *On the Phenomena of the Glacial Drift of Scotland,* by A. Geikie; 1863.

road lead over to the south end of Arran. At 800 feet, close to the road-side, ice-grooves are well marked on sandstone; they point N.E. by N. and N.E. at the shoulder of the Holy Isle, and S.W. over the col at Ireland. At this level the stream would not be influenced by the low Ayrshire coast, for 800 feet of water would sink most of the low lands. To the south of the road is a hill-top 1350 feet high. Here, on a rock which has the form of glaciation, a deep groove points N.E. by N. over the Cumbraes at Ben Lomond. In the other directions a stick nearly clears the Mull of Ceantire, and points at Antrim. At this level a stream would be free to move over Scotland and Ireland.

These marks were not made by land-glaciers, for they do not point at the high mountains beside them. They seem to belong to the hollow which crosses the south end of Arran diagonally, and to a stream which flowed through it.

In the deep glen which runs south-westward, enormous masses of drift are piled; but the drift is not arranged in conical heaps like a moraine. In the glen which runs N.E. there is less drift. Trees show the prevailing direction of the wind to be S.W., for the branches point up-stream in one glen, and down-stream in the other.

Arran, western coast.—A road coasts northwards along the back of the island. At a point called *Leaca Bhreaca* (Speckled Slabs) certain igneous rocks are much weathered, but ice-ground to a great height. At 200 feet or thereabout, grooves are distinct; they run horizontally along the hill which faces Ceantire; at this spot these contour-lines run N., S. Perched blocks and jointed tors are numerous up to the sky-line. In the lee of this point to the south are great beds of drift which contain stones of many kinds, but one pattern. After a long search no flints or Antrim chalk were found. North of this promontory, another deep glen leads to Brodick over a pass, and the coast-land is a wide flat moor. Over this a path leads to the King's Caves. Close to the sea is a fine mass of columnar basalt.

At *Machuri* the drift is arranged in terraces, which look like ancient sea-margins, but these are chiefly composed of glacial drift.

The actual sea-beach, where no ice now forms, is a good specimen of its class. It is a hollow curved slope of large stones, with ripples of coarse gravel about high-water mark, and a calm of sand below it ; but every here and there a great ice-boulder is planted in the midst of these stone-waves like a beacon amongst breakers. About *Dubhgarrie* walls are a curious study. They are made of big stones found about the sea-margin ; they were washed out of the drift-terraces by the sea, and they have been broken by men so as to show their internal structure. Some blocks are conglomerates, which contain rounded water-worn quartz boulders as big as turnips, bits of water-worn granite, gray and red sandstone, and other stones all cemented with a coarse hard reddish cement. Others are blocks of old red sandstone, which contain large pebbles of water-worn quartz with the sand packed round them, as sand is packed about pebbles on the sea-beach. Others are blocks of granite very like those which are found on the beach near the Giant's Causeway, and along the Antrim coast. There are many chips broken from Arran hills, but amongst them are no bits of Antrim flint or chalk.

At the house of *Dubhgarrie*, at the end of the longest and deepest glen in Arran, a river is crossed. It rises amongst the highest hills, 2874 feet. Here is a washed moraine with conical hillocks and terraces. A little beyond the house the road passes under a steep bank of brushwood growing on glacial drift. A few streamlets have cut scars in this face, which is about 100 feet high. The bank contains scratched and polished stones of all sizes imbedded in fine gray clay, very unlike the common drift-clay.

This then appears to be a record of the local glacier-system of Arran, a museum of Arran stones brought down to the sea, and partially arranged by the sea.

At *Iomachar* the north-western corner of the island is reached. There a sea-cliff about 150 feet high rises above a beach of rolled stones and broken crags. This is modern sea-work, but the rock-surface on the top of the cliff is ice-ground. It is so weathered and worn, that it is impossible to tell the direction with certainty. The rock is contorted slate, and on it rounded blocks of compact granite are perched at this level.

At a little more than 1000 feet, on the shoulder of a hill which makes the base of *Ben Bhanrigh* (the Queen's Hill), ice-scores are very well preserved on a smooth patch of slate, which appears from under the peat-moss. The direction at this promontory is again N. and S. A stick aims nearly at Skipness Point, and at the Mull of Ceantire, along the run of the coast. A little lower down, and further from the hill, scores upon similar rocks point N.N.E.

At *Whitefarlane*, close to the road-side, at less than 100 feet above the sea, striæ on slate are very clear. They point N.E. by N., and so do bent trees beside them. Grooves are tool-marks of ice and water-streams; trees are shaped by streams of air; the equinoctial gale followed the run of the Arctic Current, and both were driven by the same forces past this spot in opposite directions.

The Galway curve is carried over Arran at Lamlash at 1300 feet, and past the west and north-west corners of Arran at more than 1000. To account for these marks by land-ice alone, a glacier must be imagined reaching from 1350 feet to the sea-bottom, and from Ceantire to the nearest hills of equal height on the mainland of Scotland. To account for the marks by floating ice, like that which is working off Labrador in the same latitude, a change of climate and of sea-level must be assumed.

The run of the tide in the Sound corresponds to the ice-lines on the hill; the wind follows the ice-grooves along the hill 1000 feet higher. A south-westerly breeze, which soon became an equinoctial gale, and whose path along the sea was marked by blue squalls and crisp waves, swept the fringe of a low cloud of sea-mist northwards along the hill at the high level. Further up the Sound the same south-west wind curled round the hills and blew from the south-west; further up it blew from the west. In the lee of the mountain the sea-mist hung and boiled and rolled over and over. A stream of water of equal depth moving the other way would move solid floats as the wind moved clouds; surely the stream did flow here, and the floating solids have recorded the fact.

In the night, when the breeze became a storm, it was a Dutchman's hurricane, straight up and down, in the glens. It surged over the hills like great rollers on a beach, and plunged down upon the house-tops, as if to crush them; and ocean-streams must roll over sunken hills in the same way.

At *Cath-mihic-Dhuil*, which strangers have baptized Cati-kill, and at Loch Ranza, are two long glens which held glaciers, for terraced moraines are near the sea. A lofty ridge divides the glens, and the hill-top was a good point for high grooves.

Loch Ranza.—Up to 1300 feet, rocks on this ridge are ice-ground, but so weathered that the direction is hard to make out. On a shoulder at this level many large boulders of granite (some six feet long) are poised on slate saddles. The smoothest side of these slate knolls points N.N.E., the broken side S.S.W. The dip has nothing to do with the shape and fracture. These forms give the direction given by grooves at 1000 feet, and the wind which followed the grooves below blew against the fractured side of the rock here.

At 1400 feet, a deep groove in granite again pointed down wind N.E., over everything in Arran and Bute, up the Firth of Clyde, at hills about Ben Lomond.

So the Galway curve is here carried over Arran at 1400 feet.

At the top of the ridge, 1800 feet or thereabouts, several large stones had been moved a few yards from their beds towards the S.W., but here the granite is weathering fast, and has weathered so far as to obliterate all small marks.

Gravel as large as peas, scudding before a gale, was forming tiny beaches in front of every heather-bush and peat-bank; and rain-drops pattered, and splashed, and rattled against the hill, driven by the gale. It was bad weather for spooring on high grounds.

Low Marks.—In the bottom of the glen near Loch Ranza, about 200 feet above the sea, is a fine section of an ancient water-washed moraine. It is chiefly composed of granite gravel swept from the hills, and of very large granite boulders, which something stronger than wind and water must have piled there; but this is not a perfect moraine, the surface had been worn down. Lower down, stones, sand, and gravel are ranged in terraces, and packed upon a different principle. The stones are sorted in sizes, and laid in sloping beds, where the rivers shot them out during floods and low waters. These are the washings of moraines arranged by burns in the sea. At the mouth of the loch in the sea is a ridge of stones washed into another shape, and arranged on a different plan, by the ebb and flow of the tide, and by sea-waves. An old castle stands on the sea-bar to mark a date, and amongst the gravel at the point a large block of granite stands firm in the station which it took up before the castle was built. From Loch Ranza to the south end of Arran, and along the eastern coast of the island, similar large granite boulders are planted on the beach; and more boulders of the same kind are perched on the top of the Holy Isle, according to a work on the geology of Arran.*

Thus granite blocks and ice-marks, *in situ*, can be traced from the central high hills to the south end of Arran, but there are traces of two kinds of glaciation. In the glens are marks of a large local system, but high up on watersheds are marks of something larger. According to theory these high marks record the passage of the same arctic current whose traces were found at Belfast, and in Connemara; because ice-grooves point from the E. of N. to the W. of S. in this district.

Having carried the Galway curve thus far, the Westport curve must be carried a stage if possible. Having beat round Arran, and found the spoor as high as 1400 feet, and all round the coast, the next cast is northwards across the stream to Ceantire.

* *Geology in Clydesdale and Arran, embracing the Marine Zoology and the Flora of Arran, etc.* By James Bryce, M.A., LL.D., F.G.S.

This author says, at p. 15, that he had failed to discover any decided cases of glacier moraines in Arran. He mentions piles of drift at the mouth of Glen Iorsa, and at "Catacol," which are mentioned above, as moraines washed out of shape. Mr. Bryce attributes them to currents of water sweeping these glens when the area was rising from beneath the sea. At pp. 86 and 87, and elsewhere, terminal and lateral moraines are mentioned and described at higher levels in these Arran glens; and at p. 89, the combined action of local glaciers and ice-floats is suggested to account for the dispersion and placing of blocks of native granite, which are perched on distant high points in Arran, such as the Holy Isle at which high grooves above Lamlash point (see p. 66). The author has failed to notice these and other high marks which would have helped his argument. This seems to be the work of an able geologist who changed his first opinion after careful examination and due comparison with other parts of the country, so his evidence is the more valuable.

CHAPTER XXXII.

BALTIC CURRENT 5—BRITISH ISLES 5—SCOTLAND 2—WESTPORT CURVE—CEANTIRE.

BETWEEN the Galway and Westport curves is Ceantire, at which place grooves at the Giant's Causeway pointed. A steamer runs from Loch Ranza to Campbelton, and thence a road leads to the lighthouse at the Scotch Land's End. The east coast is broken and weather-beaten all the way, but the highest hills are rounded. At *Campbelton* the hills are very unlike ice-work. Not a symptom of glacial action could be traced up to the top of a hill 1100 feet high which rises south of the town. But if the sea were 1000 feet deeper, the town and the country between the two seas would be about 990 feet under water. This district has been swept and the surface destroyed by the sea.

There is no trace of old ice in the low grounds further west. A few suspicious boulders at the end of glens may possibly be remnants of moraines or drift, but these are few and far between. Within four miles of *the lighthouse*, rocks on high grounds begin to assume the familiar shape, and at a height of 700 feet, a large block is perched upon a rounded hill-shoulder to the right of the road. At 900 feet, some blocks of rounded granite peer through the moss by the road-side, and beside them are lumps of the crumpled contorted slate of the country. Fifty yards further, on the north side of the road, is a well-preserved surface. It is a miniature tor, and a deep groove on the top of it points nearly E. and W., at the notch through which the road passes.

Over the brow to the south of the road, hills rise to a height of 1260 feet, according to a barometer which passing gales made an uncertain guide for the time. All these tops have glaciated surfaces, broken short off on the Irish side; and the run of hollows and hill-sides, and of ridges of rock, nearly agrees with the opposite hollow in which Belfast Lough now ebbs and flows. But all fine lines seemed worn out of the contorted broken mica-slate. One hill-top after another was drawn blank. After a long search some very remarkable grooves were found below the brow, at the very end of the Mull. They are on a point of hard rock at 1080 feet or thereabouts. Two smooth regular deep grooves, about six feet long, run parallel to each other, so as to cut out a narrow ridge upon which a man could ride. One groove is a foot deep, and two feet wide, the other about the same size. Part of this rock has split off and fallen, and large blocks of it lie below the solid point. The fragments are deeply grooved, and these marks ran parallel to the others, before they split off.

One of these fallen grooves ends suddenly, so that the hollow would fit a man's head like a stone helmet. The grooves cut through the edge of beds in the stone, and the whole rock is rounded. In profile it has the form of a great gray leech, and Fair Head in Ireland is seen over the rounded back. A stick laid in one of the grooves points W. N. W. just outside the Rhinns of Islay, along the run of the tide, which hurries past heaving and boiling 1000 feet below. Here then a stream bearing ice once curled round the Mull, and ran, as streams now run, from Loch Fyne and the Kyles of Bute, round Skipness Point, along the Sound of Kilbrannan, and past the great Scotch rendezvous for modern storms and tides.

These smooth grooves are all the more remarkable from the shattered rocks which surround them on all sides. It remained to be seen if waves and streams make similar marks at the shore, without the help of ice, and after a close search no grooves were found. The coast-line is made up of angular forms, land-slips, rifts, riven cliffs ready to slip, and vast piles of broken fallen cliffs, amongst which a wild sea raged and roared, while the wind drove spray, cutting showers of rain, and hail scudding over land and sea. About the aiguilles of Mont Blanc (chap. xii.) similar piles of ruin are strewn; here all the power of the Atlantic has failed to obliterate high ice-marks on the brow of the Mull of Ceantire.

From Campbelton to *Glenbar* the road coasts the Atlantic for twelve miles along the north shore. The rocks about this level are all shattered and riven, and the power of ocean-waves is displayed in the grand tumbling surf which rolls in upon the sand at Machariehanish Bay. On the land side are piles of drift, which seem at first to be hills of blown sand, but the sand covers heaps of large stones. At Glenbar the mouth of a glen running north-eastwards towards Arran is passed, and there numbers of large polished and grooved blocks of hard stone, foreign to the district, had been freshly dragged from a field, and were piled along the road-side for building fences. The ice-marks on these were quite fresh. The Giant's Causeway bears S.W. by W. from this spot, and is clearly seen on a fine day. Ice-marks at the Giant's Causeway pointed N.E. by E. into Glenbar, and along the shore of Ceantire. There is no Antrim chalk at Glenbar, but there is granite in Antrim. From this glen to the mouth of West Loch Tarbert the coast gradually loses the shattered form of ocean denudation, and smooth ice-work is better preserved as the shelter is reached. Rocks are less and less broken as the mainland is approached, and as one island breakwater after another shuts out the waves. As the western surf decreases in power, and waves get smaller, rifts and geos become hollows; cliffs change to ridges and tors; patches of drift with stones appear on hill-sides, more large boulders are seen on the shore, and every rock-form points into Loch Tarbert, and the wide hollow in which it lies, as the direction from which some grinding force moved. At *Fronichean,* upon the top of an isolated hill about 200 feet high, a weathered surface is preserved, so that the direction can be determined by deep grooves and other sure marks. At this spot ice moved from N.E. towards the island of Cara.

At about 100 feet above the *clachan* the marks are fresh. The rock is smooth and rounded, and straight grooves on it, from one and a half to three inches wide, from half an inch to an inch deep, and some more than six feet long, prove that ice moved from E.N.E. at this spot. At 300 feet on the same hill the general form alone is preserved. The same rock has weathered, so that waving ribs—the edges of beds of crumpled slate—rise an inch or more above the surface. At first sight the fresh grooves would seem to be the work of a small modern glacier, which slid down a north-eastern hollow from low hills in Ceantire. The moraine seems just below the village, but the shape of the hills, deep glens, and the direction of the grooves, make a modern land-glacier impossible. One surface has been preserved at one spot by clay, and lately exposed, so it remains entire beside a bare surface spoiled by weather.

The highest hill on the road-side is opposite to *Ardpatrick,* and is 400 feet high. The surface is bare rock, ground and weathered. Deep marks here point E.N.E. up-stream, at the mouth of a pass which leads over Ceantire to Skipness, and W.S.W. past Ardpatrick at the southern point of Islay. A number of loose stones are scattered on this hill, one of which is a large block of white quartz.

At the end of *West Loch Tarbert,* Ceantire is joined to the mainland by an isthmus about half a mile wide and some thirty feet high. West Loch Tarbert lies in a deep hollow about ten miles long, which nearly corresponds to the strike of rock-beds. On either side of this large groove are hills from 1500 to 2000 feet high. Those to the south-east make the north-eastern end of Ceantire; the other side of the groove is a block of high land which ends in another large groove at the Crinan Canal, and the highest point in the district is Sliamh Gaoil (the Hill of Love), about which many songs and legends are repeated. Above the town of Tarbert, in the middle of the trench, is a long ridge about 600 feet high. On the top of this ridge are perched blocks, and, though much weathered, ice-marks abound on the hill. At one place a long narrow ridge like the back of an animal ends abruptly where it was broken off; at another a patch of hard stone ground smooth has resisted the weather, and marks are plain. The ridge itself gives the direction. A stick pointed at Dunskeg in West Loch Tarbert, points down-stream over the island of Cara at Lough Foyle in Ireland, and up-stream N.E. by E., over Cowal, past the northern shoulder of high hills near Ardkinglas; and every rock-form in the neighbourhood points the same way. With the sea at this level Ceantire would be three islands, with sounds near Skipness and at Campbelton. A stream flowing as the ebb does in Loch Fyne, would split on hills east of Tarbert. One branch would join a stream coming from the Firth of Clyde, as the ebb does at Skipness Point, and follow the direction of ice-grooves on the Arran hills; another would flow past Tarbert through two narrow sounds, and join the other streams about Clachan, where ice-grooves point at the hollow which crosses Ceantire. At higher levels similar streams would still follow these deep trenches, and flow round islands which are hills now. In walking north-westwards from Tarbert, long parallel ridges and deep troughs are passed as the hill is mounted. From Tarbert to the top of the first ridge is about 550 or 600 feet. Then comes a steep descent of about 500 feet into the next groove. Then a steep hill rises to 650 feet, and a point is reached which opens the narrow end of Loch Fyne. Ben Cruachan is seen to the north, the Ardkinglas hills to the south, and a wide hollow with hills and glens between these high points. Ridge follows ridge up to the top of Sliamh Gaoil, and the whole district seems ice-ground.

All the low hills are of one pattern. At 700 feet are perched blocks, and more can be seen higher up ; rolled stones are at the bottom of the glen, and many are foreign to the rocks on which they rest. Every bare rock in this district, even rocks below high-water mark, and under water, are grooved and rounded in the same general direction.

So, after a check at the Mull of Ceantire, the spoor which was taken up at Westport, at Clew Bay, in Ireland, is fresh on the mainland of Scotland. It lies in a wide hollow between the Jura and Arran hills ; between Cruachan and Ben Lomond further inland ; and central Scotland is right ahead. The track will be taken up there again.

On Mr. Geikie's map arrows point from N.E. to S.W. over these Argyllshire hills, and the marks are attributed to glaciers of very large dimensions sliding off Scotland. According to the marks now described, ice moved south-westwards as far as Galway and Westport, in Ireland ; if it was a glacier, it was 2000 feet thick at Shan Folagh ; it was at least sixty miles wide on this part of the Scotch coast, and it moved over the tops of hills, between 1500 and 2000 feet high, in Arran and Ceantire.

CHAPTER XXXIII.

BALTIC CURRENT 6—GALWAY AND WESTPORT CURVES—ARGYLL, ETC.

Galway Curves.—THE spoor taken up at Galway, and found at Belfast and in Arran, is fresh in Bute ; but at the low level of Bute the lines, according to Mr. Geikie's map, follow the run of the ebb tide, and curve back to the E. of S.

At Greenock a glaciated rock peers out from under a garden-wall in a footpath near the town.

So three lines taken up in Ireland are landed in three grooves which cross Scotland.

The Derry Veagh line points to the Caledonian Canal ; the Westport, Derry, and Tarbert line to Glenorchy ; the Galway, Belfast, and Arran line to the Firth of Clyde : and these must be followed.

At or near the present sea-level it is easy to trace the path which ice followed in all the lochs of western Argyll.

In crossing from Greenock to Inverary, from the Galway to the Westport curve, a series of hollows are traversed. It is plain that land-ice or sea-ice, moving at low levels, could only slide down, or float up or down, these deep grooves.

Loch Long (the Ship Loch) runs up N.E., and rocks on its shores are ground from the N.E. as far as *Tarbert,* where Ben Lomond stands sentry. A low neck of land divides Loch Long from Loch Lomond. At the level of sea-shells found about Paisley, Greenock, etc., the sea would reach Glenfalloch, and surround a large block of high land in Dumbartonshire. At Tarbert the ice-marks do not point at Ben Lomond, but turn round and point at the shoulder, and at the end of the loch, where engineers chose Glenfalloch as the lowest pass to reach Loch Tay. Ben Lomond was not the source of the ice. A great stream was moved down from Glenfalloch, leaving great stones, to which legends are attached. One is the "Stone of the Bulls." It was capsized and rolled down from the mountains during a mythical fight between two mythical bulls, and it has been used as a pulpit in later days. High up on the sky-line, on the shoulder of Ben Lomond, at least 2000 feet up, more boulders are perched, where they could not have rolled. They must surely have floated. If these be marks of ice-floats, the Glenfalloch stream split at Tarbert; one branch went S.W. down Loch Long, the other round by Dumbarton to Greenock. The proof must be sought at the head of Glenfalloch, at the watershed, and that station has not yet been made good.

At *Rowardennan*, on Loch Lomond, where steamers call, a point of rock at the water-level has deep conspicuous grooves which clearly indicate very heavy ice passing towards the Clyde, and grounding or sliding here. The only doubt is whether the ice was aground in a sea, or high and dry.

Glencrodh.—The Loch Long stream was joined by several others. A large branch can be traced from Ben Iomma to the col at "Rest-and-be-Thankful." There the level is about 800 feet, and the question is, What was the sea-level when the last glacier reached it? The marks can be followed from the col two ways; down Glencrodh (the Fold Glen) to the sea at Loch Long, and down to Ardkinglas. The question to be answered is—

Did the ice slide all the way, or did it slide part of the

way, to be launched at 2000, or 800, or any sea-level other than the present?

Loch Goil.—Loch Goil branches off from Loch Long lower down, and heavy ice came down that pass from the north. The rocks are all ground, and the weather-side is towards the pass.

At the mouth of Loch Goil, Clach-an-Turaman (the Stone of Staggering) is perched upon the sky-line about 100 feet above the level of the sea. The loch is about 250 feet deep, and the shape of the bottom is known to herring-fishers, who say that "it is all in pits and ridges." It is therefore like the shore. If this be the work of land-glaciers, the ice was at least 600 feet thick.

At the head of Loch Goil two glens branch—one to the "Rest," 800; the other to Glen Ifrinn, where the col is 630. A coach and a character convey travellers to Loch Fyne. At 200 feet, and on the top of this pass, are piles of glacial drift, and at the sea-level on both sides are conspicuous ice-marks.

But the difficulty is to account for the high drift at 730 feet. No land-glaciers met there, for there are no glens to hold them.

Loch Eck.—Lower down, a third low pass joins Loch Fyne to the Firth of Clyde, at the Holy Loch and Dunoon.

The shores of Loch Eck are strewed with large boulders, and grooved. The col is about 100 feet high, and according to Mr. Geikie's map, the ice moved towards the Clyde from Loch Fyne.

The question to be solved is the sea-level. At 100 feet there would be a strait at the Holy Loch; at 730, a second strait at Glen Ifrinn; at 800, Loch Fyne would join Loch Long in a rock-basin called Loch Restal, and it would meet

Loch Lomond at the head of Glen Chonaglas, and at the head of Glen Fyne. If the sea ever was at that level, there must be evidence of the fact somewhere, and ice-grooves on watersheds may be examined as silent witnesses.

Loch Fyne.—Loch Fyne runs nearly N.E. towards Loch Tay. Striæ are laid down on Geikie's map ; and they are very conspicuous at low levels. Everywhere along the shores from end to end, ice-marks are fresh upon rocks near the sea and awash. The direction of movement was along the run of the ebb, S.W.

The woodcut on p. 92 is copied from a photograph made by an able artist. It is a good example of the form of such rocks.

Inverary.—North of Loch Fyne, two glens—Glen Aoradh and Glen Siorrath—run northwards and eastwards towards Loch Awe. In these are piles of drift, and in branch glens which run into them are similar collections of rubbish at similar elevations, generally from 600 to 800 feet.

At a place in Glen Aoradh, called *Tullich* (mounds), are great conical heaps of scratched stones, and other glacial debris, arranged like moraines described above (chap. xxviii.) On one of these mounds courts were held in the olden time. The drift extends to the top of the col, which is about 800 feet high, level with "Rest-and-be-Thankful." There is nothing in the shape of the country to suggest a glacier ending at the head of Glen Aoradh. Ben Cruachan is beyond Loch Awe, and the drift did not come from that direction. But if the sea were 1000 feet higher, Loch Awe, Loch Fyne, and Loch Lomond would all be joined, the sea would reach the foot of the hills of central Scotland, and all these passes would be straits.*

Lorn, Cowal, and Ceantire would be ten islands added to

* For the shape of rubbish-heaps dropped from melting ice, see vol. i. p. 380.

the Hebrides, and the mainland of Scotland would be an archipelago at this sea-level.

The river Aoradh has cut sections in the drift, and it seems to have come round a hill-shoulder from hills and glens about the upper end of Loch Awe. Above a certain level, about 900 or 1000 feet, the hill-tops are bare rock, and striæ on them point in that direction.

Loch Awe.—Loch Awe runs N.E. and S.W., like the principal glens in this district. It points up to Loch Lyddich and Loch Ericht in central Scotland; and rocks along the shores of Loch Awe are ground from that direction.

The general features of the country, then, suggest the action of some powerful engine which has ground the whole district, so as to furrow it from N.E. to S.W., and cross-cut it from N.W. to S.E., leaving a few high points unground, ∧ ⌒ ‿.

Above a certain level, about 2000 feet, the tops are riven, weathered, shattered, bare rocks, as Beinn Copach ("the Jagged Hill," which Saxons call the Cobbler, and Celts Arthur's Seat); the Gray Head, and others. Lower hills are smooth rounded ridges, with the worn strata peeping through the turf to show that the glens are grooves hollowed out. They are tool-marks of some graving engine, not fractures in the earth's crust.

The shattered peaks prove that the glens are not weather-marks. River-beds prove that the glens are not simply water-marks.

Right down these smooth hill-sides small streams are sawing rough splintery trenches. They are cutting across the grain into the rounded sides of smooth grooves gouged out with some other tool.

The sea-coast proves that the glens are not the marks of

ocean-currents. Sea-waves chop like an axe at the root of a tree, or like a pickaxe at the foundation of a wall ; and the west coast is a wall of cliffs, wherever the sea has its full swing.

These west country glens seem to be large ice-grooves ; the problem is, How came the climate to change, and when did the change take place? If there were a measure for river-work, the Highland burns would give one answer. A stranger, wandering along a smooth hill-side, may see a narrow belt of brushwood meandering through the heather. On coming to the place, he will find an impassable gorge, hidden amongst the trees. Unless he knows the fords, he may wander for miles, stopped by the work of a rivulet.

Legends tell how Rob Roy took up his abode at a river-fork of this kind, and called the place his castle. The house is there still ; and, without the modern bridge, a stranger could hardly get to it, though the fords are easy, when found.

Further back, it is told that a forfeited earl and a faithful guide escaped from hostile Athole men, "who had made a stable for horses of the Castle of MacCailain." The foes got near enough to speak, but the strangers could not cross a burn whose very existence a stranger would hardly suspect.

The river-bed is a fact, if the story be too picturesque for sober history. It is a deep gash, with vertical sides, cut in the smooth rounded hollow, which was made before the rivers began to saw ; and the rivers are sawing through ice-grooves, which are as fresh as if they had just been made in the low grounds of Argyll.

Westport Curve—high marks.—In order to find out the course of a general movement in ice and water, sufficient to account for denudation on this scale, it is necessary to get out of this network of deep narrow glens. The top of the steeple is the place for the weather-cock, and hill-tops must be sought for the spoor of the Baltic Current.

Dun Chorre Bhile.—On the north side of Loch Fyne, near Inverary, is a hill which generally goes by the name of Dun Horrible ; but the name means the hill of the steep brink. It is about 950 feet high. The top is isolated, and at the end of a ridge which separates Glen Aoradh from Glen Siorrath ; Ben Cruachan is to the north, and the cols are lower than this hill-top. Loch Fyne, and hills and glens about it ; the Ceantire hills, and many other distant points, are seen from this spot. With the sea at 800 feet, it was a rock far from shore. Near the top are loose blocks which must have floated there, unless they were carried by glaciers or men. The hill itself, and rock-surfaces laid bare, have the usual rounded form.

At about 750 feet, weathered rock-tables are bare in the moor below the top. Any marks which can be found on them seem to point at Glen Siorrath and the shoulder of Beinn Buidhe, beyond which lies Loch Tay. A block of hard stone, beautifully smoothed and grooved on two sides, lies here ; and fences are made of boulders gathered on the hill. At this level, and above it, rocks to the north are ice-ground all the way to the head of Glen Aoradh, and marks there turn round the hill-shoulder into the Loch Awe groove.

These marks lead to central Scotland. But there are higher marks.

Beinn Bhreac.—The highest point on the ridge which divides Loch Awe from Loch Fyne is Beinn Bhreac (the Speckled Hill). In ascending to it from Inverary, signs of glacial action appear everywhere. Large grooved stones, enormous wandering blocks, patches of drift, contorted beds of sand, and other marks, appear in the woods, and amongst the heather. At 1200 feet, at the N.E. end of one of the numerous ridges of which the top is composed, a well-marked deep groove points N.E. by E., into a hollow to the north of Beinn Buidhe.

Up to 1350 feet, the whole ridge is ice-ground, and every rock-form points at a sea of hills in central Scotland. A spirit-level and a map show that the passes in the distance are lower than this point.

FIG. 75. TORS AND PERCHED BLOCKS AT 1600 FEET. TOP OF BEINN BHREAC. 1863.

At 1550 feet, at the end of the next ridge, weathered grooves, six feet long, run horizontally along the sides of long weathered tors, which rival those of Connemara ; and these marks all point one way at central Scotland.

From this point to the top, 1650 feet, according to a disturbed barometer, excellent specimens of roches moutonnées, with perched blocks, abound. The cut was sketched on the wood : it is reversed ; but the form was carefully copied, and it is characteristic of ice.

If the sea were at 1650 feet, there would be a clear course over Scotland by Strathspey to Scandinavia. Dalwhinny, at the end of Loch Ericht, is 1169 ; Loch Garry, 1330 ; and the highest point on the Perth and Inverness Railway is 1480 feet.

And it is to these places that horizontal marks on Beinn Bhreac point.

Looking S.W. along the supposed line of movement, there is a clear horizon between Jura and Arran along the north shore of Ceantire; and beyond the horizon is a clear way to Loch Foyle, and thence to Westport, as shown above (chap. xxx.)

Looking N.E. there is a broken horizon between the vertebræ of Scotland—between Ben Lomond and Ben Cruachan; but the way is clear at this level, all the way to the Bergen glaciers which have been described above (chaps. xiv. and xv.)

From Beinn Bhreac a magnificent panorama is seen: a wide stretch of moor and lake, with hills, islands, sounds, and the wide ocean; Arran and Ceantire are seen; Tarbert and Sliamh Ghoil; the distant smoke of Greenock beyond Cowal and Roseneath, all the Argyllshire glens and cols above mentioned; and central Scotland right ahead. From this point the evidence seems complete. These ice-marks were surely made by sea-ice, of the dimensions described by Lamont, Dufferin, Scoresby, and others; moving at this level as sea-ice moves off Labrador.*

If the other theory be taken it will not fit the facts. To arrive at the top of Beinn Bhreac from central Scotland, land-ice would have to climb for six miles along the back of a steep ridge, out of Glen Aoradh for about 800 feet, if it stuck to the col; for 1500 feet, if it came straight from Loch Awe; and there is no hill to the N.E. high enough to give the necessary pressure. The hill-top is higher than the watershed of central Scotland in passes out of which the ice must

* These high marks were first noticed by the present Duke of Argyll, who, in 1857, wrote a paper on the subject, and attributed the marks to sea-ice.—*Edin. New Phil. Journal*, new series, vol. vi., p. 153.

have come according to the marks which it made. Glaciers might slide down to the sea by Loch Awe and Loch Fyne; but they never climb if they can slide past a hill.

Supposing a solid mass 2000 feet thick to travel along parallel glens in Scotland, like a sledge in ruts. Let one runner be in the Caledonian Canal, another in Loch Awe, a third in Loch Fyne, and a fourth in the Clyde. Let the ice-tract be as large as the largest known, still even that strong supposition will not carry the ice over the top of Shan Folagh, 2000 feet up, and hundreds of miles away. Nor is there any apparent reason why such ice should move from N.E. to S.W. or thereby, from the watershed of Scotland to the west coast of Ireland.

But if ice floated at the level of the highest marks, ice in Greenland and off Newfoundland explains the puzzle.

It is easy to understand how a prevailing current may have left marks, as a prevailing wind bends trees. It is easy to watch clouds floating past those hill-tops at a well-marked level, and turn them into ice-floes and icebergs, glaciers and snow, from pictures copied by memory from books and nature.

The average annual rain-fall in this district is about six feet. If the rain were snow, as "it is whiles," and the climate a trifle colder, forty or fifty years would build a snow-heap more than 2000 feet deep, and glaciers and icebergs might resume their unfinished work in Argyll. The climate has changed, and may change again; a reason for the change is surely worth seeking. One has been sought in a rise of Lapland and a Baltic current, and so far the British spoor looks well, for it points the right way.

Tides.—If high ice-marks are attributed to ice-floats, and low marks to local glaciers and fjord ice, part of the ice-problem is solved. The powers which move these floats on the opposite coast of Labrador are ocean-currents and local tides, and their movements regulate the movements of the ice, as a stream determines the path of froth. Ever since there was fluid to be moved on the earth's surface, there must have been tides, if the laws of nature are permanent laws; so existing tides on the Scotch coast throw light upon marks made by old Scotch ice.

In the tidal chart of the British Isles, given in Keith Johnston's *Physical Atlas,* plate 15, the local wave of flood is shown travelling north-eastwards across the Atlantic from America towards the Baltic, when it runs foul of Ireland. There the wave is stopped and divided. It is high water on the south-western coast of Ireland, and the ebb begins to flow back. But the wave of flood sweeps on, and curls round till flood meets flood behind Ireland in the lee, near the Isle of Man. It is high-water in that channel, and the ebb begins there, but the wave of flood sweeps on past Cape Wrath and the Land's End, and the waves meet a second time in the lee, as waves do behind a stone in a pond. It is high-water on the eastern coast, and a third ebb begins behind Great Britain. Finally, big waves which travel westwards in pursuit of the sun and moon, and which are reflected from the shores of America back to Europe, pass eastwards to Christiania, Trondhjem, and Götheborg, where the Baltic Current flowing out meets the wave of flood and stops it in the narrow sound.

The general principle of this tidal movement is simple and easily understood, but the details are very intricate.

On the western Scotch coast it takes a lifetime to learn the tides in a small district. At one point it is said by the fishermen that seven tides meet. At another, a current swift as a mill-race pours through a small sound in one direction for about eleven hours, and after a pause, runs back for one hour. At another place Corrie Bhreacan whirls round, and can only be approached at slack water. The famous gulf is but a whirlpool like those which whirl behind stones and posts, and the piers of bridges. It is the offspring of a strong tide whirling about steep islands, and there are scores of small whirlpools in every Scotch and Scandinavian strait.

It is difficult to unravel the maze of the tides at the sea-level where sea and land are clearly defined, but it is impossible to map out all the movements of water beneath the surface. It is hopeless to attempt to follow extinct tides which flowed through passes amongst archipelagoes of hills, and at various levels from 3000 feet downwards.

Still, general movements of fossil tides may be inferred, and some high ice-marks may be referred to them.

At the level of 2000 feet, which would be shown by contour lines on a Scotch map, if one existed, the flood-tide which comes in from the S.W. would pass over low lands in Ireland, and through straits at Loch Laggan, Loch Ericht, Loch Garry, Loch Tay, etc., in central Scotland, and so on over Sweden, into the Baltic; and the ebb would return by the same direct route.

At the level of 1000 feet, Loch Garry and Loch Ericht would be closed, but Loch Laggan and Loch Tay would be open, and the tide might still pass that way.

At the level of 500 feet, the Caledonian Canal and the Forth and Clyde Canal, and Scottish Central Railway line, would still be straits, though central Scotland had become a single island.

So long as there was a direct passage the waves of flood would sweep through it as they now sweep through the Pentland Firth and the Straits of Dover.

So long as there was an ice-float to be moved by tides, the flood-tide would move it towards Scandinavia, and the ebb would drive it back towards America, as tides are supposed to move ice in sounds which cross Greenland (vol. i. p. 395.)

If, when the sea-level was at 3000, 2000, 1000, or 500 feet, there was an arctic current moving south-westward out of the Baltic, it would help the ebb to drive the floats and breed glaciers on any Scotch or Irish hills that remained above water. Now that Lapland is 1400 feet above the sea, there is no such Baltic current and no British ice.

Inverary nearly corresponds in latitude to Nukasusutok in Labrador.

Great floes, big icebergs, and fields fifty miles wide, are moving along the Labrador coast south-eastwards, driven by the reflected current which cannot escape south-westwards from the arctic basin, because the north-west passage is too narrow. The Labrador ice is moved by tides and rocked by Atlantic rollers; it whirls round islands and points and rocks, but there is a general direction of movement, and there must be a general direction of ice-marks on rocks under water.

So old Scotch floats may have recorded a general movement from N.E. to S.W., though every group of islands and every change in the level of sea and land would alter the run of local tides, change the drift of ice, and so vary the direction of low marks.

The highest marks are, therefore, best for getting at general movements. The Scilly Bishops off Scilly, the Dubh Iartach off Mull, the Mealsack off Reykjanæs in Iceland, and similar rocks in the ocean, are washed by tides, but they do not change the course of a tidal wave as Ireland does.

On Shan Folagh in Connemara, at 2000 feet; on Beinn Bhreac in Argyllshire, at 1600 feet; and on other isolated tops which were solitary rocks if the sea-level ever was so high, ice-marks do agree with the assumed direction of tides and currents. The actual path of Labrador ice coincides when copied and transferred to Britain in the map (vol. i. p. 232).

At lower levels in glens and amongst mountains, in places where hills made an archipelago, and the glens a network of sounds and firths, the marks become an intricate problem, which would cost an army of observers years to solve. To these low-level marks the attention of Scotch observers seems to have been chiefly directed hitherto ; if they will leave the beaten path and try the hill, they may work out the whole problem in time.

This at least is plain : If land rose or sea fell from 2000 feet or any high level so far as to dry glens in central Scotland, and Beinn Bhreac in Argyll, even then glaciers might flow down straths into sea-lochs in Glenfalloch, Glencroe, and Loch Long ; in Glen Fyne, Glen Siorrath, Glen Chonaglas, and Glen Aoradh ; in Glen Orchay and Loch Awe ; in Loch Etive and Glencoe ; in Loch Nevish, and in similar grooves ; while tides and currents still flowed directly past Edinburgh and Inverness, over low lands in the British Isles.

If there were glaciers on the Argyll Bowling-Green when a cold stream was in the Clyde valley, that branch of the stream might carry ice grown in Lanarkshire, Dumbarton, and Argyll, to Connemara ; while the Lochy branch carried an ice-fleet built about Ben Nevis to be wrecked on Donegal.

If this really happened, there should be ice-marks to correspond about Edinburgh and Glasgow, about Inverary and Dalwhinny, about Fort-William and Fort-Augustus, and on hills and watersheds in central Scotland ; and of these six points one is made good by Beinn Bhreac at Inverary.

At "Rest-and-be-Thankful," a weary pilgrim once sat him down and sang

> "O king! O Peter and Paul!
> There's many a stride from Rome to Lochawe."

Above this wild spot, from which a distant lowland horizon can be seen through a gap in the hills, a tall mountain rises; and on its steep ice-ground sides, fresh moraines hang where ice left them 1000 feet and more above the present sea. Where the old pilgrim sat, tides surely met since the hills took their present shape; and if they did, their way was clear along this route from Galway to Aberdeen, and to places further from Lochawe than Rome.

So now to the spoor once more with a cast southwards.

FIG. 76. *Westport Curve.*—AN ICE-MARK IN SCOTLAND.

Striæ upon a rock in Loch Fyne, about three miles south-west of Inverary. From a photograph. 1863.

CHAPTER XXXIV.

BALTIC CURRENT 7—BRITISH ISLES 6—SCOTLAND 3—GALWAY CURVE—LANARKSHIRE, EAST LOTHIAN, ETC.

THE last cast was northwards, the next is southwards into the low lands which were seen from "Rest-and-be-Thankful;" and the next point high on the Galway curve is near Glasgow.

Dechmont.—About eight miles from the town, on the south bank of the Clyde, is an isolated hill of blue whinstone, called Dechmont. It is an igneous island in a sandstone sea—an upthrow in the coal formation. Looking at this hill from the N.E., near a bridge over the Clyde, it seems to have been worn down from the eastward, at right angles to the line of sight. It is broken down to the westward. It has a rounded top; and cliffs on the west and north. In shape it resembles other hills of the same kind; for example, Stirling and Salisbury Crags in the same glen, and Bren Tor and other tors in Devonshire.

At the Clyde level, rocks are sandstones covered with beds of sand, clay, and glacial drift. Amongst stones taken from the fields are boulders of hard rock, foreign to the district, polished and grooved. Many of these are set up along the road-side, and marks are so clear on them that they can be seen from a passing carriage.

Mud in the Clyde, which is washed from this district, is of the same colour as the drift-clay to the south-west, along the Galway curve; and Lanarkshire boulders are like Irish boulders.

On the eastern shoulder of Dechmont, a large pile of stones had been newly dragged from a field by an improving farmer, in September 1863, and amongst them were large blocks of crumpled mica-slate, quartz rock, sandstone, and various kinds of whinstone. Thus glacial drift extends far up the side of this valley. On the hill-top, at 550 feet, the blue whinstone is barely covered with soil and turf. There is no drift, so this hill-top has been swept bare. Close to the keeper's house, the turf was moved in 1862, to make room for a garden, and in 1863 the rock was still exposed. Ice-marks on it are perfect; so Dechmont was ice-ground, and has not lost an eighth of an inch by weathering.

There are deep scores with finer sand-marks in them, and all these point S.E. and E.S.E., at hills on the line of the Caledonian Railway near Lanark. North-westward, the grooves aim over Glasgow, down the Clyde. Wherever the turf has been moved on this hill, marks are fresh, and point in the same direction. The hill was ground by ice moving over it from the S.E.

Bent trees on Dechmont point the old way, N.E., at right angles to these grooves. Water, according to theory, ought to have followed the track of air. But here, when the shape of the land is studied, when the mist of the coal-fields of Lanarkshire opens for a moment to show distant hills, a reason appears for a change in direction at this level.

If Dechmont were awash in a current flowing at the 550 feet level, it would be a hard rock off hard hilly islands, amongst which the Clyde now rises, and off a round-backed island on which the Kirk of Shotts now stands. If the stream came by the Firth of Tay and the Firth of Forth, over Dundee, Perth, and Stirling; North Berwick, Edinburgh, Carstairs, Lanark, etc.; the block of hard high land about Tinto would turn the stream northwards along the valley of the Clyde, as far as the next bank, where Cowal now bends the Clyde at Dunoon. Cowal sends Clydesdale water S.W., to follow the ebb N.W. round the Mull of Ceantire. On the large scale, it was the case shown at vol. i. pp. 127, 130, and illustrated by every stream of moving water and ice.

If the Dechmont marks were made by land-ice, the glacier was more than 600 feet thick; a branch slid down Clydesdale, and one side of the glacier was beyond the Edinburgh and Glasgow Railway.

The low lands of Lanarkshire now drive a busy iron trade. Coals and iron are dug from below; furnaces, coke-heaps, and engine-fires darken the air with smoke. Night and day ringing hammers, machines, and roaring blasts make a ceaseless din; and at night the very clouds glow in the light of panting fires, which flare and fade like groups of small volcanoes in full work.

Close to the most active centre of artificial igneous action, at Airdrie, arctic sea-shells have been found in drift at a higher level than the top of Dechmont. But when the sea-shells lived at Airdrie, Lanarkshire, with all its hidden treasures, was under water in a wide sea-strait, which crossed Scotland where the Edinburgh and Glasgow and Caledonian Railways now cross, and ocean-currents swung from hill-side to hill-side, as the Thames, Clyde, and Forth do from their banks.

The Airdrie bed of arctic shells makes one more link in a chain of evidence. The marks on Dechmont were made by floating sea-ice, which was moving in a fjord; or towards Galway in Ireland, in a stream which curled round islands, of which the high land about the Kirk of Shotts was one.

In mining for coal and iron the internal structure of this country is learned, and from that internal structure one original surface-form may be guessed.

It is common to find that a rounded hill consists of a pile of flat beds of rock, laid one upon the other like a heap of roofing slates. But the shape of the surface has nothing to do with the structure of the rock. If, in mining, any one of these beds is followed far enough, a fault or dyke is reached where a whole series of flat beds has been broken, and the bits displaced. One side of the fracture or the other is generally lifted or dropped many feet. In a series of 10 beds No. 1 may be opposite to No 10 ; but if No. 10 has been lifted a hundred feet up to the place of No. 1, then the side of the broken dislocated fragment ought to be a cliff a hundred feet high, with nine beds shown in section. If the broken surface of Lanarkshire were preserved entire, it would be a land of flat slopes and sandstone cliffs, like an ill-laid pavement, for the whole of this coal-basin is shattered by faults. The beds dip all manner of ways. But this broken surface has not been preserved.

Lanarkshire is a land of swelling hills and ridges. The only cliffs in the county are hard trap-cliffs like Dechmont, and river-banks where running water has done the usual work of sawing and undermining. The surface has been worn smooth, and the cliffs ground off. The edges of nine beds, to correspond to the nine which are found on one side of a vertical fault, are found by searching along the hill-top where the beds crop out. Cliffs have been denuded.

Here is another link in the chain. The whole of Lanarkshire has been ground down. The sea was up to the level of the Airdrie shells ; ice moved over the top of Dechmont, and ground the trap ; so the great valley was finished by sea-ice, though subterranean fire blocked it out, and so prepared a groove for ice and water to move in.

That seems to be the rough translation of part of the outline of the story ; the details have filled many volumes, and will probably fill many more.

Following the direction of the marks on Dechmont, the 550 feet level leads to the highest hills in the country, which are nearly 2000 feet above the sea-level about the head of Clydesdale.

Seven miles in a straight line from Dechmont, at *Dalzell* on the Clyde, a sandstone rock close to the river, 80 feet above the sea by the Ordnance Survey, is polished and striated. The direction is S. 55° E.

The Clyde here winds about in level haughs, in plains of clay, earth, and gravel ; but where this alluvial deposit was moved to make a walk in 1863, the old ice-surface was found perfectly fresh upon the hard sandstone within three feet of the surface. A line ruled on the Ordnance map points up a deep wide rock-groove which the Clyde did *not* make, because the marks of ice are there ; preserved from the water by the alluvial beds.

Leaving the Clyde groove at Dalzell, the country to the north and east rises with a gentle swell. At *Wishaw* the rise is about 350 feet, and a river has dug a **V** 90 feet deep. The sandstone cliffs are fractured, and the river-bottom is an unbroken ripple-marked bed of sandstone. In fields near *Coltness* are scratched boulders of quartz, porphyry, limestone, and other hard rocks. At the road-side are large blocks of hard igneous rock taken from the drift, some with grooves more than half an inch deep.

At Camnethan the rise is 480 feet; so the level of Dechmont is passed at a distance of about 10 miles.

Further east, at Carstairs and Cleghorn, the height is 752 by the survey, 765 by barometer. Here the drift is disposed in conical and rounded mounds, like those which result from the melting of frozen sand and gravel in water (see vol. i. p. 380.) The highest point is 918 feet by barometer, and the form of the surface on this high level is much the same. If this were first a shoal, then an isthmus, drifting ice would be apt to ground on it, and this is the place at which the Dechmont grooves point.

The Pentlands are about 1600 feet high. The rock is much weathered, and ice-marks are obliterated. A rolled quartz pebble was picked up on the highest hill in the range, and a scratched boulder was found in a wall at 1200 feet.

The range is chiefly composed of volcanic rocks, and the hill-tops are strangely like volcanic shapes in Iceland. Part of the Pentland range may, perhaps, be of later date than the Scotch glacial period ; but on many of these hills ice-marks are abundant.

Maclaren mentions other signs of glacial action on this range :—A block of mica-schist, weighing eight or ten tons, is at the east end of *Hune Hill*, the nearest rocks of the kind being fifty miles off, about Loch Vennachar or Loch Earn ; Ceantire, eighty miles westward ; or Forfarshire, seventy miles northward. But as all the ice-grooves point eastward, the block probably sailed from some land beyond the seas, together with the hills of drift which are piled up near this track.

At 800 or 900 feet, at a place called *Westwater of Dunsyne*, " dressings" were found by Maclaren.*

The direction was E. and W.

So at 1000 feet (the level of marks on the Arran hills) the Galway curve is carried over Scotland by the Caledonian

* Maclaren's *Geology*, p. 215.

Railway; the hills of Connemara and the Pentlands are joined by a curve on the map (vol. i. p. 232), and high ice-grooves correspond tolerably well all the way.

At lower levels this gap in Scotland was blocked by the high land about the Kirk of Shotts. But the way was open along the Edinburgh and Glasgow line, and ice followed that curve.

Edinburgh and Glasgow line.—Two rivers, a canal, many roads and railways, all follow the path which an ocean-current may have followed from sea to sea at and above the level of 1000 feet.

To the north of the Edinburgh and Glasgow line, as far as Castlecary, the north bank of this large groove is a range of hard hills. These have smooth tops and sides, and they are scarcely varied by glen or watercourse. The low grounds belong to the coal formation; and the surface of the low country, which was at the bottom of the sea-strait, is furrowed by ridges and hollows parallel to the roads, canals, and railways, and to the range of hills.

Ice did not slide from the hills into the plain. If it had, furrows would point at the hills; but ice made the grooves in passing along the base of the hills, and it seems as if some grinding machine had passed over the hill-tops also, for the range is but a large copy of smaller ridges in the plain below it. All the outlines are curves ⌒. All the grooves point from sea to sea.

All the hill-tops in this valley are ice-ground, according to the observations of Maclaren, his predecessors and successors. At *Binny Craig*, near Linlithgow, grooves and ridges point E. and W. *Craiglockhart Hill*, three miles S.W. from Edinburgh, is a tor pointing E. and W. It is quoted as a specimen of crag-and-tail, but the tail points E., as the tails of ice-tors do when ice comes from the E.

When a street in a populous town is paved with flags which contain hard nodules, passing feet wear the surface unequally. Ripple-marks go first, and at last an old paving stone is hollowed out and worn down, till knots of harder stuff rise like miniature hills in a rolling plain, on which puddles gather when it rains. The knots are worn and scratched by sand and hobnails, and they retain marks best, because they are hardest. The softest bits are "rock-basins."

Renfrew, Lanark, Ayr, Linlithgow, Edinburgh, and Haddington, are like the flagstones. They are worn, though not by the feet of men, and the hard knots are hills of igneous rock in softer strata, which have been ground by ice.

The low country is strewed with glacial debris everywhere, and lakes and rivers are like puddles of rain-water resting in hollows in streets. Dechmont is like a knot in the stone. At Edinburgh, Corstorphine Hill and Arthur's Seat are hard ice-ground knobs which also retain marks.

On *Corstorphine Hill* conspicuous marks are to be seen over a space of more than a square mile. Some grooves are fifteen yards long and a foot deep. Where the rock has been newly laid bare in fields, small grooves may still be copied by rubbing. The direction is E. by N., at a height of about 400 feet. Great weathered rock-tables are to be seen on all parts of this hill-top. They were noticed by Sir James Hall many years ago, as mentioned p. 214 of Maclaren's *Geology*, 1839. The direction of these grooves is confirmed by observation; but the cause formerly assigned—a deluge of water driving stones towards the east—must be abandoned. No stream of water now makes similar marks without the aid of ice. There is no sea-beach in the Western Isles, where Atlantic waves and currents have made marks which could be taken for ice-marks.

On *the Calton Hill* are grooves almost obliterated by human feet. The direction is E. and W. at about 300 feet.

On *Arthur's Seat* are three sets of marks at least.

One is about 400 feet above the sea, at the side of a steep path which leads to the hill-top from the Queen's Drive.

Here grooves dive north-eastward into the hill, at an angle of 22°. If this be an old weathered ice-surface, it has been covered by the newer igneous rock which makes the top of the hill. It may be a weathered slickenside.

A second series is lower down on a rock which was laid bare in making the Queen's Drive. At this spot the fine surface is almost perfect, and the grooves are very plain. The movement was from E. by S., S. 78° E., past the hill-side towards the castle-rock through a gap at the back of "Samson's Ribs."

Close to these ice-marks, a slickenside has been preserved. These grooves dive into the hill, and bits of crystal deposited on them still adhere to the worn surface.

A third set is at the edge of the western cliff of Salisbury Crags, at a level which would join the two seas by the Edinburgh and Glasgow line. Here two sets of cross marks are well preserved ; but the surface is beginning to split off and weather. The chief direction was from N. 65° E., or roughly N.E. by E. These grooves run to the broken edge of the cliff, where a good push would break off more of the columnar greenstone. They point over Edinburgh, along the line of the Caledonian Railway and the base of the Pentland Hills, at a low conical mound in the glen S.W. by S. The shape of the Crags alone would suggest movement in this direction ; but the marks are sure guides.

The greenstone, together with beds of sandstone which rest upon it, was at some time lifted up like the lid of a box, but since then nearly the whole of the upper sandstone layers have been rubbed off. At this spot the hard greenstone has been reached, and marked by ice passing westwards. The cross markings point from W.N.W. to E.S.E., from the low lands of Fife to the Pentlands. If this hill rose up in a current flowing from the eastward, these and the grooves in the Queen's Drive point out the junction of streams which split upon Arthur's Seat, and joined in the lee, or these are marks of heavy ice drifting backwards and forwards in the local tides.

In any case, they cannot be marks of land-ice, for they avoid high ranges, and aim over low grounds.

Here seems a fit place to quote authority in support of theory, and the authority in this case carries weight.

In his later years, Hugh Miller, that type of a Scotch peasant—the man of vigorous intellect, sturdy limbs, and strong faith—used to wander from morn till evening on the shores of the Firth of Forth, seeking to extract the secrets of the boulder-clays and brick-earths, and to unravel the old coast-lines. The result of his labours in this direction was published in 1864 by his widow. No attempt was made to account for the ice-period, or the direction in which ice moved; but Hugh Miller, as usual, saw a picture of the old ice-world of Scotland through its marks, and showed his vision to others painted in coloured words.

At page 35* is a woodcut which is not a picture, but represents a fact. It is a rough plan of a "boulder pavement;" a patch of boulder-clay washed clean by the waves of the Firth; an old ice-pressed sea-bottom of stones squeezed into clay and ground in their bed.

* *Edinburgh and its Neighbourhood*, etc., by Hugh Miller. Adam and Charles Black, 1864.

The geologist says—

"The agent was evidently the same as that which grooved and polished the rocks beneath. It was the ocean-borne icebergal cars of winter that rutted these strange subterranean pavements, compared with which, those of the buried cities of Vesuvius are as yesterday. All of them I have seen have their direction and striation east-north-east—the general direction in the district of lines and grooves of the rock below."

From ice-marks, old shells, the position of shell-beds, the shape of contour coast-lines, and other evidence, Hugh Miller concluded that a glacial period—the life of arctic sea-shells, sea-ice, and rock-grinding—coincided with a sea-level at least 1000 feet higher on Scotch hills than the present beach. From the levels of old sea-margins, from the depth of the double line of sea-caves at the Sutors of Cromarty, and such evidence, he attempted to deduce a few limits of time, and a rate of change. Of the reality of the ice-period, and the direction in which sea-ice moved, he was satisfied, and his direction corresponds to the observations above detailed.

North Berwick.—Marks on Arthur's Seat point towards North Berwick.

The Law is an isolated conical hill of igneous rock 617 feet above the sea, and at the end of this Scotch part of the Galway curve. The low country is chiefly composed of sandstones and beds of whin, and the soil is a mixture of glacial drift and volcanic debris.

The top of North Berwick Law is much weathered, but grooves are still visible on the highest point of the hill.

Looking downwards, all the small rocky islands in the Firth seem to be ice-polished from the direction of the ebb-tide, but the high grooves were probably made from the north-east. A stick laid in one of the high grooves points like a weathercock on a steeple at places from which ice came and to which it went. One end points out to sea at Scandinavia, the other towards Ireland along the ice-track which has thus been followed from Shan Folagh to North Berwick Law. The bearings in Ireland were N.E. by N., here they are E.N.E.

Because of the shape of the rock-surface there can be no doubt that ice made these high grooves, and if it was land-ice the source of the glacier may have been in Scandinavia ; it cannot have been in Scotland, because of the high marks.

Near the top of North Berwick Law is a strange old thorn which shows the force of the prevailing S.W. wind. Branches and trunk stream far away from the root, bowing towards the N.E., and every exposed tree in the neighbourhood points the same way. The equatorial current of wind sweeps over the land from Galway to North Berwick, and winds amongst the hills like any other stream. An arctic current of water surely flowed along the same curves in the opposite direction from North Berwick to Galway. Grooves and trees tell one consistent story all the way.

If the excellent Ordnance map of the Firth of Forth is set up where the general shape of the country can be seen, a curve drawn from Bergen to North Berwick passes between the Pentlands and the Lammermuir Hills. Looking down from the Pentlands this country is seen like a map, and it would be a sea-bottom at the level of ice-grooves on North Berwick Law. If a current flowed from N.E. over Scotland at the 1000 feet level, it would curve round the Fife hills, as the flood-tide now curves round the East Neuk of Fife on its way up towards Stirling. The high ice-grooves coincide with ridges and hollows laid down on the Ordnance map between the Lammermuir and Ochil Hills. If the map were laid according to its bearings on the top of North Berwick Law, the great glen of Scotland would coincide with the groove which ice made at one end of it. It seems fair to conclude that floating ice and ocean-currents—the tools which made the small groove—also made the big groove which contains so many ice-marks of so many sorts and sizes.

When the Ordnance map is studied, or when any tract in this district is seen from a high hill, the form of the wearing or denudation is seen to differ at different levels on both sides of the Firth. Down to a certain level (about 800 feet) hill-glens branch and radiate from high points and ridges. Streams which flow into the Tweed are like twigs on a branch which springs from the sea at the English border; glens in like manner radiate from the Ochils. But below a certain level, in the big hollow, all ridges and hollows run in sweeping curves like mud-banks in the Firth, which follow the run of tides which wear them. These shapes tell of water-work; the sea-shells at Airdrie prove the case, the ice-marks speak for themselves.

Streams of rain-water, which flow into the big glen from hills which make the sides, are now cutting small cross furrows to the sea, like those which older streams of water and ice cut out at the upper level. The Scotch map then seems to show two distinct forms of denudation—one due to radiating local systems, the other to a general system of movement from N.E. to S.W. The Irish map shows similar forms.

So here is another link in the chain. From Galway to North Berwick rocks have been worn and grooves made by ice; floating in an ocean-current, south-westward; but high hills have also been worn, and grooves made in their sides by land-glaciers sliding in every possible direction, downwards, into the sea, from watersheds. The sea-level was a high one when the horizontal marks were made, for they rise high.

The broad track taken up at Galway seems to be carried over one part of Scotland. If followed from North Berwick the spoor should be found about Stavanger, where it was left in chap. xvii. The next cast is northwards to seek the Newport curve which was left on the top of Beinn Bhreac in chap. xxxiii.

FIG. 77. A WATER-MARK IN ICELAND.—MERKIAR FOSS NEAR HEKLA. 5th August 1861.

CHAPTER XXXV.

BALTIC CURRENT 8—BRITISH ISLES 7—SCOTLAND 4—GALWAY CURVE—NORTH-EAST COAST.

Scotland—Galway Curve.—If one great glen in Scotland was partly hollowed out by ice, and has been so little altered by water and weather as to retain ice-marks half an inch deep, in many spots ; it is probable that other Scotch glens are but ice-grooves on a large scale, and that many of them are parts of curves which record the movements of a general glacial system whose centre is the North Pole, and whose path, like that of the present Greenland Current, was like the curve of the letter **P**, part of the figure **8** drawn on a meridian.

A glance at a map will show that the Galway curve coincides in general direction with many of the glens which cross Scotland, with rivers, firths, sounds, and main coast-lines ; denudation in Scotland as in Ireland has manifest reference to curves which cross meridians from north-east to south-west or thereby. The Galway curve was run out at North Berwick ; it can also be followed along the north-eastern coast. The tract to be searched for the Westport line found on Beinn Bhreac in Argyllshire is somewhere in central Scotland, about Loch Ericht or Loch Garry ; so the way is north.

At the level of marks found on Dechmont and North Berwick Law, the *Ochil Hills* would be a steep island cut off from central Scotland by a strait through which the Scottish Central Railway now passes to Perth.

Stirling, or Windy Gap as it is called in Gaelic, is at one end of the strait where it joins the valley which now holds the river Forth; and here a railway crosses to Loch Lomond, following the low level. On the castle-rock, Maclaren found marks of a movement from the N.W. Sir James Hall found dressings which pointed the same way; but if a current came from the E., it would bend round the foot of the Ochils.

The Carse of Stirling is an alluvial plain of rich flat land, with sweeping mounds of stratified gravel and sand rising every here and there. The stones are small and look water-worn, and the shape of the country is the shape of a dry river-bed. Canoes, the skeleton of a whale, shells, and other such marks, confirm the evidence of form. The battle of Bannockburn was fought upon an old sea-bottom.

The rock on which Stirling Castle is built, the *Abbey Craig* on which a monument is slowly rising to the memory of Wallace, and other hills in this tract, are of the same pattern as Salisbury Crags and Dechmont. They are broken knobs of hard rock, and they seem to be tors worn from the Scandinavian side, for they are broken to the westward.

The *Scottish Central* line passes northwards in the lee of *the Ochils*, and at the Bridge of Allan it leaves the plain. The cuttings are through masses of glacial drift fifty feet thick at least. The beds are not stratified; the stones are not sized and sorted; but big and little stones of many kinds are confusedly mixed with fine soil. The materials are glacial, but the surface-form is aqueous.

At Dunblane, 150 feet up; about Greenloaning, 300; and thence to the watershed, 350, where the Allan is left and water flows towards the Firth of Tay, the shape of the country is like the shape of the Carse of Stirling and the neighbourhood of Falkirk. It is a large copy of a broad west country sound when the tide ebbs. Flat fields suddenly end in hillocks, steep points, and ridges, whose slope is the slope of loose rubbish. There are piles of drift in the supposed strait which joined the Firths of Clyde and Tay, and the shape is that of the model (vol. i. p. 380). Above this drift the hills are barely covered with turf. They are rocks, but rounded to the very top.

Seen from Falkirk the Ochils slope down to Fife, but fall suddenly towards Stirling. Seen against an evening sky from hills above Dundee, the Scandinavian side of the Ochil hills has the same general outline; but the low shoulder is like a great rolling stormy sea, driven westward by a north-easter, for the larger form is repeated in miniature as ripples copy larger waves; all the low ridges slope towards the sea and are steep to the land. On the weather-side, near Fife and about Perth, there is less drift, and it is more evenly and thinly spread over the rocks. So the shape of the Ochils is like that of smaller tors on which ice-marks remain.

At Auchterarder, 200 feet up, the hills of *central Scotland* are seen. When the first snow of winter has whitened the hill-tops, and a bright sun shines through a clear frosty air, every mountain form is clearly shown by colour, light, and shade. The hills are seen to be rounded weathered masses of stratified rock, with sides furrowed by glens radiating from the watershed down to a certain level. Below that, ridges and furrows sweep along the hills. There are visible marks of vertical and of horizontal denudation on the mountains beyond Strathearn.

Weathered edges of the strata, when picked out with snow-drifts, make the great hills like coloured wooden models. They owe their convex rounded shoulders and hollow glens to carving, as models do; and their structure, like the grain in wood, has nothing to do with their surface-forms.

Amongst these distant hills are well-known well-remembered river-marks. Steep picturesque gorges, where birches wave, and heather blooms over gray crags; where mountain-streams brawl and thunder down into black boiling pools, from which they leap foaming, till they reach some quiet lake and rest. There, the broad Tay winds past Taymouth, and the Isla glides past "the Bonnie House o' Airlie;" silver threads in a carpet of green. But these are not the tools which carved these mountains, glittering like silver in the crisp frosty air. Rivers might work for millions of years, but they never could do such work. As well might an artist sculpture a bust with a hand-saw.

This work was done with other tools.

Looking north-east from Auchterarder the horizon is clear of hills, and the plain of Strathmore fades in the distance. But on either side of this level strait of rich flat land rise steep islands of rock. The Sidlaw Hills are to the right behind Perth, and the Forfarshire hills, on the left, stretch to the blue horizon. On such a day, when a wide tract is seen like a model, it is easy to fancy the horizontal snow-line to be a sea-margin, and to follow the coast along the dark line where the snow is melted.

The dark lines on a railway map show low grounds; and here railways surround two blocks of high land; they mark out the base of the Ochils and Sidlaw Hills. There is a tract of low land all the way from Aberdeen to Greenock; and if the sea were at the snow-line, tides might ebb and flow along the east coast of central Scotland and round the coasts of the islands of Ochil and Sidlaw. If the ebb did in fact pass westward, bearing vast graving-tools, and grinding hills with them, their marks should be found on the north-eastern islands, and in particular on the Sidlaw range.

Sidlaw Hills.—The next large north-eastern island, at the 500 feet level, would be the Sidlaw range, which stretches from Perth almost to Forfar about N. 30° E. The steepest ends of the hills and broken cliffs face the south and south-west, and the longest slopes are towards Forfar and Strathmore.

Strathmore, the big glen, runs parallel to the Firth of Tay, and cuts the Sidlaw range from central Scotland. A railway follows this old strait over flat land from Perth to Aberdeen now; but at the 500 feet level, Strathmore would be a strait. A stream, which rises behind Dundee at a low level, flows into Strathmore, past the northern end of the Sidlaw Hills, round by Perth, and so down the Firth of Tay past Dundee, and back to within a few miles of its source. The hills which are thus isolated are about 1000 to 1300 feet high. They are chiefly composed of sandstone and bedded trap.

The *Carse of Gowrie* to the south is a low plain of rich clay-land highly cultivated. It is very little above the present sea-level; and many marks show that it was under water at a late period. Reeds force their way up amongst the corn from bogs which are now buried. Every now and then a rude boat, an anchor, an iron ring, or some other mark, turns up a long way from the present shore.

The air above the Carse is often heavy with water, and, as the natives say, "In rimy weather, when the frost takes the air, when ye look doon frae the hills, it's just like a pond." Looking down from a height of 700 feet, on a still frosty morning, the whole Carse is hidden by a level sea of mist, above whose distant horizon peer dark islands, in Fife and Kinross. The Ochil Hills and the Fife Lomonds are the islands in this misty sea. From its depths rise sounds of busy life—barking of dogs, the crowing of cocks, the low of kine, the cawing of rooks, the rattle of carts, the buzzing of steam-ploughs, the distant roar of the train, and the near voices of men; but for all that appears to the eye, the Carse and the low lands of Scotland might be a sea-bottom a hundred fathoms down. The Carse was a sea-bottom, and deeper down, since the Sidlaw Hills took their present shape.

Behind *Rossie* are two wide straths, which at 800 feet would join Strathmore to the sea. These glens, seen from the col, seem to run N.E., but below 800 feet they are sheltered from the N.E. by hills. The glens make a kind of bay in the range. At 900 feet, at the head of these glens, and at 450 feet, at the back of the first range, are collections of drift. When a field is newly taken in, thousands of large stones are taken from the red soil. Amongst them are specimens of gray granite, white quartz, contorted gritty stone, blue limestone with white veins, whinstone, brown trap, hard gray and white quartz rock, mica-schist, porphyry, greenstone, and other hard rocks. Many of these are smoothed and grooved. Similar stones are built into walls, bridges, and houses, and they are broken up in thousands. This then was a cross sound amongst the Sidlaw Hills at 800 feet; and at 700 a sheltered corner in which drift gathered. When the col dried at 800 feet the glens were sea-lochs, dotted with islands, which are now steep hills.

The hills are all sandstone and trap. The beds dip various ways, but the dip and fracture do not accord with the shape of the hills and glens. It is plain that they were carved out; the question is—By what means?

From one col (800 feet) a steep pull leads to the foot of a

cliff of igneous rock, which seems, by its structure, to have boiled. The old igneous surface on the upper side of one layer may be seen by moving the next plate. The rock is like Icelandic lava, a hardened brown crumpled froth. The tops of "*the Giant's Hill,*" above the cliff 1350 feet, overlook Strathmore, and they are rounded knolls. The rock-surface generally is too much weathered for striæ, but some remain. They point N. 58° E.

The King's Seat is the highest point in the range, 1400 feet. The shoulder is manifestly ice-ground, but too much weathered for marks. The top is an artificial barrow of loose stones, on which the sappers and miners have built their cairn. At the foot of these hills, which were marked at 1350 feet by ice moving from the N.E., are the piles of drift above mentioned. On the hills above 1000 feet there is not a boulder to be found. But the sea of mist floated up, and settled upon the King's Seat, and then nothing was visible but a gray cloud as thick as Icelandic thoka.

At 800 feet, and some miles nearer to Forfar, a hill-top, at the head of this basin, called *Bala Hill*, was drawn blank for ice-grooves, but a polished grooved block of porphyry was found in a field near the top.

Further north, at about 900 feet above the sea, at the foot of a trap-cliff above the *Loch of Lundy*, is a long deep narrow strath which crosses the range diagonally. Through this groove distant hills about Glenartney are seen in one direction, and in the other the coast is clear to Scandinavia. At this level it would be clear to Galway also. At this spot is a bare rock-surface about 20 yards square, much weathered but deeply furrowed in the direction of the glen, N.E. by E. A steep slope of grass-grown talus 32° and 40° leads to the top of the cliff, 1150 feet, and from this point the hills of central Scotland are well seen on a clear day. Ben Ledi, Ben Vorlich, Ben Mor, Ben Lawers, Schiehalion, the Cairngorm range, and the Braes of Angus, are all seen beyond Strathmore, with its winding rivers and rich corn-land. The Fife Lomonds and the Ochils are seen beyond the Firth of Tay. On the top of *Lundy Hill*, near the edge of the cliff, the rocks are manifestly ice-ground but weathered. Near a new wire-fence, a surface newly laid bare is better preserved, and grooves on it point S. 75° E. out to sea at Denmark and Sweden. Other weathered marks seem to point E. and W. and others N.E.; but without a spade to remove the turf, fresh surfaces are hard to find. None of these high marks point directly across Strathmore at central Scotland, but they point along the Sidlaw range, and the glens in it, and join in with the line marked out by railways. Looking towards central Scotland, it is seen to be a rounded block ⌒, with conical mountains Λ rising above it. It is well named Driom Albain, the back of Scotland.

At about 900 feet, on an isolated top near a keeper's house, at a place called *Wart Well*, about four miles south of Lundy Hill, striæ on a trap surface freshly bared by the fall of a tree point N. 60° E. out to sea. These marks are nearly parallel to the general run of the tides in the Firth of Tay.

Thus, from about 1300 feet down to about 900, high grooves coincide generally with the probable run of the tides, if the sea were at these levels. At 1300 feet the Sidlaw Hills would be rocks awash, like the Bell Rock; at 900 feet they would be a straggling group of trap islands, some with caps of sandstone. At 800 feet the islands would be joined by narrow ridges. At 800 feet Denmark would be under water, and Sweden awash at places to which some grooves point.

The drift is generally below the 900 feet level. It is

foreign to the Sidlaw range, and glacial. It did not cross Strathmore, and come from central Scotland, because high ice-grooves do not point that way.

The question is: Whence did it come? and the grooves all point eastwards to Scandinavia, as similar grooves did in East Lothian. At lower levels on *the Hill of Dron*, at four stations about 850, 700, 650, and 650 feet high, and three miles apart, well-marked grooves on trap point up into glens which at 800 feet would be bays. These point N. 67° W., N. 78° W., N. 65° W., N. 65° W., round the hill-shoulder into the shelter; they point eastwards out to sea over the Firth of Tay, at Sweden and the Baltic. The flood-tide now makes a similar curve round a point close above Dundee, and the ebb returns by the same path.

It seems then that ice drifted over the Sidlaw Hills when their tops were, like the Bell Rock, awash, and that it came from the eastwards and northwards, passing along the Forfarshire hills, and grounding on Lundy Hill and the Giant's Hill at 1100 and 1300 feet.

2d, That the stream split on the Sidlaw range when the land rose, flowed down Strathmore to the Clyde, and wound about in straits amongst the Sidlaw islands, grounding floats on the Hill of Dron, at 900 feet.

3d, When that hill-top rose the stream curled round it in the lee, beside the keeper's house, and flowed up into the glens, as the tide now does at a lower level after passing Dundee.

4th, Whatever the stream did after that, there seem to have been no land-glaciers strong enough to remove the glacial drift which is piled in the glens as high as 900 feet.

5th, When ice had done its work it vanished, and streams of water sorted the upper part of the rubbish. Rossie means promontory, and Rossie church stands on a promontory of drift, at about 200 feet above the sea. The sides have the slope of rubbish-heaps sorted in water, and the materials are water-washed glacial drift. The stones were gathered at home and abroad, and piled in the mouth of the glen on whose sides are the ice-marks above mentioned.

When the cold period ended the bay in the hills probably sent a rapid ebb-tide through the glen beneath the Hill of Dron, where the burn is now cutting into the point of drift. On the point stands a cross so old that even the race who carved the sandstone are forgotten ; yet the ice-sculptures on the hill-side are fresher than the quaint figures on the cross.

The rich clay-land of the Carse of Gowrie seems to be fine glacial drift and soil washed out of coarser drift by rivers and tides, and evenly spread over rough piles of coarser drift, gravel, and big stones, which are hidden under clay and mould. The sand is washed further down about Buttonness and St. Andrews. The rock marked by ice is under the drift, and shows wherever the covering is moved.

So when the Carse of Gowrie looks "like a pond," and the Sidlaw Hills are islands in a sea of mist, this part of Scotland puts on an old winter dress for the time. When the sun shines on it a fairer landscape would be hard to find than the plains and hills which lie "atween St. Johnstone's and Bonnie Dundee."

Ice-marks then here give evidence of a rise in the land equal to 1300 feet, sufficient to account for great changes in climate, and in the course of ocean-currents.

At 500 feet a stream might flow where railways now point out the lowest ground, south-westward from Aberdeen through Strathmore, past Perth and Dunblane, to Greenock on the Firth of Clyde ; thence over Bute, past Arran, where ice-marks at 1000 and less than 500 feet point along Ceantire; thence to Belfast Lough, Galway, and Connemara.

The ice-track then has been followed from Galway to North Berwick, and to the Sidlaw Hills, and it points thence to Scandinavia, where the curves are carried into the Baltic by ice-marks, at levels higher and lower than the Hill of Lundy and the Hill of Dron, 1150 and 650 feet. At higher levels the curves must be sought on higher Scotch hills.

A fire-mark under water-marks and ice-marks in Scotland.

FIG. 78. GRANITE VEINS IN SHATTERED BEDS OF ALTERED SLATE RAILWAY CUTTING AT DALWHINNY (p. 121).

Drawn from nature on the block. Reversed.

CHAPTER XXXVI.

BALTIC CURRENT 9—BRITISH ISLES 8—SCOTLAND 5—NEWPORT LINE—CENTRAL SCOTLAND.

THE next cast is northwards to seek the Newport curve on the ridge of central Scotland.

Central Highlands.—A new mountain railway leads from Perth through the central Highlands along the line of the old Highland road. It follows and crosses a number of theoretical curves of movement shown on the map (vol. i. p. 232).

It first runs up the valley of the Tay, leaving Strathmore at Logierait.

Here a groove leads from Aberdeen along the foot of the Forfarshire hills to the west coast by way of Loch Tay, south of Schiehalion, through Glendochart to Loch Fyne.

The bottom of this groove is filled with lakes and flat alluvial plains, through which noble rivers wind. The sides are ice-ground hills, with terraces of drift along their flanks, and piles of drift opposite to each cross glen which joins the main line.

Before Scotland lifted her back, at the sea-level indicated by high grooves on Beinn Bhreac, near Inverary, and on the Sidlaw Hills, this was a strait; and according to the marks above described, ice then moved in this groove south-westwards to Tarbert in Ceantire, and the Giant's Causeway in Ireland.

Main roads follow low grounds across Scotland, and coaches and streams of tourists have succeeded ocean-currents, icebergs, and boulders; but before the flood of travellers poured into these glens, a tribe of land-glaciers perched upon the Highland hills, and slid down from the high mountains into long sea-lochs. At some sea-level this ice thoroughfare was barred by a col about the braes of Balquhidder, and thenceforth ice must have moved north-east along the course now followed by the Tay and its feeders.

But Scotch ice, grown in Balquhidder, and launched about Dundee, might still sail to Ireland through the deeper channel of the Galway curve, and join a Glenfalloch iceberg launched at Dumbarton, off Arran in the Firth of Clyde.

The railway follows a branch of the Tay to the Pass of Killiecrankie, and there, at the 600 feet level, was a sea-loch. Many of the railway cuttings are through drift, many embankments are piles of drift. In the autumn of 1863 great boulders, freshly dug from the hill-side, were scattered along the whole line. Low down, where rock-surfaces were newly uncovered, they retained their polish. High up on the skyline the hill-tops are rounded, and smooth wet rocks shine like convex mirrors amongst the grass and heather.

At Killiecrankie a second series of glens leads southwestward to the west coast, passing north of Schiehalion, by way of Rannoch and the Forest of Glenorchy to Loch Awe, where marks at 1650 feet point at these glens.

At Struan, north of Blair-Athol, the railway has passed the 600 feet level, and here is a conspicuous moraine of which a cutting gives a section.

From this point the way rises over a col to the end of Loch Garry, 1330 feet. The rocks there are ice-ground and the soil is glacial drift. Here a third set of glens lead from Driom Uachdar, the upper ridge of Scotland, and the Cairngorm range, south-westward by way of Loch Lyddoch to Loch Awe and Beinn Bhreac, where ice-marks at 1650 feet pointed N.E. by E. With the sea at perched blocks on Beinn Bhreac stones might sail upon ice from Loch Garry to Argyllshire hills. So the perched blocks on Beinn Bhreac may have come from Cairngorm, or the hill of the black pig, which Saxons call Ben Macdui.

At 1480 feet (1620 by barometer), *the watershed* is passed, and the level of perched blocks on Beinn Bhreac is 1650, or 170 feet to spare.

Water now runs north-eastward to Speymouth, and as soon as this col dried, land-ice must have slid the same way that water flows.

At this high level in central Scotland hill-tops are rounded and rocks ice-ground. Here are large piles of glacial drift, apparently the moraines of glaciers which slid down small glens on the western side of the railway. The hillocks are 200 feet high at least, and their shape contrasts with that of drift hills near Dunblane.

They consist of large boulders, gravel, and sand, and amongst the boulders are many of a fine hard gray granite. These are in such abundance that they have been used to build bridges and other railway works. There are also specimens of a very heavy tough compact red porphyry, and blocks of quartz, gneiss, and altered flags of various colours. The hills are of the latter rock, which is much shattered and veined with pink granite. No gray granite is found *in situ* on this hill.

In a railway cutting opposite to one of these piles of drift, a quartz rock surface has been laid bare. It is ground very smooth, and grooves on it point N. 38° E. down into Glen Truim, and S. 38° W. up into the glen. This spot is about 1480 feet above the sea.

A little further on a second smaller glen on the same side has a smaller pile of rubbish in the opening. This glen is about six miles long and clear of drift high up.

At *Dalwhinny*, at about 1169 feet, a fourth groove is crossed. It contains Strathspey to the north-east, Loch Ericht and Loch Awe, and the Sound of Jura, to the south-west. With ice floating at 1650 feet, central Scotland would be an archipelago intersected by narrow sounds, and this was a strait 500 feet deep.

So here is the tract in which the line marked on Beinn Bhreac is to be sought. With Monadh Liath (the hoary mountain) on one side, Monadh Ruagh (the russet range) and Cairngorm (the blue cairn) on the other; an arctic current might pick up Scotch icebergs and Scotch granite boulders and carry them along the Loch Ericht trench to Inverary, Ben Bhreac, Ben Cruachan, the Jura hills, or Derry Veagh in Ireland.

At the 600 feet level all these passes would be stopped; Strathspey would be a sea-loch ending at Grantown, and boulders would have to slide down Strathspey and sail round by Inverness and the Caledonian Canal. If there were no ice-rafts, when the land rose to any particular level, the voyages of boulders ended for the time.

A particular kind of boulder, carried to a certain height, in a particular direction, marks sea-level, movement, and a cold climate, for it is a float which ice alone can carry.

On the south side of Loch Ericht is a high ridge of gritty flags and slates traversed by veins of pink granite; it is a *spur of Driom Uachdar.*

In a rock-cutting at Dalwhinny the rock is bare; on the hill-top it crops out, and it is seen in burns at other spots, many miles apart, high and low. The hill would be an island at 1650 feet. At Dalwhinny, boulders of gray granite abound. They are foreigners who travelled on ice from some other district, and to get to the end of Loch Ericht they must have moved up hill if they travelled on land-ice. If they travelled on sea-ice they mark old sea-levels, and here they mark about 1350 feet at the end of the loch.

They mark higher levels on the spur of Driom Uachdar, which divides Loch Ericht from Loch Garry.

At 2000 feet is a round block of granite.

At 2200 is another, and from this stone the sea-horizon towards Bergen is open north-eastwards beyond Speymouth. A pass lies open to Loch Leven on the west coast. At the top of the ridge was a shallow pool made by a turf washed in between two small hillocks. At the bottom of the pool was a plain of fine soft black peat mud, and fine sand washed in by rain-water. A thrust with a stick demolished the dam and drained the pool, and changed the bottom into a working model of Glen Truim and Strathspey. Knobs of peat were the hills, peat-mud the drift; tufts of grass and gray moss were the forests; the river was a tiny rill of black water. But the water set off for Speymouth, and the forms of the alluvial plains were alike. There were terraces of stratified drift; there the river-windings, the Ys and S, the banks of small stones, high patches, long points, and steep banks of drift sweeping round steeper and harder slopes. There were glens of denudation circling round hard islands which became hills as the water drained away. All these shapes formed in the moss-hole in a few minutes, and they were all formed long ago in the big glen below. The model a few yards off, and the glen stretching to the horizon, filled the same space in the eye, and seemed alike even in size. Running water has done great work amongst the glacial drifts of Strathspey, according to the shape of the country, and the lesson taught by the model.

At 2650 feet this hill-top at the head of Strathspey, and about 1000 feet higher than the col at the western end of Loch Ericht, is strewed with big stones of gneiss and pink granite. The flat is rippled by the S.W. wind. Stones are in the trough, heather in the lee, gray moss on the weather-side of these waves; and far down below, waves driven along the surface of Loch Ericht had the same shape. Even winds leave a spoor where they pass.

This is one great thoroughfare for currents in the lower atmosphere, and a whole wood of fir-trees at the inn lean down towards Strathspey, as if driven by a strong S.W. gale. The prevailing wind is then an equatorial current moving N.E.

At 2580 feet, within sight of the Cairngorm Hills, are three large boulders—one of gray granite, one of a very coarse mica-schist with large weathered veins and nodules of white quartz, and the third is a coarse sandstone grit. The lithograph on the margin of the map (vol. i. p. 496) is roughly done from a hasty sketch made here.

At the same height—six miles from the inn and close above Loch Ericht—is another boulder of gray granite beside a rock of gritty flag, traversed by pink granite and white quartz.

At 2740 feet is another round stone of the gray granite; at 2800 another three feet long; at 2850 three more about the same size;—and all these contrast strangely with flat stones amongst which they lie.

At 3150 feet is a cairn on the top of the ridge, and at this spot is a wide view over central Scotland. Strathspey is open to the sea. Then come Cairngorm and Beinn-na-Muic-Duibhe, then a hill shoulder; and beyond the opening Beinn-y-Gloe. Then comes a wide tract of lower ground open to Fife and Stirling; then the shoulder of Ben Lawers and a lot of near hills, which shut out the distance. Then a notch through which hills near Loch Tarbert in Ceantire are seen. Then a near hill; then a wide opening at the end of Loch Ericht, with Ben Cruachan rising to the clouds. Then comes the mass of Ben Alder, with patches of last year's snow, and Ben Nevis peering over it. A glen leading down to the sea, and a col of 800 feet, divide Fort-William from Strathspey in this direction. To the north, the hills about the Caledonian Canal are overlooked, and something in a cloud seemed to be Wyvis. If boulders mark a sea-level, it is here carried to 3000 feet at least.

The hills of central Scotland, up to this level and a little higher, are all rounded tops and hog-backed ridges, above which a few conical tops rise. At this level gray granite boulders mark floating ice, which might wander amongst those peaks in any direction. A man may travel on ridges ⌒ or in hollows ◡ from N.E. to S.W. without much climbing; if he travels in any other direction, he must mount and descend from glen to glen.

A puff of cold wind and a wreath of mist blotted out the whole of this wide landscape, and Scotland disappeared behind a few drops of water, as it hid under the sea when the boulder was dropped on the top of Driom Uachdar.

Fifty feet down from the cairn are more round blocks of gray granite, and they occur all the way down the burn-side to the railway, three miles south of Dalwhinny Inn.

Now 1480 feet, the summit-level of this line, would make Loch Ericht a sea-strait; and 3100, the highest granite boulder, would make the strait about 1600 feet deep at the shallowest part. So the railway bridge is built of granite quarried somewhere, and carried by ice which floated where clouds now settle, where grouse crow, and golden plover whistle and wheel in flocks. Where dun deer and mountain hares, ptarmigan, sportsmen, keepers, and wanderers now pass to and fro, amongst green moss and gray stones, ice surely floated. The railway train passes along the bottom of a strait which crossed Scotland at Dalwhinny, because transported gray granite abounds on hill-tops to the S.W. at a far higher level than the top of the pass.

Gray granite is found *in situ* to the N.E. at higher levels.

Opposite to the end of *Loch Ericht* the drift seems to be arranged by water. A small proportion of the large stones retain scratches. They generally have water-worn or weathered surfaces. From hill-sides to the north these rubbish-heaps are seen to be terraced layers resting upon the solid rock, and sweeping down into the wide strath in points and knolls rising one above the other, like drift-terraces in Norway and Sweden, though on a smaller scale. They are the contour-lines of the country following the hollowed surface on which they rest, up to a certain line, beyond which are solitary boulders on bare rock or in heather.

It is very hard to represent these forms truly with a pencil. For that reason no woodcut is given of sketches done on the spot. The place is easy to get at and the forms are distinct. In nature they are marked out by colour, light, and shade, rather than form; and on a dull day they are lost in the distance; but when the sun shines they come out clearly. Any one who knows the Highlands knows the aspect of these dry heathery gravel hills, on which grouse delight to strut and shout their defiant chorus of "Go back, Go back, Go back, Cock Cur-r-r-r! They are "the parallel roads" of a great many Highland glens besides Glen Roy. They are the "ancient sea-margins" of Chambers, and here they rise to nearly 1400 feet. In the middle of Loch Ericht (see map, vol. i. p. 496) are two bars, similar in shape to bars which cross tideways in narrow straits; as at Roseneath, near Greenock; in Alten Fjord, in Norway; at Portland, in the south of England, etc. etc.

The ridge *north of Loch Ericht* would be an island at 1400 feet, cut off from another lower ridge about 2000 feet high by a deep glen. In the glen was a glacier. A rock-surface has been laid bare by a torrent which has washed away part of a terrace of drift; enough of gray granite to make a railway bridge is strewed below. The rock is a hard fine dark quartz with beds dipping W.N.W. 26°. Grooves on their edges are horizontal, and point east into Glen Truim. The terrace of drift is 100 feet thick at least. On the opposite side of the glen, the burn has dug into the rock, exposing a set of nearly vertical strata. This, then, is a fault; a rift which ice found and smoothed and filled with glacial drift. Lower down the hummocks of a moraine are piled in rows opposite to the glen; but 600 feet higher up, on the bare hill-top, are perched blocks of gray granite, keeping watch over Strathspey and Loch Laggan. At their level, and 600 feet lower, the high ridge north of Loch Ericht would be another long island.

At *Kingusie* another groove with a col only 800 feet high, according to late measurements, runs S.W. to Fort-William, down Glen Spean. The N.E. corner of the island beyond the fault, and opposite to Laggan Inn, is a gray granite, but not the granite of the boulders. The tops are bare and weathered, have the usual rounded form, but retain no small marks. There are many perched blocks of compact gray granite on the highest points, about 2000 feet above the sea. According to these marks the famous "parallel roads" were under water and rose, and if so they do but resemble terraces elsewhere. (See chaps. xxii.-xxvii., etc.)

While basking in the sun in the lee of one of these stones, far away from any visible sign of man, how strange it is to hear the yell of a steam-engine, and then to watch a streak skimming like a silver eel, or the mythical white dragon, through this wide strath, where an icy sea has ebbed and flowed. It is no wonder that natives stare agape, and that sheep scamper for their lives, when this fiery steam-dragon comes yelling and roaring through deer-forests where lurking stalkers used to speak in whispers.

Strathspey has seen many changes since it was hollowed out of the rock.

And this is the popular account of the matter got from a countryman of Hugh Miller, who was also a fellow craftsman of the Scotch geologist :—

"Where do you get that granite?"

"Oo, they fand a wheen o't lyin' i' the grund, eneuch to build a hail toon."

"Is there a quarry?"

"Na, there's nae quarry onyway here, jeest muckle stanes."

"What kind of rock is there here?"

"Jeest a bastard kind o' a stane."

"Well, but where did the granite stones come from?"

"Hoots, they just grew whar they lie."

Chip, chip, chip, and a look of puzzlement.

With a rising land and a rising temperature, with glaciers shrinking and melting in these Highland glens, moraine after moraine would be dropped in Strathspey, for the river, the road and the railway engineer to dig through. The last stone would be stranded high up on some lofty hill-side. In fact, the Spey winds through a flat plain of rounded stones, and the railway cuts through piles which seem to be lateral moraines re-arranged by water, while perched blocks are stranded high up on hill-sides which bound this large groove.

When this district was the birthplace of glaciers, it gave rise to those which flowed from Driom Uachdar into Glen Truim, and to six which flowed from Cairngorm and Beinn-na-Muic-Duibhe, along the valleys of the Dee, Don, Doveran, Avon, Spey, and Tummel; and each of these must have left tracks, because in Glen Truim and Strathspey they are conspicuous.*

Frothy spots of blood on heather, water oozing into the footprints of a deer, do not point out the track of a wounded stag more surely, than moraines in Strathspey map out the backward course of melting glaciers. But the low moraines are all washed out of shape.

At *Boat of Insh* station, 765 feet, the fresh wound of a new railway cutting bares the flesh of the country and its worn bones.

At the fork of two glens, glacial rubbish, sand, gravel, and great boulders, are piled as moraines are piled in beds and layers, which dip and curve all ways, and rest upon each other where they were washed off the glacier or iceberg. Beneath these rubbish-heaps are ground rocks, and behind the old moraine a shallow loch nestles in a hollow.

At *Aviemore*, 692 feet (700 by observation), the drift is flat and terraced, as it is elsewhere, at this level. When the moraine was whole there was a larger lake behind the dam, in the flat country which fills the glen higher up.

The grand hills whence this drift may have come tower up-

* Glacial phenomena about Balmoral have been described by an able local geologist. They seem to prove the existence of land-glaciers on the side of Strathmore, etc.

wards to the mist, with sun and shower, light and shade, and glorious colours of purple and gold, playing on their furrowed sides. The works of ice in the plain are now arrayed in forests of yellow birch and dark-green pine ; but whoever has seen ice at work must know these tool-marks and these chips. On an autumn day, a single snow-patch gleaming through a cloud is enough to call up a vision of the Alps, the Folge Fond, or the great ice-floods which hem in Sprengisandr in Iceland. But the sea-level of the mental landscape rises on the hill flanks.

At Grantown, 731 feet (800 feet up on the hill-side, by observation), the new line leaves Strathspey and crosses a ridge 1000 feet high to the Moray Firth.

It cuts through hills of glacial drift which rest on contorted ice-ground slates, and other rocks. Woods glowing with rich autumnal tints; purple heather, yellow corn, and blue hills, far away beyond the rich strath; the warm rosy colours of a Scotch moor lit up by the sun—contrast strangely with the cold gray desolation of the picture which ice-marks recal so vividly. And yet these Scotch landscapes were like the hills of Iceland, and the weather and the river Spey have done little to alter the land since ice and sea left it bare for plants to clothe.

In descending from the ridge to the sea-level, the whole character of this country changes. Glens and wide straths, moraines, and other marks of river-glaciers, are left in the Spey-groove.

The train approaches a north-eastern corner, and it is like others in the British Isles. Seen from Wyvis, it has a regular slope ⌒. If land-ice grew here, it slid north-west into the Moray Firth, in a wide sheet like that which covers parts of Iceland at Ball Jökull, Lang Jökull, etc. (chap. xxv.) The whole of the Morayshire side of the Firth is one ridge from 1000 feet to the sea-level, from the Spey to Inverness. Above that level, a few Λ hills—such as the Knock of Brae-Moray—rise, but they are exceptions. The soil is still drift; but the coating of loose debris is more evenly and thinly spread, and more regularly packed. Layers of sand and gravel are sorted, sized, and generally laid flat one upon the other above the sandstone rock. The Findhorn, and other rivers, have cut deep gashes in this rock. If land-ice had moved in the same direction, it would surely have dug grooves ⌣.

At Rafford station, 169 feet above the sea, drift is arranged in knolls and mounds, and layers dip many ways. Most of the stones look washed and rolled, and large boulders are rare. At Forres, the flat plains of Morayshire are only 26 feet above the sea; and thence to Inverness the whole of the low country bears marks of water-work. But it was not water-work done by shallow unfrozen seas, for the beach at Inverness and the shores of Scotland are not arranged like the hummocky drift-hills and points which rise up in this low tract. Drift-ice might do work of the kind; and plenty of glaciers to make icebergs grew between Perth and Inverness in central Scotland, and on the opposite coast in Norway.

The evidence in this tract seems to prove that *central Scotland* was crossed by narrow sounds, through which ice-floats drifted, as they now do through the straits of Belleisle; that the land rose gradually; and that glaciers on shore have not been lower than the two moraines near Dalwhinny, since the sea packed terraces about the end of these moraines.

If after the land had risen to this level (about 1400 feet), central Scotland was an island with a sound passing westward at Stirling, another sound passed westward at Inverness, and ice-grooves at 1100 feet near Derry Veagh in Ireland pointed in this direction, as shown above (p. 57).

The Galway and Westport curves have both been carried over Scotland; the spooring must go northwards again, if the Glenveagh marks are to be found on the Scotch mainland.

INVERNESS AND PERTH JUNCTION RAILWAY.

List of Stations, showing their respective Heights above the Sea-level, High-water Mark, ordinary Spring-tides (rising 14 feet at Inverness.)

	Feet.		Feet.
Forres . . .	26	Newtonmore . .	764
Rafford . . .	169	Dalwhinny . .	1169
Dunphail . .	614	Summit of Drumochter .	1480
Foot of Knock of Brae Moray,		Loch Garry . .	1330
about . . .	1000	Struan . . .	615
Grantown . .	731	Blair Athole . .	421
Broomhill . .	656	Pitlochry . .	334
Boat of Garten . .	706	Ballinluig . .	202
Aviemore . .	692	Guay . . .	186
Boat of Insh . .	765	Dalguise . .	179
Kingussie . .	740	Dunkeld .	212 ft. 4 in.

The heights estimated by the pocket aneroid barometer agreed pretty well with these heights, which were kindly furnished by a director of this railway.

CHAPTER XXXVII.

BALTIC CURRENT 10—BRITISH ISLES 9—SCOTLAND 6—DERRY VEAGH CURVE—CALEDONIAN CANAL AND NORTHERN SCOTLAND.

INVERNESS stands at the north-eastern end of a large groove which crosses Scotland. At 100 feet level the glen which now holds the Caledonian Canal would be a sea-strait; at the 500 feet level it would be a deep narrow strait through which a rapid tide would flow, like that which now boils and seethes through Kyle Akin, between Skye and the mainland. North of Inverness the rocks are a coarse conglomerate. Up to 400 feet great banks of sand, shingle, and large stones, are confusedly piled on the hill-side. This drift contains stones of many sorts and sizes, granites of various colours, and hard igneous rocks, mica-schists, and various kinds of quartz. They have the shape of stones in glacial drift, but the surface of waterworn stones. They look like stones on the beach near Galway, which have been rolled by sea-waves after falling out of the clay bank, in which similar stones retain their grooved surface (p. 20). This seems to be water-worn glacial drift at the end of the old strait. The plain below is of like materials, spread out and laid flat, and a conical pile of loose stones is left in the middle like the mounds which workmen leave in a cutting to mark the original level of the surface from which they have dug. At the head of many a Scotch glen, at about 600 or 700 feet, a like plain of rolled drift remains. If rapid tides ebbed and flowed over Inverness, they would dig away Tom-na-Shirich, and the rest of the drift ; but a watershed 100 feet high stops the tide, and the Ness can do little in such heavy ground. Wherever they came from, these mounds of large stones were carried, and they are piled upon ice-ground rocks. The hills have the usual shape, and enormous fragments of conglomerate have been moved and dropped where they stand, amongst heather and trees, 800 feet up, clear of the terraces of rolled drift.

In Geikie's map, lines are marked about the watershed of this groove. The whole country is glaciated ; and it is manifest that ice can only have moved N.E. or S.W. along this deep groove, whether it was land-ice or sea-ice.

The next great groove which crosses Scotland from N.E. to S.W., runs from the Dornoch Firth to Loch Carron.

The intervening district is a large block of high land, deeply furrowed by glens. On the eastern side, the northern shore of the Moray Firth is low land in the Black Isle of Cromarty, and this district is thickly strewed with drift. It seems to be glacial and waterworn.

Beyond the Black Isle is the Firth of Cromarty, which ends at Dingwall, below Beinn Uaish or Wyvis, which is a great block of high ground, with a rolling plateau on the top.

Beyond the Cromarty Firth is a long low tract of drift, which ends eastward at Tarbert Ness, and beyond that is the Firth of Dornoch.

Lines of existing and projected railways mark the division between hill and plain from Inverness to Dornoch.

From the Firth of Forth to Duncansby Head, the map of the eastern coast is like the teeth of a blunted saw. The lines run alternately westward and south-westward, and hills inland correspond to the coast-line. Railway lines, in like manner, run westward and south-westward in pursuit of low levels. Roads which follow low levels cross this district in similar directions. Beyond Dornoch, the low coast-land becomes a narrow strip in Sutherland, which comes to an end at the Ord of Caithness, where the sea washes a line of eastern cliffs.

The hills now trend northward to Thurso, and westward to Cape Wrath ; and Caithness is flat land, with a soil of drift.

If the north-eastern corners of Caithness and Berwickshire were not blunted teeth, St. Abb's Head, Kinnaird's Head, and Duncansby Head, would be points of land of the same pattern as Tarbert Ness and Fife Ness. The whole east coast is a repetition of the same pattern on different scales, and it is repeated in miniature in every firth where the tides are wearing the coast. It seems fair to conclude that the shape of the Scotch coast results from the wearing action of water-streams, which flow on a fixed principle, and in certain directions. Here the points aim N.E. and the bays S.W.

In the northern division there are glens to correspond to notches in the coast-line, and glens which are prolongations of bays. Deep grooves run up westward at Glengarry, Glenmoriston, Strathaffaric, Lovat's Forest, and Strath Conan ; and, after passing the watershed, glens run westward down to the coast about the Sound of Sleat, in Knoydart, Glenelg, Loch Alsh, Kintail, etc.

Further north glens in Sutherland turn north-westwards, and on the eastern coast they curve north. No map of Scotland gives the true shape of these hills and glens. Black's road and railway map gives some of the main features, and it shows that the main hollows and passes which cross Scotland all converge upon the Næs of Norway and the Skagerrak. Any geological map will show that these forms of denudation bear no reference to the geology of Scotland. The grooves have nothing to do with dip, or strike, or subterranean disturbance. Most of these Scotch glens are tool-marks of some denuding engine, and the study of their shape is a part of "superficial geology." Conspicuous ice-marks are in all these glens, and in all their branches, so far as they are known to the writer. They all seem to have held river-glaciers of large size, which followed the present run of water from the watershed to the low land.

With the sea at the 1000 feet level, this tract would be crossed by sounds, and the main coast-lines would generally trend N.E., E. by N., or thereby, as coasts and sounds do in the Hebrides, at the present level of sea and land.

At 1500 feet there would be ample room for the tide to flow over the low land of Sleat, through Loch Carron and Strath Bran north of Wyvis, and so along the Sutherland coast to the Ord of Caithness. The ebb and a north-eastern arctic current might flow the other way along the same path as the flood-tide and the Gulf Stream now flow together outside of the Hebrides northwards, and the marks should remain.

The most likely place for sea-marks is on the watershed in passes. Drift accumulates in shallow sounds; and low tracts in the Scotch and Scandinavian islands, which join high hills, are generally composed of terraced drift with recent shells. If the backbone of Scotland rose from the sea, the watershed of each glen would be first a shallow sound, and then a "tarbert," with raised sea-margins. But if the rise were gradual and general in Scotland, passes would dry in their order of height; so the highest terrace is the oldest.

The col at Dalwhinny is at 1480 feet; so, on this supposition, it was dry when the Forest of Gairloch was an island, and Strath Bran a strait 850 feet deep about Achnasheen. There the barometer marks 630 feet at an ancient sea-margin.

When there was a tarbert at the head of Glen Dochart, where the barometer marks 800, there was still a strait 680 feet deep at Glengarry on the Caledonian Canal, and there was deep water above Lanarkshire, where sea-shells have been found in drift at Airdrie. When the sea was at "Drumochter," the Parallel Roads of Glenroy, about which so much has been written, were sunk 324 feet; for the highest of that series is only 1156 feet above the sea.*

The ancient sea-margins of the British Isles have been examined and described by Robert Chambers, and they lead to the conclusion that the last rise was general, for terraces of shingle are found at corresponding levels at many distant points in Britain. A terrace of stratified gravel is a sea-mark which could not resist a land-glacier; it would be swept away by the force which sweeps moraines before it, and grinds solid rocks; it is therefore a kind of thermometer, and it is easily distinguished from glacial drift.

Where a terrace is found resting on glacial drift, beneath which rocks are marked by ice, there is a series of records.

1. Ice ground the solid rocks and made the marks.
2. Ice dropped the great stones which floated on it, and which now rest upon the marked rock.
3. Water packed loose gravel in horizontal layers upon the moraines or drift.
4. Streams cut through the terraces, washed the gravel, and arranged the mud in hollows lower down.

These records, then, give relative dates for the last glacial period, and elevation of land.

There has been no land-glacier at the place where a terrace of stratified gravel remains, since the terrace was arranged by water upon glacial drift. There has been no glacier since

* Antiquity of Man, p. 253.

the moraine was stranded in the glen. So the highest terrace of sea-gravel marks a sea-level at which the land stood after glaciers had disappeared, and the highest Scotch terraces of washed drift known to the writer are at Dalwhinny, 1169 feet, in Loch Ericht (?), and near the summit level of the new railway, which is at 1480 feet.

Assuming that this argument is well founded, the record in Strath Bran proves that the water-level has been at 700 feet since the Scotch hills were clear of ice, and that there have been no large glaciers since that time in Strath Bran.

For the same reason, because the rubbish at Dalwhinny is terraced, there has been no land-glacier in Glen Truim since the water-level was at 1400 feet; but there were land-glaciers as low as 1600 feet near Dalwhinny, and their moraines have not been washed out of shape.

But if so, and if the rise of land was general in Western Europe, then the end of the glacial period coincided in level with the rise of the low isthmus which now joins Scandinavia to Russia, 1400 feet, and the last cold period in Scotland coincided with the level which allowed the Arctic Current to flow down the Gulf of Bothnia (see map, vol. i. p. 232).

Horizontal ice-marks on hill-sides and tops, and on watersheds in passes above 1400 feet, were probably made by floating ice, at a time when only the highest Scotch hills were above the sea and smothered in ice.

The nature and direction of ice-marks at high levels is the foundation on which this theory rests; and the shape of hills of drift is another stone on the cairn.

One of the most beautiful of all the Scotch lochs is Loch Maree in Wester Ross. It lies in a deep trench which runs north-west along the foot of a block of high land, which makes the Forest of Gairloch. To the north are lofty hills—

Slioch, Beinn-araidh-char, and others—which rise to nearly 4000 feet. In the loch are rocky islands on which natural woods of Scotch fir still survive; and in deep glens and corries which furrow the hill-sides, gaunt trees toss their twisted arms, like the last giants of a departed race. On a still morning when the eastern sun peeps over the hills and under the mist, it sends a flood of yellow light and heat streaming westwards, into the level glen at the head of Loch Maree. Blue peat-reek, which before sunrise followed the run of the stream down every hollow, turns to a golden haze, and it eddies and curls upwards as the air answers the sun-power and rises. East and west, north and south, the smoke of scattered farms sweeps towards the spot where the light falls and warms the ground, and the chill breath of the hills comes down the hill-sides like a stream of cold water. Heat and cold stir the air, and the smoke and the sunlight show the currents which a ray of sunlight sets in motion. On such a morning the hills are like great cones of lapis lazzuli set in glens of gold and lakes of quicksilver. As the day wears on the mists rise up and creep slowly round the highest peaks, till they rise upwards and float away in shining clouds. Then the blue cones change; bare white quartz glitters in the sun like snow, and Ben Eith looks as if it were "ice" in truth.

To a height of about 2000 feet these hills are ice-ground. It needs but a glance to know the shape, but here all marks are clear and distinct.

At the bottom of the glen, at Kinloch Ewe, at 200 feet, ice-grooves run towards Loch Maree, N. 30° W. These might be marks of a local glacier.

Thence, for 700 feet up the western side, the rock is broken. At 900 feet glaciation begins. At 1100 feet, at the edge of the glen on the west side, a large hollow groove three feet wide, and as smooth as polished marble, contains striæ of all sizes, down to fine sand-marks. They point a little more to the west, N. 40° W. At a higher level than the watershed of the glen, which is also the watershed of Scotland, and 800 feet high at Glen Dochart, a tract begins which is not easily matched. The rock is a very hard stratified quartz—gray yellow, white, and pale pink—and for several square miles the rock is bare. It is weathered in some places, and there fossils rise up half an inch from the surface. The stone looks like a sugared cake, with chips of almonds stuck into it. Other beds are weathered into a pattern of round flat lumps, like small ivory shirt-buttons laid close ; others have larger shapes ; concentric rings an inch across, which wear away, leaving concentric ridges and hollows. But the greater part of this rock is either freshly broken, or ground perfectly smooth. At 1350 feet, on the top of a ridge high enough to clear most of the cols which join Scotch hills, and close to the foot of Beinn-a-Ghuis, the marks are perfect. They point N. 20° W.

In that direction they aim over lower hills about the river Ewe, twenty miles away, and over the sea outside of the Butt of Lewes ; in the other direction they aim over the head of Glen Dochart (800 feet), over Strath Bran at a big hill supposed to be Sgur-a-Mhulin, but found to be further south. There is no apparent source for land-ice within reach of this spot, except the high peaks beside it, and the grooves aim past these hills, which are some of the highest in Scotland.

They were not made by land-ice.

At the same level, 1350 feet, a mile nearer to the foot of these hills, and opposite to a glen which seems made to be the home of a glacier, the grooves point N. 56° W., and here is a tiny moraine, still perfect in shape. It is bare and looks like piles of broken white sugar poured out across the glen. Here, near the level of moraines near Dalwhinny, a similar form tells the same tale. The sea has not been here since the glaciers melted. At 1800 feet, close to the foot of Beinn-a-Ghuis, the marks point N. 25° W. The sea must have been here when the marks were made. So the glacial period seems to have ended when the sea was at the terminal moraines on the side of Beinn-a-Ghuis at about 1400 feet, and on the side of Driom Uachdar at about 1400 feet also.

At still greater heights the rocks have the same ground shape (see cut, p. 17, and map, vol. i. p. 496), but time would not admit of a closer examination.

It seems to be proved by marks on hills on one side of Loch Maree, that ice crossed Scotland from the east to the west at a level of more than 2000 feet. Above that line the Gairloch hills seem to be conical piles of broken quartz talus leaning against jagged cliffs and peaks. The shape is ⌒ up to one level, Λ above it.

If a stream came from the eastward and split on these high hills it would sweep off north-westwards, as ice did according to these marks.

There can be no doubt of the direction. For 100 yards in length, and 20 in breadth, one great waving sheet of white quartz is smoothed and grooved on one side, and fractured on the other, and for several miles rock-surfaces of the same kind abound. A few blocks of dark trap are scattered about at this level, but on this exposed shoulder there are few perched blocks. Looking inland from the Gairloch Forest, an open gap in the hills about Loch Fannich bears E. by N., and there is nothing in that direction to stop ice floating at 1800 feet.

Looking through that gap the first land of equal height is

in Scandinavia ; so this path, too, is clear, for in Scandinavia there are grooves on the watershed which point N.E. at about 2000 feet above the sea near Trondhjem (see vol. i. pp. 103, 234).

The next point on this line is on the opposite side of the glen, where a ridge 2100 feet high is cut off from all neighbouring hills by deep glens. It is cut off from Slioch by Glen Bianastle ; from the Forest by Kinloch Ewe ; and a wide deep strath divides it from Ben Dearg to the north-east. It is called *Beinn Mhonaidh.*

If a stream at this level came from the east by way of Fannich it would split on the side of Slioch, which is about 4000 feet high, and run foul of the place last described.

In the bottom of the glen at *Kinloch Ewe* drift is arranged in flat terraces up to the 300 feet level. The river is digging into these banks, and it is building a new set in the loch three miles down. This is stratified water-work done since the ice disappeared. But the gravel banks rest in an ice-groove, for the marks show as soon as the drift is cleared.

At the 1000 feet level the hill-top is above the level of the col at Glen Dochart, which would make Strath Bran and Loch Fannich sea-straits.

At 1200 feet the groove which holds Loch Maree is seen to be a short transverse rut, for the big groove which runs from sea to sea E. by N. is open between Beinn More and Fin Beinn. A few large perched blocks of gneiss are scattered on the tops at this level, and the wide hollow and the shape of hills and knolls in it, all indicate movement from the east towards the high hills beyond Loch Maree.

At 1200 feet some weathered grooves on gneiss point E. by N. The rocks are much weathered, but their shape is clear. At 1620 feet is a perched block 9 × 9 × 9 feet, and many smaller angular blocks of veined gneiss and granite are balanced upon rounded knobs of gneiss near a small tarn.

At 2150, on the top of the ridge, are perched blocks and grooves pointing N. 65° E. These are almost obliterated, but they can be made out.

From this point the opposite quartz hills are well seen.

Unless central Scotland was one vast snow-dome, there is no possible source from which land-ice could reach this spot. Deep glens surround Beinn Mhonaidh, and the shortest way to sea from the hills at which the grooves point is behind Slioch, three or four miles away, and 1500 feet lower down, where the water runs. At the same level, and a little higher, the very same kind of rock-surface, and the very same pattern of smooth hills, are seen in every direction; but a little above this 2000 feet level, hill-tops are jagged, conical, weathered, fantastic peaks, fit rivals to the Lofoten hills, which have been likened to the teeth of a shark.

On an autumn day when the air is clear, a grander scene is not to be found in all Scotland.

When yellow lights, purple shadows, and showers are chasing each other from hill to hill, rainbows and windgalls, bright clouds and blue sky, make this wild tract a scene of wondrous beauty. It is a picture to look at and remember. But it is easy to map out the glaciers from other pictures stored in the same memory. Through a gap in the hills is the way to Bergen. There stand peaks of the pattern of Bodals Kaabe and Åreskutan; below is a long rounded swell like the Norwegian Fjeld. Deep down from the rift of Glen Bianastle comes the distant hushing sound of a mountain-torrent. It is in the path which ice must have followed if it came from Scandinavia through Glen Fannich, and ran foul of Slioch. It is easy to fill in the whites in this picture, and it is easy to test its truth when finished.

At the head of *Glen Bianastle,* at 1450 feet, the rock is the same quartz which makes the opposite hill-tops in the forest. The beds dip the same way, and some are weathered and some polished. At the very edge of the cliff a set of perfect grooves point from N. 65° E. to S. 65° W. over Loch Maree.

At the same level, thirty yards off, similar grooves on gray quartz point N. 60° E.

In the glen below the cliff at 1200 feet the marks are quite perfect. Long white ridges and grooves are "for all the world like a marble chimney-piece," as an astonished native of Dingwall remarked. Striæ point from N. 50° E.

From this point down to Loch Maree are similar marks wherever the bed of quartz is the surface.

But at the bottom of the glen a bed of sandstone is smoothed by water in the burns, and on the side of Slioch, where strata nearly vertical meet the edge of the sandstone beds, the hill-side is deeply furrowed by rain. These ruts aim at the peak, the others run horizontally past the hill.

The burn has cut a rock-trench twenty or thirty feet deep, but though all this weathering has taken place, many quartz surfaces have not lost the thickness of a sheet of paper since ice left them bare.

At 700 feet is a bed of flat drift apparently arranged by water amongst old moraine stuff.

At 700 feet the rock is bare, and marks point at right angles to the shore of the lake. Here a quartz cliff about 1000 feet high is ice-ground to the top, and the opposite hills, ground to the level of 2000 feet, tower up beyond the lake. At 150 feet the shore of Loch Maree is a river-delta forming on a moraine, which has lost the characteristic shape, and the lake as usual is said to have no bottom. It is very deep and a true rock-basin, for the Ewe escapes through a channel of rock.

So, looking on these great hills as stones in a stream, ice-marks at the high level indicate a current flowing through sounds, and splitting upon blocks of high land as streams do on posts; the floats must have been ice of large dimensions, but not necessarily larger than drift-ice, in the same latitude.

The plan laid down at the beginning was to follow ice-marks wherever they might lead. Marks on the top of Beinn Mhonaidh pointed at quartz hills on the opposite side of Loch Maree, and they were followed. Marks at the head of Glen Bianastle led down to the shore of Loch Maree, marks at the bottom of the glen pointed down the stream; on the shoulder of Ben-a-Ghuis, opposite to Beinn Mhonaidh, at about 1800 feet, the arrow (see cut, p. 17), carried 55 miles, to the visible horizon of the highest spot, aimed about Stornoway in Lewes. The ice-lines were found to wind about the hills, and finally aim over two blocks of isolated hills 15 or 20 miles off. This spoor has been followed, and it is very plain on these distant hills.

The *Hill of Groban*, over which the arrow passes in the woodcut, is between the post-road to Gairloch and the shore of Loch Maree. The highest knob of the central eminence in the midst of this group of small hills is about 1200 feet high. It is all ice-ground, but weathered. On the S.W. shoulder, at 800 feet, is a shelving rock of great extent; from which rubbings were taken, first by a gamekeeper and afterwards by a gentleman who was kind enough to follow the instructions given at page 15. Allowing 20° for magnetic variation, the direction is from S. 83° E. at a height of 800 feet.

Thus, after a flight of nearly 15 miles, the arrow curves westward 48° (A). At a point about 350 feet above the sea, behind Flowerdale, and near the post-road, marks have the same direction. These are in the bottom of a hollow, and cross it diagonally from S. 43° E. (B).

On the other side of the hollow, in the bottom of a wide shallow valley, which runs nearly north and south, the marks point from S. 40° E. (F). They do not aim at the hills. These three spots, A B F, are in the middle, and to one side of the large glen, which is split by the Hill of Groban, 20 miles from the watershed at Glen Dochart. At the northern extremity of the block, beside the road which leads from Gairloch to Pool Ewe, the marks point at the sea from S. 60° E. (C), which is the direction of the watershed.

Further north, and further from the hills, and out of the jaws of the glen, another set of marks, perfectly preserved, give two cross directions—from S. 85° E., and from S. 35° E.

Still further north, and quite beyond the glen, is Meall Mor, a hill 600 or 700 feet high, on the north point of Gairloch, isolated; and near the western coast-line of this part of Scotland, a rock on the N.E. shoulder is clearly marked, and the rubbing shows two distinct movements—from S. 85° E., and from N. 35° E. (allowing 20° for variation) (D).

Thus the arrow is carried over the watershed of Scotland, at about 2000 feet, with the direction N. 65° E., which might bring it from Scandinavia along the coast of Sutherland. It is turned aside on the shoulder of Beinn-a-Ghuis, at the same level; and is made to glance northwards from S. 25° E., down a wide and deep groove. Followed for more than 20 miles, it is found bending gradually southwards, and left aiming from east to west and from N. 35° E. to S. 35° W. near a coast where currents flow various ways, according to the state of the tide. Tides close at hand do in fact flow in directions which correspond to marks upon this last isolated hill.

All this seems to point at floating glaciers, grown in sea-lochs, and amongst small islands, moving in currents and tides.

For a perpendicular height of nearly 2000 feet, for a length of about 25 miles, and a breadth of five or six at least, rocks are marked on one plan. Perpendicular cliffs, the bottoms of grooves, the tops of ridges, the tops of hills, all are marked alike : all the smooth sides are towards the watershed, all the broken faces towards the sea. All the grooves have a manifest relation to each other till they get clear of the glen. It seems plain that this big groove was full of heavy ice. But there is no great extent of higher ground at the watershed, and there horizontal grooves 1200 feet higher than the watershed aim past the higher peaks from which alone glaciers could slide.

If the other direction is taken, and the grooves followed, the same thing appears. From the watershed striæ lead down to the eastern coast, winding seawards in the grooves, and they are found on hill-sides far above the bottom of the glen. But at the watershed there is no possible source for a land-glacier, and no apparent reason why land-ice of any dimensions should move horizontally over Scotland at 1200 feet above the watershed of glens which isolate the hill. It must be remembered that similar marks pass over Scandinavia at about the same level, and in a similar direction, and that similar marks are found upon American hills. If these be marks of land-ice it was unlike any which now exists. If they be marks of sea-ice, the Arctic Current explains the puzzle.*

The head of *Glen Dochart* is four miles from Kinloch

* While this sheet was passing through the press a new work on this subject appeared—*The Physical Geology and Geography of Great Britain*, etc., by A. C. Ramsay, F.R.S.: London, Stanford, June 1864. The opinions of the author are well known, and have been adopted by several eminent geologists; in particular by the authors of the *Geology of Canada*, 1863; and by Mr. Geikie, author of an excellent pamphlet on the *Phenomena of the Glacial Drift of Scotland.* The theory assumes a period of intense cold, which prevailed throughout all high latitudes, and in all elevated regions of the earth, simultaneously; and which caused an enormous growth of ice during one or more geological periods. But no attempt is made to account for this cold period. The theory which this volume is intended to illustrate is that the present time is the "glacial period;" and that an explanation of ice-marks is to be found in the present condition of other parts of the globe. The marks in Scandinavia suggest glaciers on the scale of glaciers in Greenland; the marks in Great Britain suggest sea-ice on the scale of Labrador ice; the change of climate at one place is accounted for by a change in the course of an ocean-current, caused by a change in the level of sea and of land. All are agreed as to the facts; the questions left for argument are the cause of the change which has surely taken place, the nature of the ice which made the spoor, and the amount of work which this engine has done.

Mr. Ramsay attributes many rock-basins and their lakes to glaciation, and few agree with him; these volumes go further, and attribute these and many of the main lines of denudation in Northern Europe and elsewhere to glaciation, *combined with ocean-currents.* Mr. Geikie and other observers attribute marks in Ross-shire to land-ice. Their difficulty is how to get their glaciers over watersheds, and account for the cold of the exceptional glacial period. Mr. Ramsay appears to have proved that glaciation coincided with the deposition of certain breccias of Permian age in Britain. The stones are glaciated stones, that is certain; their position rests on good authority. If the glacial period began soon after the coal formation, and has endured till now, the acknowledged work of denudation gains the aid of an engine which works faster than streams and waves do. If arctic currents are now to be added to the list, they are bigger and stronger tools than land-glaciers, and may have helped to do the work, which has certainly been done somehow.

Ewe and 800 feet above the sea. Here the rocks are brittle and broken, and there are no marks.

Loch Roisg is 630 feet up, and from the head of it to the S.W. the Applecross hills are seen at the end of a wide strath. Here is a high col, and here at the head of Loch Roisg are heaps of drift.

Five miles off, at the lower end of the lake, near Achnasheen, are flat terraces of stratified water-worn gravel and sand, resting on a large lateral moraine, and the moraine is on grooved rock. Beyond the glen towers Sgur-a-Mhulin, and a range of high hills. The grooves point along *Strath Bran* at Ben Wyvis and Loch Carron, so ice did not come from the high hills.

The terraces stretch far up along the road which leads to Torridon, and they are very large.

Tides surely flowed through this strait at about 700 feet, for no small streams could do such heavy work.

The glacier-work was finished, and the drift left, before the gravel was packed over it. And the river is now winding along a plain of fine sand and mud which it washes out of older water-work, and packs away in lakes in Strath Bran.

The lateral moraine or the glacial sea-margin, which begins about Loch Roisg, is followed by the road for about twenty-five miles to *Garve* from 630 to 350 feet. Here the road descends from the high glen and turns away from Ben Wyvis into the valley of the Blackwater.

The grooves are well marked on rocks all the way from Achnasheen to the lower end of Loch Garve.

At 630 feet near *Achnasheen* grooves on gneiss point N. 65° E.

At 530 feet, at the junction of two glens near *Loch Liochart,* and the junction of the river which drains Loch Fannich, grooves on gneiss point N. 85° E.

Lower down, at *Loch Liochart,* at about the same level, 550 feet, weathered grooves on gneiss point N. 82° E.

About this level the high glen ends suddenly in a transverse glen. The drift in the upper groove is arranged in layers which slope down-hill towards the W.S.W. at an angle of about 35°. This is like the packing of silt by the ebb (vol. i. p. 339).

Above the inn at Garve, at about 600 feet, grooves on a rib of white quartz turn with the glen. They do not point at Wyvis or up into Strath Bran. They coast round a hill-side, carefully avoiding the high hills, as rivers do at the lower level. They point S. 45° E.

At the end of Loch Garve, beside the road, grooves on contorted gneiss take another turn with the glen. At about 150 feet above the sea, the marks point N. 70° E., and aim at the shoulder of Wyvis, which bars the way. On this hill-side are piles of drift, and it seems as though a glacier had ploughed down to the sea-level through the bed of the Blackwater. Near Contin inn the rocks disappear under plains of rolled drift.

Now, if these marks were made by a land-glacier, it was twenty-five miles long at least, and it must have had a large moraine. That mark ought to be found somewhere about the foot of Wyvis, or about Brahan, or Conan. But there is no large moraine with conical hills. There is glacial drift in profusion, but the moraine shape is not there.

If Strath Bran held a glacier which flowed north and east towards Ben Wyvis, stones left by it ought to be blocks of white and gray quartz and gneiss, fragments of rocks in Strath Bran, and near it. But there is no such collection of native drift here. If ever there were true land-glaciers in this district, they were launched at a high level, in a sea like that which is now passing Cape Farewell, near the same latitude, and which now carries "heavy drift ice" and "northern drift" southwards and westwards in sweeping curves.

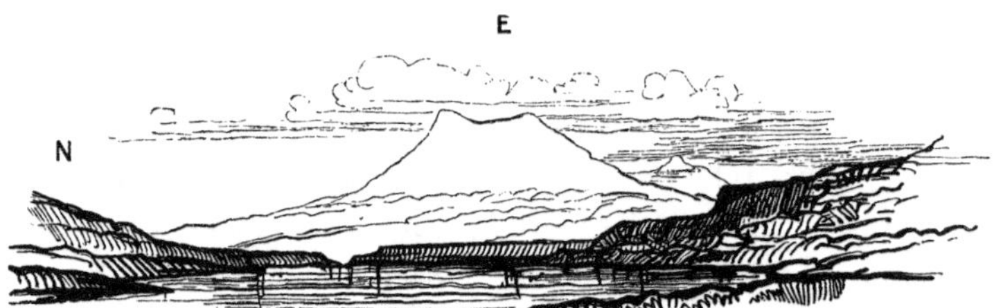

∧ ⌒ — Weathered hill, ground hill, and flat drift. Terraces of water-worn gravel and sand at the foot of Loch Roisg, near Achnasheen, at about 650 or 700 feet above the sea. Sgur-a-Mhulin, beyond Strath Bran. Ice-marks run north-eastward to the left along Strath Bran to Ben Wyvis.

FIG. 79.

CHAPTER XXXVIII.

BALTIC CURRENT 11—BRITISH ISLES 10—SCOTLAND 7—STRATH BRAN, BEINN UAISH, SUTHERLAND, ETC.

Beinn Uaish.—In travelling down Strath Bran, the end of the groove seems barred by the great mountain mass of Wyvis or Beinn Uaish. The highest point of the hill is nearly 4000 feet above the sea, and the base covers a very wide tract. Seen from Morayshire, and from the new railway near Inverness, it is a great block ⌒ with a rolling plateau on the top, and on this high base lofty clouds rest when neighbouring hills are clear.

From the bridge over the Conan, the movements of floats of white froth may be studied in the black peat water. The floats move as the water moves, past the piers of the bridge; and such curves described by froth are roughly drawn at page 127 and at the end of vol. i. On Conan Bridge, as on any sloping road, marks made by streams of water flowing past a stone may be seen. The forms agree with the movement of floats. In walking up Wyvis from the south-east, the course of a supposed north-eastern current, which came down the western shore of Scandinavia, is crossed. These large forms should resemble the miniature glens on the bridge, if they are in any way the work of ocean-currents. The shape of the land about Wyvis corresponds to hollows made by rain on sand, and to the curves drawn by froth on the Conan; and the floats in the Arctic Current in this latitude are large floes and deep icebergs loaded with boulders. Here boulders, like the hill-forms, seem to record the passage of ice-floats south westward at a high level.

Above Dingwall, in the woods behind Tulloch, are numerous boulders of a peculiar kind of pink granite. They are not common angular blocks, but large rounded blocks, like those which abound on the northern shores of the Baltic (see vol. i. pp. 297, 322).

At 540 feet is one 27 feet round and 8 feet high ; it is rounded on all sides, and a big tree beside it has bent round it in struggling to grow upright. Near it are others of the same kind, and these rest upon a foundation of brittle slaty sandstone (p. 167).

At 600 feet (the level of Achnasheen) is a flat block of gneiss of the same colour and composition as the granite ; and this block is scored on the upper surface. It is 9 feet long by 6 broad.

At 800 feet (the level of the col at Glen Dochart) are three large rounded masses of the same granite.

At 950 feet is another, and at this level the top of Brahan Hill and Torachilty are overlooked.

At 1100 feet, on the top of this hill, are more large granite boulders on a wide heathery moor ; and from this spot a deep ⌣ groove is seen crossing the ridge of Scotland W. by S. It is Strath Bran. If these boulders mark a sea-level, then the seaway was open over the watershed of Scotland.

A corresponding groove runs N.E. along the foot of Wyvis. At the same height, four miles inland, is another granite boulder at the head of Strath Peffer, opposite a notch in the shoulder of Wyvis, which opens Strath Conan above Contin inn, and Strath Bran behind Torachilty. The water in the glen behind Tulloch runs into the Cromarty Firth ; but at this level the tides would flow in from the Firth of Dornoch.

At 750 feet, the burn has cut through a pile of terraced drift level with terraces at Achnasheen. The bank is a cliff of gray clay, which contains numerous scratched stones, chiefly gray slaty blocks of various sizes, amongst which are specimens of granite. In the bed of the stream, where the largest stones are washed clear of rubbish, many large boulders of granite are mixed with slaty blocks. But there is no granite hereabouts *in situ.*

At 1000 feet, up the side of Wyvis, the rock is laid bare in a small burn. It is a soft slate dipping 10° south, or thereabouts.

Thus the shape of Wyvis ⌒ has nothing to do with the structure of the rock, but is due to denudation, and ice has done part of the work so far. There are blocks of granite on the hill, and a moraine in the glen. Great part of the moraine seems to have come from the flanks of Wyvis ; and the corrie in which the glacier moved is seen on the hill-side ◡ . But granite is foreign.

At 1650 feet is a conical hill called Cioch Mor. It is a lump of hard coarse conglomerate left standing in the groove. The sides are scored ; the greatest length corresponds to the run of the groove ; the steepest end is down-stream towards the west ; it is a large tor. In the supposed lee are large blocks of mica-schist, bits of gray quartz rock, and a big boulder of gneiss.

At 2600 feet, the sea-horizon is open through a groove to the north-east.

At 3000 feet, the ground on a shoulder of Wyvis is smooth, flat, and covered with a velvet carpet of yellow-green moss, over which mountain-hares have traced a pattern of footpaths. The rock shows in the edge of the deep corrie which was seen from below. It is a coarse gritty sandstone which splits into thin flags; it dips about S.W. On this high shoulder are blocks of gneiss, weathering and splitting to bits.

The view over the central district of Scotland is very fine. All the low hills are seen to have one even slope to a certain height ⌒, and above that the tops are of a different pattern Λ. The Knock of Brae-Moray is a cone planted upon this upper level, as Cioch is on the shoulder of Wyvis. The high hills about the head of Strathspey are steep conical hills, and the way over the Toridon hills is open. It is a groove ‿ ; and, as shown above, it is ice-ground and terraced.

At 2600 feet, on the shoulder, is a rounded boulder of the Dovre Fjeld and Finmark pattern, ten feet long, and made of gneiss. It is visible from Dingwall; and it must have floated to the shoulder of Wyvis, unless it flew, or slid upon ice all the way from the parent rock.

The seaway to Scandinavia along the coast of Sutherland is clear from this point at this level. Not so the top of Wyvis, which was hidden in mist.

At 2100 feet rock-surfaces are bare on this side facing the south. They are rounded but much weathered.

At 2000 feet and lower down glaciated surfaces abound, but they are all weathered. At this level the steep side of the hill ends, and the base has a longer slope to the head of Strath Peffer.

At 1100 feet are many granite boulders. And on the top of a sandstone quarry by the road-side near Dingwall, at the end of the Cromarty Firth, is a cap of glacial drift which contains large smoothed scored blocks of granite, and many other hard igneous rocks.

In the low grounds the whole country is covered by masses of similar stones washed and rolled. It is hard to find one with ice-marks amongst those which have been moved in railway-making and other works. This seems to be the case of the Galway drift repeated. The boulder-clay has been disturbed and repacked by water, without the help of sea-ice, below a certain level, and the scratched boulders are water-worn in the plain.

From Beinn Slioch to Wyvis the way to Norway is open, and floats are stranded at 3000 feet. There are no small ice-grooves left on Wyvis to point out the way, but glens and hills are but larger grooves and tors, and here they all point up the coast of Sutherland towards Molde and Trondhjem, where the coast-line takes a sweep and curves northwards as far as the Lofoten Islands beyond the Arctic Circle.

Still following the marks on Wyvis, the *Sutherland coast* trends N. 48° E., and there are no Scotch hills from which the Wyvis boulders could have floated at 3000 feet.

At the *mound* near Dunrobin Castle is a high bluff of coarse conglomerate, on which small ice-marks cannot be seen, but there larger grooves are remarkably distinct. The whole hill-face has been scored horizontally from top to bottom. The grinding force appears to have come along the coast from the N.E. as the flood does now. But it may also have come from the opposite direction with the flood, if tides ebbed and flowed over this part of Scotland, as they are supposed to do now over part of Greenland.

The woods of Dunrobin, as far as the river Brora, grow on vast terraced piles of boulders which do not seem to be moraines. They rest upon the sides of ice-ground hills above the sea, as if they belonged to a system far larger than any land-glaciers which now exist even in Iceland. They may be marks of the "ice-foot."

These terraced heaps are like the terraces of Northern Scandinavia, and they are probably effects of the same cause. The stones are of the Scandinavian pattern, and some, at least, may be of Scandinavian origin. To decide that point special knowledge is required. If Scotland held together and sunk and rose as Scotchmen are said to do, in a mass, this coast was under water when Wyvis and the Gairloch hills were islands, and Caithness at the bottom of the sea. The terraces appear to be horizontal.

Leaving Scotland and following the curve of the Scotch coast up to *Scandinavia,* the same forms recur all the way to the North Cape. If summer lost the aid of the Gulf Stream, winter and his fleets of ice would reign in spite of the midnight sun of Scandinavia. But if there were Greenland weather in Norway, there would be a wintry crop in Northern Scotland, and Sutherland might grow icebergs instead of wheat and dun deer.

Thus starting at Beinn Eith and Beinn Mhonaidh, on the western coast of Scotland, ice-marks at a level of 2000 feet lead across Scotland to Wyvis. There boulders mark a sea-level of 2600 or 3000 feet, and the shape of the country and of the east coast, existing tides, and other marks, all point one way. When the line is run out at the North Cape, it coincides with an equatorial current, which is continually flowing into the arctic basin, along the north-western coast of Norway. If an arctic current flowed out here, and the Gulf Stream passed westwards by Panama, the climates of these northern regions would change.

This curve passes very near Trondhjem where a road crosses to Sweden. Chambers estimated the height of the col at or below 2000 feet. He found ice-grooves perfectly preserved on this watershed, and they pointed N.E. and S.W.*

North-east from this spot there is no land of equal height

* Edinburgh Journal, vol. xii. p. 75.

now, unless it be in Novaya Zemlya, or about the North Pole. So the boulder on Wyvis may have sailed over Norway.

If it came on land-ice, the névé must have been somewhere beyond Scandinavia, the terminal moraine somewhere beyond Galway; and a glacier moved in the same direction, in similar latitudes, in North America, up the valley of the St. Lawrence, according to marks there. A Baltic current is easier to swallow, though it is a large draught.

Central Sutherland is a wide rolling plateau, with a few tall conical hills rising above the moor.

On the *west coast* the hills are higher, and they are quoted by the most eminent geologists as proofs of enormous denudation. On all the bare hills ice-marks are conspicuous.

The sketch copied in the woodcut was made from a yacht 25th September 1848, on a clear calm day with a transparent atmosphere, and the outlines are tolerably accurate, though each hill was sketched from a different point, as the yacht came opposite to it. The shape of the surface in the central districts of Sutherland is like that of the upper plateau which divides the Gulf of Bothnia from the arctic basin ⌒. The shapes of the hills on the west coast are like those of hills which now rise through glaciers in Iceland Λ.

The sharp angular peaks in Sutherland are like weathered hills elsewhere. Talus-heaps rest below the cliffs from which stones fall in every frost, and after every fall of rain rivers and mountain-streams add to the heaps, and carry part of them a stage down-hill. But the low grounds in Sutherland, Scandinavia, and Iceland, are not weathered but ground, and they all have one characteristic shape.

In Iceland there is a tract of ice nearly as large as Sutherland, in which névé and ice cover the whole land like a white pall, but the fringe is a black scolloped border of hills, and

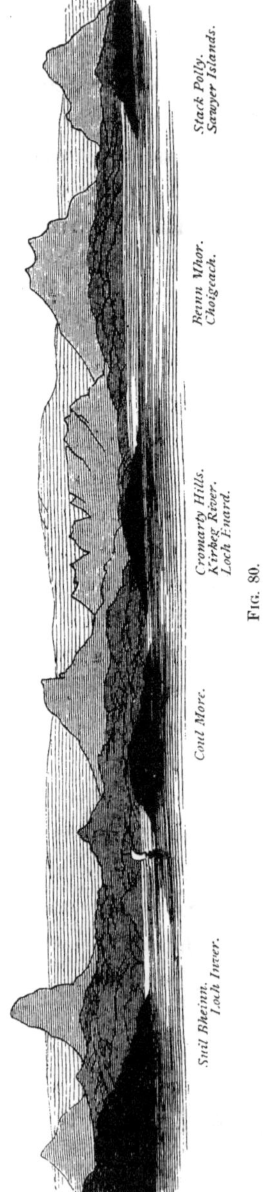

FIG. 80.

some of these are like the hills of Western Sutherland.

The ice flows into the central hollow of Iceland, but it melts before two broad streams meet. For a distance equal to that which the woodcut includes, two great banks of ice hem in Sprengisandr, and the outline of one is like that of the dark shadow in the sky of the woodcut. The ice-banks are advancing towards the sand, as if towards the sea-coast of Sutherland. But where a bit of harder rock has pierced the ice-crust, it stands up as a long ridge, a steep rock-spur in the round white ice-mountain ⌒. It is a tor ⌒.

One of these hills in Iceland has the shape of Suil Bheinn, in the woodcut of Sutherland. The ice-stream is splitting at the col, flowing along the sides, and meeting again in the lee. One glance is enough to show the movement, and the hill retains ice-marks high above the present ice-level. This hill is a great ice-tor, which the Arnefells Jökull has hewed and is still hewing out of bedded igneous rock. Suil Bheinn is another of the same size and pattern, and the same marks are on both, though one is igneous, and the other sedimentary rock.

They are long ridges pointing up-stream, tors on a large scale, mounds left in a rock-cutting, by which to measure the work done ; and the tool-marks are those of ice.

In the woodcut, Suil Bheinn is seen end on, and it looks like a pillar.* When the hill is seen from the side, it is a long steep ridge which ends in a knife-edge, and there are not many places where it can be scaled. The strata of which it is made are nearly horizontal, and the same beds recur in hills to the right, beyond the gaps which are valleys of denudation.

According to Geikie and other geologists, who have explored this district in more detail, the direction of high ice-grooves coincides with that of passes and main glens, which run from south-east to north-west, north of Loch Maree (see woodcut, p. 17).

About the same latitude, on the opposite side of the Atlantic, the Arctic Current, after flowing south-west along the coast of Greenland, eddies round Cape Farewell, and flows north-west, with all its train of ice-floats. It whirls round again further north, and flows down to Newfoundland, along the curve transferred to the map (vol. i. p. 232). A very slight modification of that curve would make it fit the glens of Sutherland and Caithness, and ice-marks on high passes in this district. The curve would then represent an eddy in the North Sea, and such an eddy might well result from a rise of land in the path of a Baltic current sweeping round the point of Norway, as the Arctic Current now sweeps round Cape Farewell. It is easy to test this theory, by building clay maps of this part of Europe in any shallow pool with a running stream.

When the land rose, land-glaciers would follow the present river-courses, till they melted and became rivers. And this

* *Sular*, Icelandic for pillar.

seems to have been the order of change all the way from Galway to North Berwick, from Malin Head to Cape Wrath and John o' Groat's House.

First, cold ocean-currents working denudation on a large scale ; then local denudation worked by minor causes acting from watersheds downwards.

From the sea the north coast of Sutherland appears to be ice-ground, but the sea has dug into the rock, and wild cliffs overhang a wild sea.

All down the *west coast* forms of glaciation recur below a certain level, above which are forms of weathering, and the sea-cliff is forming at the sea-level.

In the islands it is still the same. In the low island of Lewes ; in the low lands of Harris ; near the high mountains of the south end of Harris ; in North Uist, Benbecula (Beinn-e-Mhaoil), South Uist, Barra, Skye, Mull, Tyree, Jura, Islay, and in scores of smaller islands, similar forms recur in rocks of every description.

In the *Long Island,* for instance, looking from the north end of South Uist, the low grounds of Benbecula and North Uist are spread out like a map. There is a wide plain of peat and sand, salt and fresh water, through which low hummocks of gray rock and piles of boulders appear. In the midst of this half-drowned land rise two hills of the same pattern. They slope to the eastward, and are steep to the westward, and they are ground and rounded from top to bottom. Memory and rough sketches are enough to show that these hills are but large tors, of the pattern of Bren Tor in Devonshire, and hills in Lapland, with the same bearings. A small depression would make them islands, like those which are scattered broadcast along the Scotch and Norwegian coasts.

If there be striæ on these hills, they will point towards

the Lofoten Islands, which they resemble; but they were not examined for high grooves.

Outside of *Harris* grooves point N.E. and S.W. along the western coast near the shore beside a road.

In *Skye*, at Loch Corrie Uisge, marks of ice can be traced to a great height, and down to the sea, as clearly as in Romsdal or Justedal, or in a Swiss or Icelandic glen, where ice is working. This district has been described by Forbes; it was first seen by the writer in 1845, while the impression left by the Alps was fresh, and the work was then attributed to land-ice.

In *Rona*, near the lighthouse at the north-eastern end, the hills seem ground from the north-east, and thence a sea-way is open to the North Atlantic.

In *Raasay*, according to Geikie, all the hills are ice-ground, as he supposes from the south-west by ice sliding from Skye.

If the grinding resulted from the alternate movements of tides, the opposite ends of these two long islands may well show opposite movements. The uttermost rock of Scotland, the *Dubh Iartach*, has a long reef to the south-west.

In *Coll and Tyree* are perched blocks. In Mull, Colonsay, Oronsay, Jura, and in Islay, are all the marks attributed to ice; and drift-terraces abound.

The Scaur of *Eig*, that strangest of all the Western Islands, is a great wall of trap, with notched sides built upon a pyramidal base of stratified rocks, and one layer in this masonry contains fossil wood, immediately under the trap-wall. The island is another case of denudation; it is a tor in the sea; and it points up into the Sound of Sleat N.E. at Strath Bran and the coast of Sutherland. South-west of it are *Muck*, *Coll*, *Tyree*, and the *Sgeire Mhor* reef;

and breakers are beyond. This is a long ridge partly sunk, and aiming S.W. outside of Islay and Ireland.

The whole of these islands—all the small ones, and the main ranges of hills and glens in the large ones—have one general N.E. and S.W. trend.

Any good map shows the form of the coast. There is no good map of the hills, but when the Ordnance map appears, it will show that all these island-forms bear reference to grooves crossing meridians diagonally south-westwards, like the chief passes on the mainland, which no map shows.

Further north, the low *Shetlands* seem all to be ice-ground rocks.

In *Orkney*, farmers find their land full of great loose stones, and the general shape of the low rocks towards the north is rounded. At the southern end, the coast-lines are chiefly cliffs of great height, which the sea is undermining.

So the general shape of this country on a map; the general shape of the hills as seen from a distance, minute details on shore; the general shape of Western Europe, and of the whole northern hemisphere,—all seem to point to symmetrical denudation, and to the action of ice on shore and afloat.

Taking the curves of the Arctic Current from Spitzbergen to Cape Farewell as a natural curve of motion which might be repeated elsewhere, and is extended south of Newfoundland, the curve can be applied to the British Isles, as shown roughly in the map (vol. i. p. 232).

A S.W. curve, which comes out of West Fjörd in Norway, passes between the Shetlands and the Farö Islands to Rockall.

Curves which start about the watershed of Lapland, near Kautokeino, etc., skirt the Norwegian coast, pass over the Shetlands and Hebrides, and coincide with ice-grooves on the outside of the Island of Harris.

South-west curves drawn from south-west ice-grooves on the watershed of Scandinavia beyond Trondhjem, skirt the Norwegian coast, and the Scotch coast from the Ord of Caithness, to ice-grooves on hills at Dunrobin in Sutherland; thence Strath Bran and small ice-grooves carry the curves over Scotland, into the Sound of Sleat. The curve passes Coll and Tyree and the Sgeire Mhor, into the Atlantic, and even under water sunken hills and hollows stretch further in the same direction.

The same curve, begun about boulders on the Dovre Fjeld, passes seaward with ice-grooves out of Romsdal, and enters the Moray Firth. The Caledonian Canal, the Muckle Glen and ice-grooves in it, carry the line over Scotland into Loch Linne, and it passes Colonsay and Oronsay, which are ice-ground. There, again, sunken rocks extend in long broken ridges south-westward into the Atlantic. Strong tides and wild seas work in the hollows, which hold sounds, amongst these islands. If the sea were cumbered with heavy ice, as it is off Labrador, there is water-power enough and to spare in this region, to work the floating ice-engine which, according to Kane, " rubs rocks."

Curves begun at the head of the Sogne Fjord, at the foot of the highest hills in Norway, follow ice-grooves to the sea, and pass by several local glacier-systems near Bergen. They fall into a series of deep grooves which cross central Scotland, and in these the curves coincide with ice-marks which cross the watershed, and touch hill-tops in Argyle ; they recur in Glen Veagh, Donegal, etc., in Ireland.

Curves drawn from boulders on the Fille Fjeld in Norway fall in with boulders about Aberdeen, skirt the Sidlaw Hills, where they coincide with marks on the rock ; pass Perth and Stirling and Glasgow ; Argyll, Arran, and Ceantire ; the Giant's

Causeway, Sligo, and Westport; and there are ice-marks all the way which seem to correspond to a general movement in that direction, at a high level.

Curves begun about the Hardanger glaciers run with ice-marks for a hundred miles in Scandinavia; join an ice-mark on North Berwick Law, and wind their way across Scotland and Ireland to Connemara and Galway, where the spoor is lost in the sea. It is there as perfect as if made yesterday, on limestone rocks laid bare in making a railway near the coast, and on the top of a quartz hill 2000 feet high.

All these several lines have not been followed expressly to study ice-marks; but some have, and the rest are pretty well known to one who has wandered amongst the hills whenever he could. There is scarcely a Scotch hill or glen, in island or in mainland, which does not bear some conspicuous mark of glacial denudation. The low marks seem generally to bear reference to local glacier systems. The high marks, from 3000 and 2000 feet down to the sea-level in low passes, appear to bear reference to a general system of horizontal movement in water and floating ice, like that which is now going on further west.

These theories, founded upon observation of glacial action in Switzerland, Scandinavia, and Iceland, and of ice-marks on rocks at home and abroad, during twenty-two years, are thus far supported by facts gathered from books and stated above. They are also propped up by facts observed and gathered by the latest writers on this cold subject.

They gain strength from facts stated by geologists in the *Geological Survey of Canada*, 1863; by Sir Charles Lyell in his great work on the *Antiquity of Man*, 1863; by Professor Ramsay in numerous papers; by Mr. Geikie in his work on the *Glacial Drift of Scotland*, 1863, which is perhaps the best book of its class which has yet appeared. All these authorities, and a host of witnesses whom they quote, are agreed that the British Isles are ice-ground, and that the land has been submerged to a height which would only leave a few hill-tops above water. The facts are beyond cavil; they seem to lead to the following conclusions:—

1st, Because raised terraces and sea-margins are nearly parallel to the plane of the sea, it is probable that the last rise of land in Ireland, Scotland, and Scandinavia, was a general swelling movement, which included a very large area of upheaval.

2d, That the last cold period in this area, and in particular in Ireland and Scotland, coincided with a sea-level at least as high as the highest erratics yet found in Scotland (on Wyvis and Driom Uachdar at 3000 feet); and with the highest horizontal ice-grooves, which are at about 2000 feet on Shan Folagh in Ireland, and 2000 feet on hills about Loch Maree. They may yet be found higher.

3d, That the cold period also coincided with the sea-level, which is marked by the highest Scotch terrace of glacial drift. The highest known to the writer is near Dalwhinny, at about 1400 feet.

4th, That ice-marks may have been made in deep water by ice-floats grounding in 1800 feet, while an "ice-foot" packed drift in terraces at the sea-level; because these operations are now going on further west in similar latitudes.

5th, That the last Scotch glaciers which reached the sea passed away after the land had risen to the level of the lowest *perfect* terminal moraine. The lowest of these yet found by the writer are opposite to glens north and south of Loch Ericht near Dalwhinny, at about 1400 feet. All lower moraines seem to be washed out of shape.

6*th*, That this level of 1400 feet, and all other levels marked above that plane, coincided with a general movement of cold water from the arctic basin south-westwards, which was varied by tides and impediments, so as to make eddies like those drawn on the map, vol. i. p. 496.

7*th*, That this general movement, varied by local tides and eddies, continued while there was a strait left open in Britain; now continues in the Straits of Dover and in the Pentland Firth; and in the Arctic Current and Gulf Stream, which alter climate in similar latitudes on opposite coasts.

8*th*, That the end of the last cold period in Scotland nearly coincided with the sea-level of 1400 feet, which is marked by a moraine of conical mounds at Dalwhinny, and by a terrace of glacial drift, partially water-worn, beside the moraine.

9*th*, That this change also coincided with the closing of a strait by the rise of land in Lapland, which is now 1500 feet above the sea, according to Von Buch's measurement.

10*th*, That a gradual subsidence in the same tract would let in the current by opening the strait, and would bring back the period of cold to Scotland when land had sunk about 1500 feet to the north of the Baltic.

11*th*, That many similar changes of equal amount, produced by the same causes, may have taken place; and that the present shape of Scotland, Ireland, and Scandinavia chiefly results from denudation by currents of air and water, which still circulate. These are driven by mechanical powers which still work the engine, and guided by laws which produce regular movements.

12*th*, Because these laws seem to govern all known quantities and dimensions, small quantities of earth and water, and streams which men can see and guide, serve to help them to comprehend movements which they cannot control or see ; or even comprehend without hard thinking.

13*th*, Because Scotch and Irish rocks, exposed to the weather at 2000 feet above the sea, and at the sea-level, still retain sand-marks which are perfectly fresh, and less weathered than Egyptian sculpture 4000 years old, the time which has elapsed since the end of the last British glacial period must be short. The occupation of the British Isles by the ancestors of races who still dwell there may have coincided with the existence of glaciers on Scotch hills, and traditions may be dim recollections of these geological facts.

In the course of this journey from Galway to Dingwall, from Malin Head to Cape Wrath, the Baltic Current theory has gained strength. Another cast southwards will try the hobby ; if he is sound after that run, he may be trotted out and started, to try his chance with other hobbies.

FIG. 81. ROUNDED GRANITE BOULDER, IN A WOOD BEHIND TULLOCH, RESTING ON SLATE, 540 feet above the sea (p. 152).

CHAPTER XXXIX.

BALTIC CURRENT 12—BRITISH ISLES 11—ISLE OF MAN.

A KNOWING old pointer quarters his ground on system, and his system is worthy of imitation by all who search.

Turned loose on the brown moor on a fine breezy morning, he capers soberly, and shakes his velvet ears, and licks his slobbering lips, to express his intense enjoyment of freedom and fresh air; and then, with quivering nose breast high, and wavering tail in full play, he settles steadily to his work. He takes his line and tacks steadily to windward, crossing and recrossing the straight line which the human sportsman draws in the wind's eye. When one beat is finished, a wave of the keeper's hand conveys the order, and the eloquent tail and ears tell that their owner knows what to do. Up goes the head, off goes the pointer down wind at score, that he may beat to windward again. Having beat the northern half of the ground on the pointer's zigzag plan, let the middle of the moor have a turn. The S.W. curve drawn from high grounds at the head of Sætarsdal, past Stavanger, runs over an ice-ground country in Norway, passes Berwick, the Solway Firth, the Cumberland hills, the Isle of Man, Drogheda, and Dublin, and passes out by the Shannon. If one leg of a pair of compasses be placed on the Isle of Man, a large circle, described about that point, nearly touches Duncansby Head, Cape Wrath, the Butt of Lewes, Cape Clear, the Scilly Isles, the mouth of the Thames, and Kinnaird Head. The lighthouse on the Calf of Man is near the centre of the British Isles, and the island may be taken as a miniature of the whole group.

The *Isle of Man* is about thirty miles long and twelve broad; and the highest point is about 2000 feet above the sea. The long axis bears about N.E. by N.

The north-eastern end of the hill country is rounded; the south-western is broken. To the north-east a long low tract stretches about eight miles from the hills to the point of Ayre. At the other end the sea has so undermined the hills, that cliffs are 350 feet high at Brada Head and elsewhere. Exposed trees point about N.E., so the prevailing wind is from the S.W. The flood-tide comes from the same direction. Drift timber, like that which the Gulf Stream lands elsewhere on the British Isles, is sometimes stranded about the Calf of Man. So the Mull hills, Brada Head, and the south-western coasts of the Isle of Man, are exposed to wind, and tide, and ocean-currents, and to large Atlantic waves, which roll up channel. The point of Ayre, on the contrary, is sheltered.

Denudation and deposition are still going on; air and water are at work; and the form of the work is conspicuous. Speaking generally, the coast-line is a shelf quarried out of contorted silurian and other strata, most of which dip at a high angle. A vertical cliff, and a shattered plain below it, form an L notch between high and low water mark. On this shelf the sea packs chips which it digs from the cliff.

At the sheltered north-eastern end the beach is made of gravel, fine sand, and clay, and it shelves gradually. The outline of the coast is smooth, like that of a mud-bank in a mill-stream. At the battered end the coast-line is jagged, and beaches are steep and narrow, and generally made of large egg-shaped boulders, some as big as a man's head. These are tools with which waves quarry cliffs, and they bear marks of work. The general shape of sea-worn boulders is curved; but their smooth surface is dinted and pitted by small hollows. Forty or fifty go to a square inch, and each pit records a blow. The water-line at the foot of the cliff is also worn smooth by the rolling of smooth pebbles at some places; but generally the rock is jagged, torn, and broken by the storm of boulders, with which heavy rollers, driven by strong winds, pelt the cliffs.

If the island has risen from an open sea, there should be beach-marks of this kind on the hills.

On a clear fine morning, after a slight fall of snow and a strong wind, the shape of the ground is picked out in lines of black and white; and on such a day hills in the Isle of Man, seen from Douglas Bay, appear to be ruled horizontally up to a height of about 1200 feet. Low down at least three notches can be made out on the hills which make the horns of the bay. The lighthouse is perched on one of these shelves. At about 150 feet above the sea, at the road-side, on the hill to the N.E. of Douglas, a quarry was open in March 1864. The rock is silurian slate, dipping at a high angle, the same as the jagged rocks which form the present sea-beach below the hill. The cap of the quarry is a thick bed of compact clay, showing signs of deposition in water. It is arranged in thin beds where it touches the rock, and it contains ice-ground stones, which may be contrasted with boulders carried from the beach. The rock-surface is not broken, but shorn across the edges of the strata, so that the boundary-line between rock and clay is an even convex curve ⌢. When this rock-surface is laid bare and washed clean, it is found to be smoothed, grooved, and striated from E.N.E.

So ice had a share in hewing out these hills and marking

these beach-lines, and it was not ice sliding from the tops, but ice moving horizontally along the coast, which made these marks at Douglas, at 150 feet above the present sea-level.

At about 450 feet above the sea, the road from Douglas to Laxey passes over the ridge in a groove which runs along the hills from N.E. to S.W., crossing glens in which the drainage of the country now flows.

On the Mull hills, at the south-western end, at least three shelves can be distinguished on hill-sides and cliff-faces. These occur at about the same levels wherever they are visible, on promontories, etc., according to very rough observations hurriedly made. To get at the full meaning of these "terraces of erosion," a careful survey should be made.

There are large boulders, at about 450 feet, at the top of the ridge, between Douglas and Laxey, and also at Brada Head, at about 450 feet, which seems to be the level of one of these rock-shelves which surround the whole island. There is evidence of an ice-laden sea up to this level at least. At Laxey are two deep glens which run to the watershed. They have the shape of glacier-glens, and they contain large boulders. The marks of a large glacier will probably be found in these rock-grooves when they are examined.

A depression of 500 feet would make the Isle of Man a row of small conical islands, stretching from N.E. to S.W. North Barule, 1842 feet, would be at one angle; the point of Ayre would be under water; Cronck Irey na Lahaa (the hill of the rise of day, 1445 feet, fifteen miles S.W.) would be at the other end of an archipelago of twelve islands. At lower levels, cliffs would still be washed by Atlantic waves, but Laxey Glen would be a long sea-loch.

The top of Snæfell (2024 feet according to maps, a little more according to observation) is conical but rounded, like all the other hills in the island. It is strewed with large slabs of broken slate and blocks of white quartz, apparently native rocks. Except the shape of the hill itself, there is no indication of glacial action at the surface near the top, unless the large quartz blocks are foreign. The hill is joined to Mullagh Oure (Dun Top) by a col which is about 1400 feet above the sea, and near about the level of a contour-line, which is seen from Douglas Bay. In March 1864, a gravel-pit made for a new road gave a section of the surface-beds. They consist of blue clay with broken angular slate and grooved stones, covered by a bed of peat and some washings from the hill. The rock foundation was hidden. The grooved stones prove that ice moved at this level on this col. The new road winds along the hill-sides for several miles, keeping near the watershed where streams part. The cutting along the road-way, and numerous gravel-pits, show that the cap consists chiefly of angular stones broken out of the hills, but these are mingled with numerous blocks carried from some distant place. Large angular weathered blocks of granular quartz rock are the most numerous ; specimens of yellow and red sandstone and of schorl were found in a day's walk, and some of the boulders were finely polished and grooved.

At the height of about 1100 feet, on a shelf which is visible from Douglas Harbour, large rounded boulders are common in fields, in cottage walls, and elsewhere. Though the surface has been destroyed by weathering and frosts, there is still evidence to show that ice floated over the cols where sandstone was dropped. If the sea were now to rise fifty feet, it would cut off the Mull hills at Port Erin. If it rose 500 feet, it would sink half the island and make a strait at Douglas. If it were to rise to 1400 feet, where a foreign boulder now marks an ancient sea-level, little of the island would remain above water except eleven hill-tops and two long ridges. If the rise were general in the British Isles, nearly the whole of England would be sunk, and the nearest sandstone island left above water would be in Cumberland.

At the south-western end of the hill country, granite and other boulders are strewed on the hills from Peel up to the verge of the cliff at Brada Head. There are various kinds, and as Manx granite appears at the surface in two places only, some of these must be wandering blocks. They are found at 400 feet and at higher levels. The people say that some of these were carried by Phynnodree, or Hairy Breek, an outcast fairy with shaggy goat's hair and cloven feet, of whom many curious Manx tales are told. One block, according to popular history, was hurled by Goddard Crovan at his scolding wife. Fin MacCool and his warriors, giants, and Druids, and other mysterious people, get credit for moving these mysterious stones.

The country about Castletown is to the south-west of the hill country, and would be sheltered from a north-eastern current. It is well described by an able local geologist.*

It has the outward form of a plain of drift packed in water. According to Mr. Cumming, it is a bed of drift containing bits of insular rock, fragments of the coal-measures of Cumberland, stones from the south of Scotland, and chalk-flints which may have travelled from Antrim, but which may also have come from Denmark.

This bed of glacial drift rests upon limestone, which is striated from the magnetic E., say E. by S. Trains of boulders and other marks indicate an ice-laden current moving

* *The Isle of Man, its History, Physical, Ecclesiastical, Civil, and Legendary.* By the Rev. George Cumming. London : John Van Voorst, Paternoster Row, 1848.

from the Solway Firth. To this Mr. Cumming attributes the "drift," and the ice-marks in the Isle of Man. He adds, "The origin of such a current is at present a mere matter of speculation." He suggests that the chief carrying and grinding agent which worked on these low grounds was floating ice; shore-ice, land-ice, and icebergs moved by tides like those which now pour through the sound of Kitterland. If the low grounds about Castletown were sunk, and the sea up to the highest notch on the Mull hills, the same tides which now flow north and south in the main channel, and east and west in the small cross sound, would flow east and west over Port Erin and the limestone district of Castletown. But if the sea were up to 1400 feet, the Solway Firth would be an open strait, and a deep sea-way would be open through Ireland along the curve which leads from Stavanger to Shannon. The tidal wave which now splits on Ireland would pass directly to Norway over the British Isles, and ice-floats would move in the direction of ice-marks, if icebergs moved seaward with the ebb or south-westward with an ocean-current from the Baltic past Cumberland and the Hill of Dawn in the Isle of Man.

A cast up-stream leads to the Cumberland hills. Boulders abound by the way-side, along the railway line which crosses this tract. The mountains are very much ice-ground, according to those who have examined them, and in all probability a local glacier-system once radiated from the watershed of this tract.

In the lower grounds, between Carlisle and Berwick, drift and ice-marks abound. The trough which holds the two main rivers in this tract follows the S.W. curve, and in Geikie's map a red arrow points about N.E. When hill-sides are examined at about 1000 and 1500 feet above the sea, the arrows will probably point the other way.

A sweep northwards brings the line to that curious set of curves which are seen in the low lands south of the Pentlands, from the top of these hills, and which are well shown upon the Ordnance map.

A sweep southwards brings the line round to Morpeth. The clay which covers the rock near Morpeth and Newcastle is about ten yards thick, and full of scratched boulders. In making new coal-pits the rock-surface is laid bare, and it is said to be scored. A promised rubbing has not appeared, but in all probability the marks at low levels point south on the east coast. At high levels they ought to point south-west or thereby, through gaps in the hills, but this point has not been made good.

On the other side, down-stream, the whole physical geography of Ireland is based upon grooves and ridges, rivers, lakes, points, and sea-lochs, pointing south-westward. According to Jukes (*Manual of Geology,* p. 680)—

> "The rocks of many parts of Ireland, especially those of the south-west corner of it, exhibit in great perfection that rounding and polishing which glaciers communicate to the rocks over which they glide. So perfectly indeed are all, even the hardest rocks, rounded and smoothed, that the very universality of the process prevents its striking an eye not instructed in the nature of the phenomenon." . . .
>
> "The surface of the rocks on the slopes and tops of the hills are traversed also by glacial striæ." . . .

The author shows that Ireland may have been elevated during the glacial period, so as to be within the climate of land-glaciers, but that it certainly was submerged during the glacial period, so as to admit of the passage of ice-floats amongst a group of Irish islands. "At 2000 feet below the present level, a few small islets only would be left."

It has been shown above that ice moved in a south-westerly direction, over the tops of hills in Connemara, one of which is 2000 feet high. The map of Ireland, reduced from the Ordnance Survey, shows that the whole island is grooved in the same direction, and the shape of it corresponds to the shape of the Isle of Man.

So a cast round the centre of the British Isles helps to swell the bag of facts, and feed the Baltic Current with a heavy feast of hard stones, tough facts, and fossil floods of iced-water.

CHAPTER XL.

BALTIC CURRENT 13—BRITISH ISLES 12—YORKSHIRE AND WALES, ETC.

A CURVE begun in Novaya Zemlya, and drawn over Lapland, near the head of the Gulf of Kandalaksha in the White Sea, passes near Torneå, runs down the Swedish coast to Sundsvall, touches Christiania and Christiansand, and lands at Whitby. It crosses Yorkshire, passes Manchester and Liverpool, and passes behind Snowdon into Cardigan Bay, skirting the coast of Ireland from Wexford to Cape Clear.

Part of the country has been described above (chap. xiv. to xx.), and there ice-marks point to a current moving south-westwards. In Lyell's *Antiquity of Man*, p. 270, glacial phenomena in Ireland are described, and the geological survey and former writers are quoted.

Signs of glaciation have been traced to elevations of 2500 feet in the Killarney district. Marine shells have rarely been met with higher than 600 feet above the sea, and that chiefly in gravel clay and sand in Wicklow and Wexford. Above 2500 feet, rocks are rough, below that elevation smooth, and "drift" has been traced as high as 1500 feet on hills which reach to 3400 feet. Taking the symbols used above, the form $\wedge$ characteristic of weathering, is characteristic of Irish hills down to a level of 2500 feet. Below that level the characteristic form is $\frown$. At 1500 feet drift is deposited ;

at 600 feet are sea-shells of arctic type in beds of gravel. Except in a few cases, the transport of erratics is southwards and westwards, and the prevailing trend of mountain-ranges is south-westwards. Sir C. Lyell's map, p. 278, is the best of its kind, and it shows that currents moving through the British Isles at a level of 600 feet, and governed by the same laws which affect the present run of tides, might pass along part of the curves which have been followed thus far.

At 1500 feet, Lapland would be under water, and the way open from Novaya Zemlya to Wicklow, if the submergence were general in this tract of Europe. Keith Johnston's map (plate 10, *Physical Atlas*) shows that volcanic disturbance has affected areas of equal size in modern times.

If the climate was cold when the districts above mentioned were under water; if glaciers grew in Scotland, Ireland, and the Isle of Man; then it is probable that climate in England was cold at the same time, and English hills ought to retain ice-marks.

In *Yorkshire* is a hilly tract where the highest points are about 2000 feet above the sea.

The country is composed of beds of sandstone, shale, carboniferous limestone, and suchlike rocks; disposed horizontally, but broken and shattered and bent, dislocated and upheaved in many places. Where a stream of running water has made a bed in the rocks, it has generally cut a deep trench with steep or perpendicular sides, or the banks have fallen so as to leave a slope of talus under a cliff. But the whole district is furrowed by deep glens whose rounded form bears no sort of resemblance to the beds of streams and torrents which flow through them, or fall into them. A section across one of the Yorkshire dales is like a section of an Icelandic glen—a sweeping curve, not a steep trench—and the sides are terraced ; each terrace corresponding to a bed of rock. The dales are deep grooves winding in long sweeping curves, like dales which now contain glaciers elsewhere ; the hills are rounded ⌒, the glens grooves ‿ ; the terraced sides are like coasts represented in Parry's *Voyages to Baffin's Sea.* These, also, are composed of beds which are nearly horizontal, and are now undergoing denudation by weathering and ice, and there glaciers flow through glens with terraced sides.

No small ice-grooves were found in a rapid journey through the Yorkshire hills, but sandstone and limestone weather so fast that fine tool-marks speedily wear out. The dales themselves remain, and they are full of patches of drift,—of ridges, mounds, banks, and hills of foreign boulders, sand, and clay.

In some glens, as in Wharfdale, small terraces like those which occur at Melar in Iceland sweep along the hill-sides. They are not horizontal, so they are not beaches or water-marks ; they are not the edges of strata, like terraces above them ; they are about the size of vine-terraces, which are made on hill-sides near the Rhine, and they sweep round hollows and promontories in green fields, like works of art. Where a river has cut through them, their section shows loose gravel, sand, clay, and stones, disposed like broad steps upon the rocky foundation of the hollowed dale.

If a local system of land-glaciers filled upper glens, and a general system of currents worked in from the north-east—while tides floated field-ice, land-ice, and icebergs up and down, pushing gravel along the bottom—the forms of these glens, and of small terraces in them, might be explained by the known effects of ice elsewhere.

These dales were hollowed out by some wearing process ; for beds of stone can be followed from glen to glen, and from hill to hill, round, and even through the hills in the mines.

They are not the work of rivers; for denudation by running water is very well exemplified at the lead-washing floors, and the work differs.

In one process lead-ore and vein-stone are crushed to powder, and washed by a stream through a funnel into the centre of a shallow pit. A machine revolves in the pit, sweeping the surface of the fallen mud with a heavy coarse cloth, so as to give it time to separate according to comparative weight. Heavy lead-ore sinks first and fastest; lighter minerals roll further, and sink slower; and when the operation is finished, there remains a stratified convex mound, whose outline is a regular curve ⌒. When water is poured upon the top of this dome, it cuts miniature glens in the sides of the hillock of sediment, as rivers do through hills of sandstone; and each glen has its delta. If rivers dug out the Yorkshire dales, their forms ought to agree with these. The miniature glens are, in fact, very like the beds of torrents in the country; but they are wholly unlike the dales in which the torrents flow.

Form asserts the agency of glaciers and ocean-currents, and denies the agency of rivers in the large denudation of the Yorkshire dales. The tool-marks are like those of frost elsewhere. As shown above, a theoretical curve leads near Christiania, and there the long groove of Gulbrandsdal runs up to the watershed of Norway at the Dovre Fjeld. The general shape of the big Norwegian dale is very like that of the smaller dales of Yorkshire.

Stoke.—About Stoke, the English watershed is 370 or 400 feet above the sea. The rocks belong to the coal-formation, but a few granite boulders are strewed about the fields. No other ice-marks were found; but the country is thickly peopled and highly cultivated; the rock buried under beds of clay and sand. Minton makes china and encaustic tiles of glacial chips, while coals and iron are dug from beds 1200 feet below the sea-level, where the temperature is 68° in the coal, and the temperature outside about 49°.

This land was above water when the coals were plants growing in air; it was under water when sand was poured over the bed of peat; it has been up and down while 1500 feet of coal-formation beds were deposited. The whole series of rocks has been hardened and tilted bodily up and broken; and the broken surface has been worn smooth and furrowed. The worn surface was surely under water when the drift and clay were dropped there; and the granite boulder records the passage of ice at this point on the curve.

The railway gives the line of lowest level, and here Bradshaw's *Railway Guide* and a net of iron roads carry the curve in any direction; for there are no hills about Stoke.

Manchester and Liverpool.—At a late meeting of the Manchester Geological Society, glaciated rocks were described.

These occur on Bidston Hill and elsewhere near Liverpool, at a level of about 200 feet. The direction was N. and S., E. and W., N.W. and S.E. Amongst these low hills, currents might flow in any direction, as tides do amongst the banks off Liverpool, at various states of the tide.

Cheshire—The railway map gives a very intricate pattern in Cheshire. The country is high and varied by round hills. Hartford station is about 270 feet above London. The low grounds are covered with water-worn drift, in which sea-shells are found. Amongst the stones are granite, chalk-flints, greenstones, and various hard rocks. Large blocks of granite, with fresh ice-marks on them, are found, and many are broken up and used.

The village of *Eaton* stands on a hill of bare rock, which is new red sandstone disposed in horizontal beds. Several large blocks of granite and greenstone are placed by the road-side, near wells, and at corners. On some of them the polish is well preserved, and grooves are fresh. On the top of the hill, in a sandy lane, a small boulder of green porphyry was found. It was about the size of a small turnip, subangular, and with a perfect surface grooved on three sides. The shape of the rounded sandstone hills bears no relation to dip, fracture, or bedding. They are carved out by some engine, and ice certainly passed over the hills at Eaton. The top of the hill is 340 feet above Oulton. Hollows seem to run E. and W. The cap of the quarry consists of broken flags and sand. Other boulders of granite and gray quartz with perfect surfaces were found in a garden ; and this was the owner's account of them :—

"Them is what we call marble stones ; they grow in the yearth, especially in places where they are bringing in new ground. You see the yearth produces all sorts of things for the good of man. The top produces all manner of vegetables, and underneath there's all sorts of mines and minerals for the good of man, and these stones grow in the yearth amongst the sand."

So spoke the village sage.

The sand seems to tell of cold tides flowing in the Vale of Chester, for sand-pits show mounds of contorted sand-beds, whose foldings are hard to unravel, unless they were frozen and melted like the sand-heap mentioned above (vol. i. p. 380). A fringe of crystal ice hung in a sandstone quarry, and a brittle crust of thin flat ice on the mill-dam, was all that remained of Cheshire ice; but mental eyes looked over the water to Hamilton Inlet, and saw the pictures which other men have drawn.

At *Northwich* numerous boulders of large size, specimens of granite, greenstones, and other hard rocks, are set up in the town. In fields near the town heaps of small boulders occur.

The whole town is sinking from the constant waste of the brine springs. About a million of tons of salt pay canal dues every year. In one dry mine the salt is quarried for a depth of thirteen feet, in an area of twenty-three acres.

The temperature is 51° at all seasons. The heat of the earth below, and the weight of cold air above, together produce a constant movement of air. It rises up one shaft and falls down another. A greater difference of temperature evaporates water in the salt-pans. Steam rises and water falls. Steam in the boiler lifts the piston of the steam-engine which pumps up the brine, and lifts and lowers the miners and their millions of tons of salt. The same heat-power, set to lift Cheshire and evaporate the sea; the same weight-power, set to condense steam and lower the earth's crust; the same natural powers which men chain to their wheels—seem strong enough to work the natural engine which ground and polished granite boulders, and carried them to Northwich.

It is plain that ice travelled here, it is equally plain that low ice-marks will not unravel the ice-problem. The Cheshire boulders did not come from Wales or Yorkshire. They may have come out of Cumberland, but it is possible that they came from Sweden or Lapland, because zircon syenite was found in Galloway by Jameson, and at Christiania and in Lapland by Von Buch, and because boulders are on the watershed of England, about Stoke.

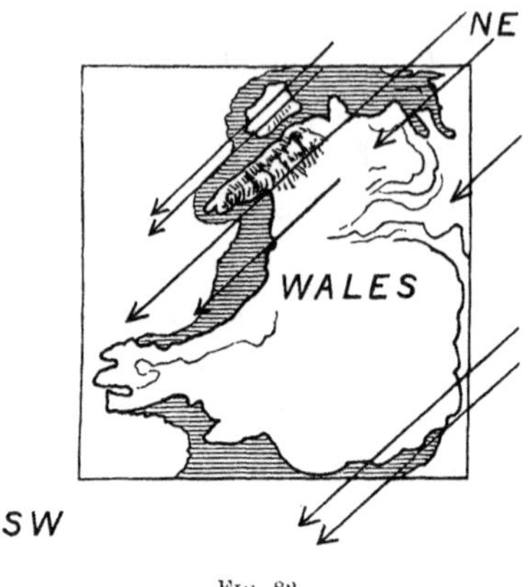

Fig. 82.

has been mapped by the Ordnance, and surveyed by geologists; it is the scene of Sir Roderick Murchison's discoveries, and classic ground. In the book of the Alpine Club* the glacial phenomena of Wales are described by Professor Ramsay, who states his own views, which coincided with those of the best modern geologists.

It seems to be admitted that sea-ice stranded drift amongst the Welsh hills at a height of about 2300 feet, that local land-glaciers ploughed out the drift when the land rose; but no attempt seems yet to have been made to account for the change of climate which destroyed the Welsh glaciers and turned winter to spring. If England were submerged 2300 feet, then the nearest land to the north-east would be Scandinavia, and a way open for the curve whose direction is shown on the woodcut.

The Principality is an oblong block of high land whose four sides face the cardinal points. The corner next Liverpool faces the north-east, the point from which an arctic current now flows in the same latitudes beyond the sea.

The corner near Milford Haven faces the south-west, the point from which the tides come now; from which the equatorial Gulf Stream flows towards our coast, and from which it is assumed that a prevailing equatorial current of air has blown ever since there was an atmosphere, and will

* Peaks, Passes, and Glaciers of the High Alps. Longman, 1857.

continue to blow till the state of the atmosphere and the laws which govern its movements are changed.

The north-western corner of the block is Anglesea, and the south-eastern is rounded off by the Severn valley.

If a north-east stream flowed from the Cumberland and Yorkshire hills, it would cross two corners diagonally as south-western gales do. If the wearing power moved from the south-west, then the soft rocks of South Wales ought to bear the strongest marks of abrasion.

In fact the coal-beds are most ground away at the north-east side of Wales.

From the western side of the block the hollow of Cardigan Bay seems at first sight to have been scooped out in a north-easterly direction by south-west waves. In looking at a map where land only is marked, we are apt to forget that the sea is but land covered with water. A sea-coast line is therefore commonly mentioned as a form resulting from marine denudation, a curved line produced by sea-waves acting unequally upon rocks of various hardness. It seems to be assumed that a hollow curve like Cardigan Bay was very slowly scooped out of the edge of a block of high land by the great rollers which still sweep in from the south-west. If Cardigan Bay were simply ocean-work of this kind, the whole coast-line would retain the tool-marks of waves. The rocks would be steep, broken, and angular, like the precipice which overhangs the sea at Aberystwith. There would be heaps of fallen debris and beaches of rolled stones beneath a bold coast-line, for sea-waves can only act between wind and water.

The sea does wear away this land, but it works as a pond does, by undermining and breaking down its banks.

The form of Cardigan Bay is not wholly due to the slow

action of Atlantic waves, for the coast is not generally precipitous. The coast-line is due to the surface-form of the land, whose valleys and ridges stretch out under the sea, and Cardigan Bay is part of a large hollow. The surface of denudation has been broken through by sea-waves at many places at the sea-level, and there are many sea-cliffs; but the rock-surface has been preserved elsewhere, and the bottom of Cardigan Bay is but a continuation of the rocks of Wales. In particular, at the head of Cardigan Bay a series of deep glens are continued under water; and if the fifteen-fathom line were the coast-line, there would still be a long fjord off Portmadoc, running N.E. and S.W. as the glens do on shore.

Tradition.—Modern geologists are rapidly nearing a conclusion at which many have arrived. It is held that men, and certain large animals which no longer exist—great hairy elephants, rhinoceroses, elks, cave-bears, and other such creatures—existed together in parts of Great Britain and in France, at a time when the climate of these countries was at least as cold as it is now in the same latitudes on the Labrador coast.

The oldest of the races who now inhabit Western France and the British Isles are admitted to be Lapps, Basques, Celts, and Cymri. If geologists are right, the ancestors of these races may possibly have lived in the end of the cold period where their descendants now live; or they may have found older races there, whose ancestors had hunted hairy elephants and wild bulls amongst glaciers in Scotland, Ireland, and Wales. The race may have witnessed great changes in sea and land. Lapps have traditions about giants and big beasts. About Basque traditions little has been published, and that little does not bear upon this subject.

There are several collections of Celtic traditions. Sir Charles Lyell quotes some British stories in his *Principles of*

Geology, and another geologist is about to publish a collection of Cornish tales. In Cornwall Celtic traditions, which seem to record changes of sea-level, abound. Celtic and Scandinavian traditions, as the oldest of western traditions yet collected, may bear upon late geological changes in the west.

Charts which give the depth of the sea, such as Keith Johnston's (plate 6), show that a very slight rise or fall of land or sea would now alter the outline of Wales very materially. If the land were to sink ninety feet, Aberystwith would be under water, and the church-steeple awash in the middle of a fjord ten or twelve miles long. If the sinking were general, the majority of Welshmen and Welsh towns would share the same fate; and if the land has in fact sunk that much, the evidence has sunk with it.

If the land were now to rise ninety feet, so as to make the line of fifteen fathoms the coast-line, great part of the land now under water in Cardigan Bay would become dry land, and rounded rocky islands and points which now slope away beneath the water-line would be rocky knolls and ridges, like those which rise up through drift and peat-moss in every Welsh glen.

If like changes were now to take place in Brittany, the coast-line would alter as much or more in that region. When land has risen from the sea, the evidence remains for those who will accept it; and in Wales the evidence shows that land has risen about 2300 feet since Snowdon was a mountain. Sea-shells have been found in the loose soil at a height of 1392 feet, according to Professor Ramsay; and at 1630 feet, according to Keith Johnston's Atlas; and, according to Sir C. Lyell, stratified drift-beds exist still higher. If these great changes of level took place suddenly, rapidly, or even gradually, by fits and starts, at a time when there were ancient Britons and ancient Gauls, memorable disasters might result, which tradition may yet vaguely remember.

In Wales and in Brittany there are, in fact, many traditions which seem to point to such geological changes as a sinking of land ; to great disasters, and to the existence of animals which have passed away ; and in all works on geology evidence is given to support these traditions.

In Wales it is told that Cardigan Bay covers a land which was thickly peopled by a wicked race who were overwhelmed by the sea, and sunken forests are at the sea-margin in Ireland.

In Brittany, according to the popular tale,* the wicked Princess Dahut, the daughter of King Grallon, and all her court, were overwhelmed in the city of Keris, near Quimper, which stood "where now you see the Bay of Douarnénèz," near Brest. King Grallon was a good man, and he was saved by a saint, whom he had made a bishop. The author of the *Foyer Breton* maintains in a note that the ruins of a town yet exist under water between the Cap de la Chèvre and the Pointe du Raz.

In Normandy it is told that the tenure by which a certain abbot held his land was the service of laying a plank for his superior to walk over from Jersey to the mainland of France. Mont St. Michel, it is said, was in a great forest when its owner went to the wars; when he returned, he found it a rock in a wide plain of sea-sand. The church on the top saved the rock from the destruction which overwhelmed the wicked plain. There appears to be some geological evidence for the existence of the drowned forest.

In England there is a tradition that merchandise was carried on horseback from Winchester to Puckaster Cove in

* Foyer Breton, vol. i. p. 232.

the Isle of Wight. But there is good evidence to prove that no great change of sea-level has taken place since the Roman invasion.

In Ireland the good O'Donoghue rises once a year, in May morning, and rides in procession along the smooth surface of the Lake of Killarney; but there is no evidence to support him.

Near the Isle of Man, Fin MacCool and his sunken country rise once in seven years to the surface, and sink down again; but if any one could cast a Bible on the land, the good old times of Fin and his heroes would return, and his land would remain above water. Geologists suppose that the channel was in fact dry when big elks lived in the Isle of Man, where skeletons have been found entire.

In Scotland there are endless traditions of the same kind. Tales of castles, towns, and houses sunk beneath the waves, and visible in calm weather; of islands which appear upon the western horizon, and sink down again; of lands where no land is, discovered in a thick fog by sailors, who find grand-looking stalwart men drinking ale from vast cups. They are the ancient mythical heroes in the "land of youth," and the "green isle," and the "land under the waves;" and who rise from time to time to show what men used to be, and what they still are in "Flathinnis," the abode of heroes.

In Ireland, as in every Celtic country, the same tales of land rising and sinking abound in endless variety; and they prevailed in the days of Queen Elizabeth, for they are recorded by Giraldus Cambrensis as facts.

In Scandinavia, the wicked city is not *drowned*, but seven parishes are smothered under snow and ice, and the church-bells may still be heard ringing under the glaciers of the Folge Fond.

Similar traditions of ancient kings—Barbarossa, Arthur,

etc.—enchanted, with all their warriors, ready to come forth to battle when summoned, prevail all over Europe, wherever popular tales have been collected. These myths seem to resolve themselves into a belief in a spirit-land; and many incidents seem to be borrowed from Holy Writ. But popular imagination has dressed the model in picturesque drapery, and the figures are often placed in landscapes painted from nature at home.

The inhabitants of central Europe, and Teutonic races who came late to England, place their mythical heroes under ground in caves, in vaults beneath enchanted castles, or in mounds which rise up and open, and show their buried inhabitants alive and busy about the avocations of earthly men. They find their heroes where they placed their bodies—under ground.

The Celtic races who came early to the west, and to the coast-line, place Arthur and Fionn, Merlin and Ossian, and all their following of bards and warriors, and those who have inherited their attributes, in islands, in lakes, or in a land beneath the waves of the sea. Perhaps they find them where they lost them or placed their bodies.*

In Morayshire, the buried race are supposed to be under the sandhills, as they are in some parts of Brittany; and as a matter of fact, marks of ancient cultivation constantly appear in the trough of the sand-waves of Moray. Where the adjuncts of a myth fit the country and the facts in so many known ways, they probably fit equally well in the matter of unknown change in a coast-line.

If Wales sunk ninety feet, after men had taken possession of it, the line of fifteen fathoms marks off a tract of low

* The savage inhabitants of Tierra del Fuego sink their dead in deep water, according to Admiral Fitzroy.

country more than twenty miles wide, which was drowned in Cardigan Bay, as Welsh tradition relates. If France went down as much after a town was built at the end of a valley near Brest, the town was drowned as Aberystwith would be, and the valley became a bay as the Breton tale describes. If ocean-currents change places, and climates are transferred for a time, flourishing valleys and mountain pastures might become the beds of glaciers and snow-heaps, as the Scandinavians tell. The Justedal glaciers have in fact advanced and retired again a short distance, and Swiss glaciers have done the same in modern times.

All these mythical disasters may be, and very probably are, records of real events, witnessed by men, and related by generation to generation; though the wickedness of the people, the miracles, the marvels, and the religious features of the story as now told, may have been invented or added when Christianity was first taught to a rude people. If Wales were to sink ninety feet now, the survivors on the mountains would be apt to quote the destruction of the "cities of the plain" as a parallel to the destruction of Welsh watering-places, where the majority of the inhabitants are strangers who cannot speak Welsh.

In the case of extinct animals, tradition may be true also.

There is a widely-spread popular tale, common to Ireland and Scotland, and told with many variations. The gist of it is, that in the days of Fionn there were deer and birds far larger than any which now exist.

Ossian, it is said, when old and blind, lived in the house of his father-in-law, or in the house of St. Patrick, and they were busily writing down all he had to tell them of the history of the Feinne. But no one would believe what he said about the strength of the men, and the size of the deer, the birds, the leaves, and the rolls of butter, that there were in the "Feinne," the country and age of Fionn.

To convince the unbelievers, the last of the old race prayed that he might have one more day's hunting, and his prayer was heard. A boy and a dog, the worst of their class, came to him in the night, and with them he went to some unknown glen.* There, with many strange incidents, it is told how they found a whistle and a store of arms, and a great caldron, and how the blind hero collected deer and birds by sounding his whistle, or horn, or "dord." Deer came as big as houses, or birds as big as oxen. Guided by the boy his hand drew the bow and slew the quarry, and when the chase was done they dined as heroes used to dine. A hind-quarter was brought home, and the bone of an ox went round about in the marrow-hole of the shank of the creature which Ossian had brought from the "Feinne." With endless variations, this story is told all over Ireland and Scotland; and it is firmly believed by a very large class of her Majesty's Celtic subjects in Ireland, Scotland, and Wales, that there were giants and monstrous animals in the days of King Arthur and of Fionn. There is no geological evidence yet for gigantic men, but peat-bogs, gravel, and caves, are full of the bones of beasts as big as a small haystack; and the word used in the tale, "Con," means "Elk" as well as bird.

In beds of superficial drift, in caves, in peat, clay, and gravel, near Torquay, in Wales, in the Isle of Man, in Ireland and in Scotland, bones of big British beasts have been found. Amongst them are—cave-bears larger than any living species, tigers twice the size of those of Bengal, elephants twice as large as those commonly found in Africa

* The glen is pointed out in Sutherland, near Dupplin, and at intermediate spots.

and Ceylon, two large species of rhinoceros, hippopotami as bulky as those of Africa, great cave-hyænas and lions, elk as tall as horses, gigantic oxen, reindeer of the ordinary size, and big red-deer with horns like wapiti. Did these or some or all of them live within the memory of human tradition?

Tradition seems to remember big beasts and ice-clad mountains, philosophy finds human bones so placed as to support tradition. The ruins of a drowned town support the Breton tale which describes its destruction. Thus legends rest upon piles of old bones; tradition and geology support each other, and point the same way. Two separate and very different routes lead back to a time when men and elephants were drowned by changes in the level of sea and land, in countries now inhabited by Celts and Cymri, and the last discovery in France brings men who could carve good pictures of reindeer, and bones of reindeer of large size, into one place, where bones and works of human art are enclosed in slabs of stalagmite.

If the block of land which is now Wales has been up and down, under water, awash and high and dry; if arctic and equatorial streams have spent their force upon it, the surface must bear their marks.

Supposing an arctic current to break upon the north-eastern corner of Wales, that corner ought to be worn away to a slope facing the current, and beds of rock should be broken short off to form precipices on the south-western side, if heavy ice was driven over the hills towards the S.W.

It is so in the small scale in all valleys where glaciers have slid downwards. It is so in the valley of Gwynant near Beddgelert, and similar action would produce like form on any scale (see cut, p. 6).

Standing upon *Little Ormes Head* and looking south-east,

the north-eastern corner of Wales is seen in profile, and the general outline of the country has the form of small rocks worn down by ice which moved from N.E. to S.W.

To a practised eye the Welsh hills seem to tell their story of movement from the N.E. as clearly as Welsh trees do of movement from the S.W. (see vol. i. p. 59).

Looking south-west from the same point, the end of the ridge, of which Snowdon is the highest point, is seen over a foreground of bare rocks about 700 feet high, and it is manifest that the outline of the distant ridge of high hills seen in this direction is something wholly different from the foreground, which is like the rounded hills about Mold and Wrexham. These can be seen by looking S.E.

Looking W. and N.W. the outline of Anglesea is something different from them all. When that island is crossed it is like a worn grooved slab of stone. From Ormes Head it seems to be a low undulating line nearly parallel to the horizon.

If after seeing hills in profile the observer could fly over them, he would gain a better notion of their shape.

In the case of Wales the country has been so admirably mapped by the Ordnance Survey that to look down upon a map is almost as instructive as to sail over the country in a balloon. In the Ordnance map of this district, the high hills and the low country are seen to have a totally different configuration.

The Snowdon ridge, 3570 feet high, extends N.E. and S.W., and great valleys and corries seem to have been gouged out of it in every possible direction. But on both sides of the ridge the country is furrowed by long grooves, which run N.E. and S.W. In the deepest of these is the Menai Strait. Another runs into Cardigan Bay. The north-eastern corner of the block has in fact been worn down by some force acting from the N.E., and the north-western corner has been furrowed diagonally in the same direction.

To one used to the look of ice-ground hills, the whole of North Wales, except the Snowdon range, appears to have been first ice-ground in one direction, and then further ice-ground in all possible directions, by local river-glaciers of great size, which hewed out glens.

The low hills at Little Ormes Head and Llandudno are much weathered, but they retain their general form. They are very bare, so that their form can be well seen, but here and there patches of drift, clay, and boulders, and big perched blocks, occur near the top of the hills.

The broad low isthmus which joins Great Ormes Head to the mainland seems to be chiefly composed of rounded boulders of all sorts and sizes. It is probably an old moraine arranged by the sea, and it contains specimens of many kinds of rock which are not found in the immediate neighbourhood.

Looking down from the ruined battlements of Conway Castle on a fine evening, after a strong northerly breeze has nearly blown itself out, the forms of the miniature waves on the river, and of larger solid wave-marks made at high tide upon the sandbanks, by larger water-waves, may be seen and compared. They are almost identical : one set is moving, the other is at rest ; but the wave-mark shows how a wave moved, and copies it. Looking up to the hill-sides where the trees are exposed, their form tells of a prevailing wind which bends them towards the north-east. Looking to the hills themselves, they have the form of wave-marks, caused by a north-east wind ; for they have been swept by the force which carried perched blocks, and arranged the boulders about Llandudno. There is no known force but ice which could so grind rocks and carry such stones.

At Chester, Llangollen, Wrexham, Mold, Holywell, Rhyll, Abergele, high up and low down, the north-eastern corner of Wales looks like a block worn down from the N.E.

The hills are much weathered, but they all retain a general form. Patches of sand, clay, and boulders rest in hollows; and on hill-tops perched blocks rest at all elevations from the sea, to about 1000 feet.

About *Maes-y-Safn*, and this north-eastern corner of Wales generally, it is hopeless to search for high striæ upon the limestone rocks; for they are so weathered as to leave delicate fossils projecting far above the surface. Rain-water seems to dissolve limestone like salt. It is vain to search for striæ on grits and sandstones, which crumble at a touch; but the whole of these hills have their longest slope towards the N.E.; in which direction the beds also dip at a higher angle. The steepest side is generally towards the S.W.

Sometimes the beds are broken, so as to leave precipitous faces of mountain limestone. Sometimes these edges are rounded off.

Glens are rounded grooves, and seem to be gouged out of the rock without reference to bedding; and every shape in the country seems to tell of some great mass moving over the surface of the land, and grinding it down.

There are three stages—first, a low alluvial plain, but little raised above the sea-level, which stretches far up into the glens; for example, at Rhyll. This seems to consist of transported materials. The next stage is a rolling rock-plateau, about 1000 feet above the sea. It is steep towards the N., and slopes gradually towards the E. and N.E. In the low grounds to the east, and on this plateau, are beds of drift and boulders. The hills at the 1000 feet level are all rounded. Even though the slope of the low hills and the dip of the strata are much the same in direction, the slope has nothing to do with the dip. Near Rhyll, the hills slope from the N.E. at an angle of about 9°, but the dip is about 45°.

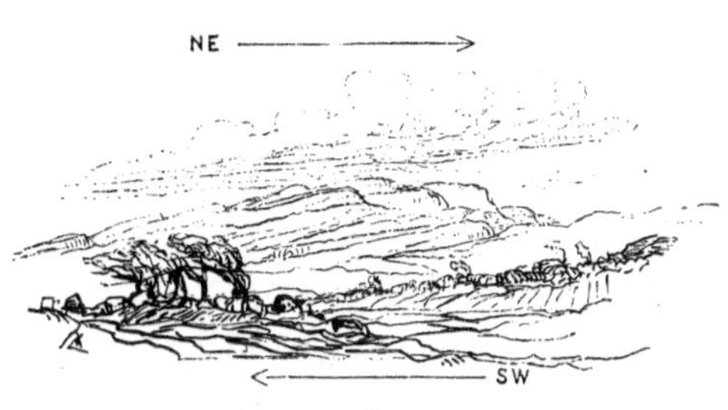

FIG. 83. N. E. CORNER OF WALES.

Above this upper level, hill-tops are weathered peaks, and mountain-glens radiate from them, cutting through the upper plateau from the watershed to the sea.

In the *Snowdon range* the rocks are harder, and striæ abound. The valley of the Conway is a great groove, which runs nearly N. and S., and which certainly contained a large glacier, or heavy fjord ice. The road to Llanberis follows its course to the foot of Snowdon. The bottom of the groove is filled with beds of gravel, sand, clay, and peat, in which large trees are buried. It is a flat plain, through which the salmon-stream winds to the estuary, where it meets the tide; trees, green fields, and neat houses abound; a railway train screams and rattles over the plain, and up the glen; but there was a big glacier there nevertheless. The railway cutting has uncovered a rock about twenty-five feet above the sea-level, and near a ferry above Conway; and glacial striæ are as freshly marked upon the slate as if they had just been made.

Above ground, the rocks are weathered and broken down. Many forests have sprung up and died since the ice was there; but under the beds of drift the original surface of glacial denudation is unmistakeably clear. If there was a glacier at Conway, there may have been others in other Welsh glens.

Leaving the valley at *Llanrwst,* a path leads up the Snowdon side of the valley, past Gwydr House, to Coed Mawr Pwll mine. There are numerous ice-marks, boulders, and suchlike, all the way.

To the left of the path rises a hill called *Coed Mawr,* from which a wide view is obtained. It is the Rhigi to this range, a kind of outlier, a flat-topped ridge separated from the main ridge by a hollow, and cut off from the rest of Wales by deep valleys. At the height of about 1100 feet above the sea, and on the top of this outlier, the ground is strewed with loose boulders.

The rocks are well marked with striæ, and their direction corresponds to no existing feature of the country. They neither point down-hill, nor from the ridge, nor along the run of any valley or river near them; they point north-east over Rhyll, and south-west over Traeth Bach in Cardigan Bay; parallel to the Menai Strait, to the ridge of Snowdon, and to the run of the great sound which would cut through Carnarvonshire between Moel Siabod (2865 feet high) and Moel Wynn (2529), and so join Cardigan Bay at the two strands "Traeth Mawr" and "Traeth Bach," near Portmadoc, if the sea were at this level of 1100 feet. A glance at the Ordnance map shows that the ground in this direction has the form of an estuary of glaciers passing south-west into Cardigan Bay.

This mark joins in with the curve which has been followed from Yorkshire, for no land-ice could well move N.E. or S.W. at Coed Mawr now, unless the névé was about the Pole.

Two hundred feet lower down, in the valley between Coed Mawr (1100) and Carned Llewellyn (3482), between the main range and the isolated hill, at a height of about 900 feet, a small lake, *Llyn Pencarreg*, has been drained close to a lead-mine. It was in a rock-basin, for they had to cut through rock to drain it into the branch of the Conway which comes from Snowdon. The bottom is filled with peat, and where the peat has been removed glacial striations are fresh and perfect. These point E.N.E. and W.S.W., out into the valley, through the hollow where the drain was cut. If ice were now sliding from Carned Llewellyn it might be caught in the trench and split on the watershed. Part of it might slide northwards into the Conway valley, along the line of the path to Llanrwst, and the rest would swirl round and move W.S.W. towards Capel Cureg, where it would meet the Snowdon stream, turn back to Bettws-y-Coed, and so flow on to Llanrwst by a circuitous path along the river-course.

If a Carned Llewellyn glacier were so large as to overflow the top of Coed Mawr, it would evidently flow S.E. into the Conway valley; but the marks upon Coed Mawr are at right angles to this direction—they point S.W. Moreover there appear to be a series of shelves higher up which correspond to the striæ, not to the present watershed.

If the Conway glacier, which must have had a source about Moel Wynn, were large enough to overflow the whole country, it might possibly move north-east, over Coed Mawr, but it would have to cross a glen 500 feet deep, at right angles at Bettws-y-Coed, and then move along a hill-side at a higher level than the opposite side of the Conway valley, about Llanrwst, which seems impossible. Making every allowance for land-ice of enormous thickness, it is still very difficult to explain the striæ at Coed Mawr without the agency of floating ice.

But if ice floated above 1100 feet, then the Snowdon range was an archipelago when this mark was made, and Moel Wynn was an island. But as sea-shells are found 500 feet higher up, and stratified drift 400 feet above the shells, icebergs may have floated along the Snowdon islands so as to mark sunken rocks 900 feet below the sea-level. Of 3570 feet of Snowdon there would still remain 1570 above water to form a base for the land-glaciers which Ramsay describes. When the land rose the Conway glacier might flow down to the present sea-level; ice certainly did move in this trench.

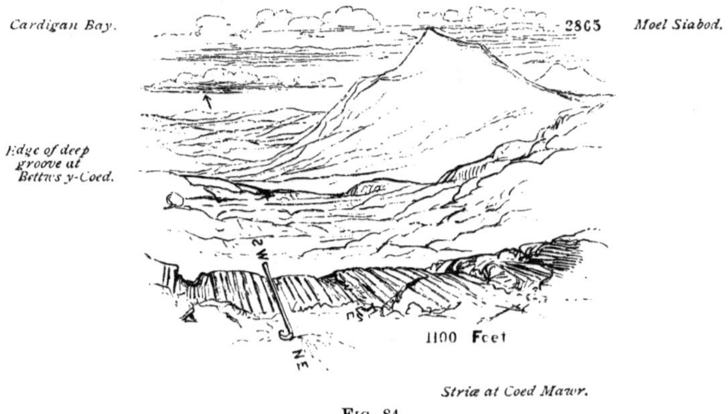

Fig. 84.

On this supposition the striæ on Coed Mawr are older than those which are seen from the train, about 1075 feet lower down, and those which remain in the lake 200 feet below the ridge at Coed Mawr. They look far older, and in this respect resemble others of their class. Looking south-westward along the line indicated by the striæ, there is a great hollow between Moel Siabod and Moel Wynn, beyond which is Cardigan Bay and its great strand Traeth Mawr.

When a great smooth Atlantic roller, moving steadily on, encounters an isolated rock, some twenty or thirty feet higher than high-water mark, the glassy surface of the wave breaks, and a torrent of boiling foam, green water, and glittering white spray, rushes over the stone with a hoarse roar. If water then left marks they would be parallel to each other, and to the direction of movement. If a stone or any other loose object stands upon the rock, it is driven on by the torrent, and follows the wave till it sinks. But when the crest of the wave has passed, the rock seems to rise up like a whale, or some other black monster of the deep. Then for a time the direction of movement changes—green torrents, streaked with snowy foam, stream down the black sides of the rock, and brown sea-weeds flutter and wave in rivulets which radiate outwards and downwards from the highest point of the rock in every direction. If these left marks they would radiate as the streams do. The rivulets would make furrows, and flow in them while there was any water left to flow. But they leave no such marks. The Dubh Iartach, the outermost rock off the west of Scotland, has a rough jagged surface, though it rises twenty feet above the sea where waves are as large as any in the whole world.

When river-ice drifting down-stream meets a stone, the ice-surface, like the smooth wave, breaks. It pushes on, up and over the stone in the direction of the stream which moves it, but it slides off in many ways. If heavy enough it would mark the stone.

If ice is moved by a falling tide, a time comes when it no longer slides over the stone, but splits upon it, and slips past it, and meets behind it with the stream.

And then if a shower falls the water streams down the sides of the stone in every direction, while the stream flows past as before. If snow falls it caps the stone, and when the tide has ebbed the bed of the stream retains marks of the current, while the snow is left to tell its own story.

If the blocks of stone which Welshmen call Plynlimmon, Y Wyddfa, and Cader Idris, were rising stones in the falling tide of an ice-laden ocean-current, like that which now overruns sunken islands off Labrador, they would retain the marks, for heavy ice does record its movements upon stone, and stone preserves the record.

The high Welsh hills do retain ice-marks, and they seem to record that the hills rose up in an icy sea which moved ice towards the south-west for untold ages, and that glaciers streamed from their sides when the cold tide fell, and continued to flow on, until a long age of winter gradually passed away, after the bed of the cold stream was crossed by Lapland.

The hills about the head of Cardigan Bay seem to record that the stream poured out that way, and that the coast-line is a result, not of waves acting at the present sea-level from the south-west, but of ocean-streams pouring towards the south-west, from the arctic basin into the Atlantic.

The deep trench in the fifteen-fathom line tells the same story. It seems to carry the south-westerly curve over England and Wales, and to launch it in the Irish Channel.

The hobby seems none the worse for this rapid burst. The story told by Scandinavian and Scotch hills is confirmed by hills in Yorkshire, by stones at Stoke and in Cheshire, by geologists and their books, by popular tradition, by the map of Ireland, and by high ice-marks on Snowdonia.

CHAPTER XLI.

BALTIC CURRENT 14—BRITISH ISLES 13—WALES 2.

ARCTIC sea-shells found in loose drift at a height of 1392 feet, and boulders, perched blocks, and drift at a height of 2300 feet,* prove that a cold sea has been as high on the flanks of Snowdon, since rock was ground into something like the present shape of Wales. High horizontal ice-marks on a hill-shoulder at 1100 feet seem to prove that the cold sea which rose so high was cumbered with ice and moved from north-east to south-west, when the way was last open. If land and temperature rose together gradually, and the cold period passed away from Wales when rising land reached a certain point; then marks on watersheds at various elevations ought to record the changes and their order.

Glacial drift, arctic shells, and horizontal ice-grooves, record the high sea-level and cold weather. Glacial drift partially waterworn, and packed in forms characteristic of sea-margins, at lower levels amongst the hills, seems to mark an ebbing sea and warmer weather, a state of things more like the present state of the beach at Galway (p. 21). Water-worn drifts at a lower level, terraces, and sea-shells, speak for themselves. It seems reasonable to assume that during a gradual change of climate, dwindling glaciers flowed in

* *On the Superficial Accumulations and Surface Markings of North Wales.* By Professor A. C. Ramsay, F.R.S., F.G.S. March 26, 1851.

rising glens, long after the greatest cold had risen off the sea.

A series of terminal moraines, entirely made of native rocks, and laid in hollows, mark the retreat of dwindling glaciers, shrinking upwards ; while the cold shell of air-temperature and land rose together ; and in Scotland the lowest perfect moraine seen is at about 1400 feet, the level of the Welsh shells.

Old striæ at Snæfell point up to, and converge upon, the high point from which smaller glaciers now diverge (vol. i. p. 432) ; and the same series of events appear to have followed each other in like order in Wales and in Iceland.

Marks made in the bottom of deep glens near the present sea-level may be marks of comparatively modern glaciers, which continued to flow into the sea long after hill-shoulders, with old scars, had risen far beyond the reach of the battle between sea-water, sea-ice, and Welsh stone, or they may be marks of fjord ice like that which now works with the tide in Hamilton Inlet in Labrador.

The old local glacier-system of the Snowdon range has been well described by abler pens.

Buckland, Darwin, Lyell, Murchison, Ramsay, and a host of famous men, have piled up a mountain of facts which would be harder to get over than Y Wyddfa. The former existence of Welsh glaciers is proved beyond dispute ; and to a practised eye the record seems patent.

At *Capel Cureg* ice-ground rocks abound. At the head of the pass, where the water sheds towards Cardigan Bay, at a place lower than Coed Mawr, ice-marks rise high, between Moel Siabod and Snowdon. If ice floated at 1100 feet, this was a sea-strait, and these may be marks of heavy drift-ice moving in a groove like the Menai Strait. Two ice-streams here split. One reached Conway by the road and railway; the other went to Beddgelert and Portmadoc. Whether both reached the present sea-level remains to be proved. It is certain that the ice was of large size, and it reached Conway.

At the col at the head of the Pass of Llanberis, about 1300 feet above the sea, a cross strait divided the Snowdon range when shells and drift were deposited upon the hill-sides at 1392 and 2300 feet.* According to the ice-marks, two glaciers met in this trench, and parted, as glaciers part now at the Col de Géant. One ice-stream probably split lower down, and went to Conway and Portmadoc; the other stream went towards the Menai Strait, for the marks are plain in this direction for many miles. Above this col, Ramsay has tracked old moraines, almost to the peak of Snowdon. One system thus tracked from Conway to the highest peak of Wales, the map of the country gives the shape of the local system. It must have been a herring-bone pattern of ice, for the glens all radiate like ribs from the backbone of North Wales.

It has been shown above (vol. i. p. 157) that rocks upon the snowshed of the Alps, on the Strahlek, at 11,000 feet, and in the midst of land-glaciers, are not ground, but riven and shattered. It is also shown (vol. i. p. 167) that rocks on the snowshed of Mont Blanc, on the Col de Géant, at 11,146 feet, and at the source of the largest of European glaciers, are equally shattered; although the snow-dome of Mont Blanc, 15,744 feet high, rises 4598 feet immediately over this pass.

From the top of Mont Blanc the Glacier de Boissons flows continuously down 12,300 feet to a level only 3444 feet above the sea. This glacier descends 3902 feet below the

* According to Professor Ramsay's paper above quoted, the drift overhangs this pass.

level of the Grimsel Col, which is 7346 above the sea. According to De Charpentier and Elie de Beaumont, one, and the highest known, superior limit of the erratic formation is at the Grimsel Col. There, at the Furca, and on similar passes in the Alps, at about this level, rocks are rounded. The top of the Stelvio (9272 feet) is not shattered but ground (vol. i. p. 144). The inferior limits of the erratic formation of the Alps are far beyond the Rhine on one side, and near Turin and Milan on the other; and the question is whether these stones were carried from the watersheds of the Alps all that distance upon land-ice, or part of the way on land-ice, and the rest of it on ice-floats (vol. i. p. 169). If the Snowdon ice-marks were made by land-glaciers, which grew in consequence of a great elevation of land (which is one theory suggested to account for them), they ought all to point up-stream, to and towards some snowshed; and the snowshed ought to be shattered when it is narrow, because the Strahlek and Col de Géant are shattered. According to this theory the snowshed at Llanberis, which is very narrow, ought to be shattered.

The top of the col is in fact rounded.

The highest grooves close to the head of the glen are as deep as grooves made in places where the heaviest glaciers press hardest, and they seem to be nearly horizontal. If the ice-work in this district is sea-work—a result of a cold period caused, not by great elevation, but by a small depression of land—the marks agree with the present state of things on the opposite coast.

If the col at Llanberis was first a deep strait, then a shallow sound, and then a "tarbert" at the end of a sea-loch open to the ocean on the west, heavy drift 1000 feet deep might grind the deep strait; lighter drift, 250 feet, as at Belleisle, might pass through the shallow sound; and heavy fjord-ice move horizontally in the sea-loch, as fjord-ice now does in Hamilton Inlet (chap. xxvi.)

It is certain that this col was a sea-strait 1000 feet deep when drift was packed in terraces 1000 feet above the pass, and that it was a sound at least 92 feet deep, when sea-shells were buried in drift, where Mr. Trimmer found them at 1392 feet.

It may have been a "tarbert" 300 feet high, when shells were buried where Professor Ramsay found them at 1000 feet on Snowdonia.

So far no one has yet found shells in drift on the high Alps ; no one seems to have sought them ; but judging from form alone, it seems probable that arctic shells may yet be found in superficial deposits at higher levels than the Stelvio (9000 feet), but not above the level at which cols and peaks are all shattered—namely, about 11,000 feet.

It seems possible that rounded Alpine passes were sea-straits when they were rounded, and that land-glaciers may have been launched from Alpine peaks which were 6672 feet above water when the Stelvio was a "tarbert," and the Ortles Spitz a tall "stack" in a European ocean whose arctic current passed Snowdonia.

According to the Baltic Current theory, such a current did pass this way, and did all the work ; according to other theories, the whole of the northern hemisphere must have been covered with one vast sheet of ice during the glacial period.

When the gorge of Llanberis is passed westwards, a wide plateau begins, where the chief product of the country seems to be glaciated boulders, but rolled and waterworn. Walls are made of them, roads are broken boulders, streams run amongst boulders, and the soil is clay. At this level, about 300 feet above the present sea, most of Anglesea would be under the sea which helped to roll these stones.

The boulder-land ends in a series of steps and a steep terrace, which makes one side of the big groove, over which the tubular bridge has been thrown. These steps and terraces, and the groove which holds the Menai Strait, cross the course of the old Llanberis glacier at right angles. If the Snowdon glaciers reached the sea at the level of 300 or 400 or 500 feet, the present tides might move icebergs and land-ice N.E. and S.W. along the coast.

Anglesea.—The geological structure of Anglesea includes igneous rocks and sedimentary beds, from the lower silurian to the coal-measures. In the mines, these beds are seen to be fractured, twisted, dislocated, and roasted; the surface consists of rocks of every degree of hardness, of beds dipping everyway and at all angles, of minerals which fracture, wear, and weather into all manner of shapes; but the whole surface of the country has one prevailing form. The hills and the rocks, wherever they appear through drift and peat, have the same form as the hills and rocks of low ice-ground Scandinavian islands; and they too are ice-ground.

Boulders and clay are everywhere. Travelling at express speed in the railway train, driving or walking, the marks of ice are manifest. "Tyr Von" is like a slab of variegated marble roughly ground flat, well scratched, and ill washed.

The direction of movement was N.E. and S.W., that of the tide in the strait, which now looks like a big river shrunk in its bed; the grinding-machines were probably icebergs and sea-ice worked by tides and the Arctic Current, with boulders for polishing-powder (see chap. xxvi.)

All the rocks seem to have their longest slopes and smoothest sides towards the N.E., so the machines worked most from that direction, and the sea-level was probably more than 300 feet higher than now, about the level of the boulder plain, when the ice vanished.*

Looking south-east, the side of the Snowdon range whose end is seen from Llandudno, appears as a long ridge most worn at the north-eastern end, and furrowed by deep glens which cross the ridge at right angles. Generally this north-western corner with its bent trees must leave the impression of something now swept by a powerful S.W. wind, and formerly ground by some force which acted from the N.E.

It repeats the story of the north-eastern corner of Wales, but in a more legible form. It surely was like the corner of Iceland (chap. xxv.), or Jan Mayen (chap. xxiv.), or Bear Island (chap. xxiii.), or islands about Hamilton Inlet over the way (chap. xxvi.)

From Carnarvon the road to Beddgelert first passes through a boulder country and over terraces, then up the course of an old glacier, which left notable marks. At Beddgelert the course of the Portmadoc and Snowdon glacier is crossed, and thence all the way to Tan-y-Bwlch, the road crosses a series of large furrows running north-east and south-west.

In some places the surfaces are beautifully preserved low down. Many ice-streams seem to have converged here. Traeth Mawr is seen to the westward, and Moel Wynn is to the eastward, and there seem to have been large glaciers on both sides of Moel Wynn which met here. The marshy plain is probably a heap of drift and glacial debris, a whole collection of ruined moraines arranged by the sea, like the plain on which Llandudno stands.

* According to Professor Ramsay, striæ in Anglesea were made by floating ice; they generally point E. 30° N., and are quite unconnected with those of glaciers in Caernarvonshire.—*Paper read March* 26, 1851.

From *Tan-y-Bwlch* the road rises into a valley, which is strewed with large stones at the height of 700 or 800 feet. The walls are of boulders, many of which are grooved, and the rocks and low hills are all rounded to the very top. Above a certain level, the hills are steep and broken, and furrowed with larger corries. At the level of the Coed Mawr striæ (1100 feet), this glen would be a strait. On the map this inland country seems to have been swept southwards, as if a N.E. current had split on Diphwys, a range 2050 feet high. The glen may afterwards have been filled by a Mer de Glace which was fed from both sides, and overflowed two ways to Tan-y-Bwlch and to Dolgelley.

The deep glens which meet at *Dolgelley* all have the form of glacier-glens, and above Dolgelley at the pass of *Bwlch-llyn-Dach*, about 1000 feet above the sea, ice set off southwards, and left a large moraine of crumbled slate, to mark the spot where it finally expired, below Cader Idris. This is not a perfect moraine, but is washed or weathered out of shape. Tradition narrates that a giant called Idris sat on the Cader, his seat, and strode from side to side of this gap. He was one of "Hyrm Thyrsar," the frost giants of Norse mythology, and he has turned to mist; for he was ice, and he has melted away.

Thence all the way to Aberystwith, the hills and glens have the same general rounded forms, and wherever a quarry or a broken stone appears, it shows that the form is different from any which could be produced by weathering or upheaval. It is neither the form of bedding, jointing, cleavage, nor fracture. It is the form of glacial denudation.

At the *Devil's Bridge*, some fourteen miles from Aberystwith, a river has made a mark in a slate rock, which proves that water could never wear slate into the form of Welsh glens. A stream working at the bottom of a curved hollow has cut its own breadth straight down for ninety feet, and is cutting backwards for some hundreds more lower down. The rock is too hard to weather or break easily, and it has not fallen, so the river-mark is perfectly preserved. Further down, the valley retains its glaciated form, and higher up, wherever a valley is left, the upper level of the country is seen to have one uniform slope from Plynlimmon to the sea ⌒.

There is the general form of denudation upon the largest scale in the outline of the country, and in the glens which run north-east and south-west; next the form of denudation by local glaciers, or glacial currents, which scooped out broad concave glens; and lastly, a steep straight ditch cut by running water at the bottom of the old ice-groove.

There is no room for doubt as to the tool which made this drain; the marks are seen from the water-level up to the foot of the bridge, and there is no joint or vein in the rock, for the rock is smooth and polished, and the slate beds are unbroken in the bed of the stream. At the bottom of the trench, which the stream has dug ninety feet through slate, there is not a chink in the stone.

If the rate of wearing could be got at here, it would be a chronometer. It is not likely that the river worked thus under ice; it certainly did not work below the sea, so it began to dig after the spot had risen. It is now 750 feet above the sea. The stream was about its present size when it began at the ninety feet, for the trench is no wider at the top than it is below. The question then is, How much slate does this river wash off in a year? By anchoring stones in the river, and weighing them from time to time, this question might be solved, and then the upheaval of Wales might be calculated from the river-mark.

At Borth is a large beach, which crosses a rock-hollow, like a sea-dam.

Behind the dam peat and silt-beds have gathered ; in front of it a bed of yellow sea-sand is smoothed by Atlantic rollers; and the mound itself is a blue ridge of slate pebbles and boulders rolled by the sea. These were probably carried from their parent rocks by the Plynlimmon and Machynlleth branch glacier from the Plynlimmon and Cemmis junction, where it joined the Severn valley ice-line, at the watershed.

From *Borth* near Aberystwith, a railway has been made across Wales to Shrewsbury, and the cutting has not yet (1863) been overgrown with turf. Travelling on this line is like studying a geological section. The hills and valleys are all of one pattern outside, but they are composed of beds which dip in many directions, and at many angles, and which are of various kinds. The rock is often covered with glacial debris, beds of clay, generally yellow, enclosing angular and rounded blocks of stone of many kinds. There are grits, white quartz, igneous rocks, and slates. Near Carno, about 700 feet above the sea, these are well seen.

At the height of 1100 feet, this would be a sea-strait. It may afterwards have been the bed of glaciers which came from Plynlimmon, split on the watershed, and worked their way to Shrewsbury and Cardigan Bay.

With the well-marked glacial phenomena of the high mountains of North Wales fresh in the mind, a rapid journey along this line is like reading the history of a glacier. Bare rocks get covered ; stones get more rounded as the train descends ; the colour of the clay changes ; confused heaps of loose rubbish are better sorted where they have been washed in hollows ; there is more variety in the materials after a greater number of beds have been passed ; and finally, when the low plains are reached, the whole is hidden under alluvial soil. The work of ice is covered by the work of water and air, and a green cloak of vegetation is thrown over all.

Then comes the plain, and the town, and archæology, and man's history recorded by his works; old houses, old glass, old churches—a museum of antiquities. Old English, Norman, Saxon, Roman, Celtic, and unknown remains—all records of a series of events, which began here after the other ended. And yet the sculptured marks of ice which moved between Snowdon and Conway, and passed over Coed Mawr and Anglesea at 1000 feet, and at the sea-level from N.E. to S.W., are better preserved than Roman sculptures from Uriconium; and there are boulders near the Stiper Stones, which tell their story at least as well as the ruined gable of an old house.

The geological sections of Wales, which have just been finished, confirm what has been said above.

On the western side of Cader Idris boulder-clay is marked at 1100 feet; at 1000 on the western side of Snowdon, and at 1700 feet at Mauchlyn Mawr.

On the eastern side of the hills drift is not marked, but drift exists in patches everywhere. If the movement was south-westwards drift ought to be found to the westward of the high grounds, under the lee of islands which are now mountains. Sea-waves would tend to wash the drift from the south-west end, where it abounds most.

The structure of the country shows trap, felspathic ash, fossiliferous and non-fossiliferous slates, grits, lime, shales, and coal-fields. There is evidence of fracture, disturbance, and bending of strata, upon a very large scale, and of volcanic eruptions. The mines show that the shattered crust has grated its broken edges to make smooth grooved sides in the cracks. Bits as broad as a parish and of unknown thickness have risen, or fallen, or moved horizontally; and every bit has moved, for there are slickensides in every mine. The surface must often have been rough and jagged like that of a broken flagstone laid upon a soft bed and trodden awry. Some of the cracks are filled with clay and boulders, so they were open when ice was here. But some great force has now ground off all the corners. The geological section gives the same lines which can be seen in every Welsh quarry, and in many quarries the surface of glacial denudation yet remains.

The geological map shows no granite in Wales. Granite boulders are found in Cheshire to the north-east, and the nearest English granite hill is further to the north and east than the Cheshire boulders.

If the assumed curve is followed up-stream it joins Wales, Cheshire, the Skagerrak, and a Scandinavian district where granite abounds, and where ice-marks are conspicuous at high levels.

So the block of land which we call Wales seems to have been ground down by an arctic current and by local glaciers, which gradually disappeared after the land had risen to a certain level, and of which the last traces are to be found in the highest part of the highest glens. Whether any of these traces coincide with any record of man, is the geological question of the day.

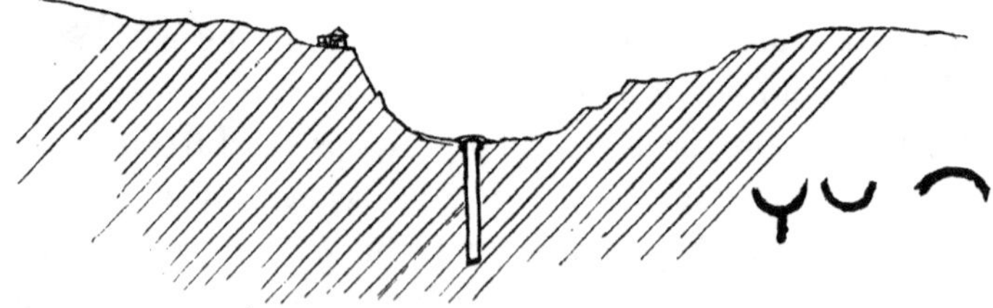

Fig. 85. Devil's Bridge.

CHAPTER XLII.

BALTIC CURRENT 15—BRITISH ISLES 14—ENGLAND (SOUTH).

A SET of curves, like the rest, drawn from Novaya Zemlya proper, pass over Russian Lapland and the White Sea; Finland, the Gulf of Bothnia, and the Baltic; the low rocks of Sweden; the drift of Denmark, Hanover, Holland, Belgium, and part of France. In England, curves pass from Whitby to Snowdon; from the Wash to the Bristol Channel; from the Thames to the Isle of Wight; and from Heligoland past Dover, down the English Channel, and out to sea.

It has been shown above that there is reason to believe that ice travelled south-westward over Sweden and Finland (chaps. xviii. xix. xx.) A succinct account of the superficial geology of Denmark is given by Sir C. Lyell in the second chapter of his last great work. Means of temperature and limits of vegetation have been mapped, and a series is published in Keith Johnston's *Physical Atlas.* From facts taken from these stores, and from personal knowledge, it appears that the present mean annual temperature in Denmark is about 46° and 48°, and the forests chiefly beech. In the upper beds of peat the trees which are preserved are chiefly beech; and in this layer human remains are associated with weapons of iron and other metals. In the next layer the trees are oak, and works of human art older and chiefly bronze. In the next the trees are Scotch fir and birch, and human implements far ruder, and chiefly stone.

Beneath all these are layers of glacial drift, clay, and scratched boulders. These several layers seem to indicate a gradual change of temperature from cold to warm; thus—

In Bear Island, Greenland, and the north of Labrador, a mean temperature of 28° now coincides with the deposition of glacial drift in the sea, and with the polishing of rocks by land and sea ice.

About the North Cape, Western Iceland, and the south of Labrador, a mean temperature of 32° now coincides with the growth of fir-trees and birches on shore, and with the deposition of glacial drift in the neighbouring seas.

About Stockholm, Christiania, Cape Race, and Nova Scotia, a mean temperature of 41° now corresponds to the growth of oaks, pines, and other forest trees, and of heavy winter-ice on shore and afloat.

Lastly, about Copenhagen an isothermal curve of 44° passes north of Scotland and south of Nova Scotia, where sea-ice now marks rocks, deposits drift, and moves south-west about lat. 45° in the Bay of Fundy.

If the climate of Europe were now like that of America there would be ice-floats on the northern coast of Spain in winter; the cold of Copenhagen and Halifax would reach Bordeaux; while the cold of Labrador, Cape Farewell, and the North Cape of Norway, would reach Copenhagen.

The glacial drift of Denmark seems to prove that the present climate of Labrador did in fact exist about Jutland when that spot was under water, and geologists are agreed that Jutland was an archipelago at no distant date. The Danish stone, bronze, and iron periods, with their vegetations, so far prove a change of climate during the human period, after the land rose.

According to the Baltic Current theory, the blocking up

of a northern strait by a rise of land was the first step in a gradual change which is still in progress, for the last Norwegian glaciers are now dwindling away.

Rivers of all dimensions have deltas ; ocean-streams, especially when laden with ice, ought also to build submarine deltas ; the Banks of Newfoundland, about lat. 50°, seem to represent the "northern glacial drift" of the present day : if so, Denmark, the Dogger Bank, and the drift districts of eastern England, may be parts of the submarine delta of the Baltic Current. The direction of striæ, shells, and the nature of the drift on shore, are the only guides.

The same high authority who states the order of superficial deposits in Denmark also describes the eastern coast of England (chap. xii., *Antiquity of Man*). The "series of documents" which lie next below the glacial drift in Norfolk and Suffolk read thus, according to Sir Charles Lyell's translation of the rocks :—

> "The fossil-shells of the deposits in question clearly point to a gradual refrigeration of climate from a temperature somewhat warmer than that now prevailing in our latitudes, to one of intense cold."

According to the Baltic Current theory, the opening of a northern strait, by the sinking of land, let in the cold climate, which is now transferred to Labrador, by the close of the strait.

The English documents, as read by Lyell, record many successive changes in the relative level of sea and land in Norfolk, Suffolk, and Essex. Forest-land has sunk, for beds of shells are spread above the upright stumps of fir-trees identical in species with firs now growing ; the sea-bottom has risen, for trees now grow above the shells, and men spread shell-marl in the fields, on the top of the English cliffs.

Through these old buried English fir-woods, elephants, rhinoceroses, and other big brutes roamed; whales, narwhals, and sea-horses swam over the same spot when it sank; and then came an ice-chapter, which the best of modern geologists thus translates :—

"Erratics of Scandinavian origin occur chiefly in the lower portions of the till. I came to the conclusion in 1834 that they had really come from Norway and Sweden, after having in that year traced the course of a continuous stream of such blocks from those countries to Denmark, and across the Elbe, through Westphalia, to the borders of Holland. It is not surprising that they should then reappear on the eastern coast between the Tweed and the Thames, regions not half so remote from parts of Norway as are many Russian erratics from the sources whence they came."—*Antiquity of Man*, p. 218.

The Baltic Current theory is thus propped up by a strong buttress of facts, stated by a great authority to prove something else. The northern strait, which is supposed to be the source of change in English climate, is at the head of the Baltic. When land was sunk in England and in Denmark, a cold sea carried boulders from Scandinavia to England, in the direction of the curves above shown (vol. i. p. 232); but when the land rose higher, the transport of Scandinavian stones was stopped, and soon after that clause in the ice-chapter was recorded in the till, the glacial period began gradually to pass from Europe. It is argued that it went to America.

Sir Charles himself suggests, that the "glacial period" may be nothing but a transfer of existing climates, by causes now active, but other causes than a Baltic Current.

One more fact may be taken from this storehouse.

At the end of the glacial period, eastern British drift came, not from Scandinavia, but apparently from the north of England.

Sir C. Lyell says—

"Patches of the northern drift, at about 200 feet above the Thames, occur in the neighbourhood of London, as at Muswell Hill near Highgate. In this drift, blocks of granite, syenite, greenstone, coal-measure sandstone with its fossils, and other palæozoic rocks, and the wreck of chalk and oolite, occur confusedly mixed together. The same glacial formation is also found capping some of the Essex hills further to the east, and extending some way down their southern slopes towards the valley of the Thames."—*Antiquity of Man*, p. 160.

Many of these fragments are not Scandinavian, and may be of native growth, and the deposition of this drift is supposed to have taken place at a time when nearly the whole of the low grounds of England were at least 200 feet under the sea.

According to theory, Scandinavian drift gave place to English drift when the stream and the local tides changed their direction, after the way from the polar basin to Muswell Hill was blocked by Lapland, now 1200 feet higher, which rose and sent the cold westward, to the place where the glacial period has now perched, to feed on rocks in Greenland.

Passing S.W. from Norwich, glacial drift is said to be found near the railway between Gloucester and Bristol, and that line leads to Devonshire. It is vain for a single hand to attempt to follow drift through all England, so it is best to get to the hills once more.

Dartmoor is an upthrow of horse-tooth granite of a peculiar character, which has upheaved and altered surrounding stratified rocks. The granite and the altered rocks are traversed by numerous veins and faults, in which mines of iron, lead, copper, tin, etc., are worked. There are numerous dykes of greenstone and other igneous rocks, which fill up breaches in the earth's crust; and there are "cross-courses," which are great cracks filled up with angular fragments of broken rock and other materials of small value. The crust has been much broken and shaken at various times, for more "heaves" and "slides," "faults," "upthrows," and "downthrows," are known in Devonshire and Cornwall than are to be seen in the cliffs of Iceland.

There are other evidences of subterranean heat and fire. There are so-called "hot lodes," where a thermometer marks 90° or 100°. The deepest mines in the district are the hottest, and volcanic products, carbonic acid gas, and such-like, sometimes escape from veins into the mines.

There are hot springs at Bath still. There is evidence of upheaval by the agency of heat-force in the geology of the country, and in the temperature under ground. There is evidence of denudation by ice above ground.

The hills are about 2000 feet high.

The upper part of Dartmoor is strewed with large blocks of granite, many of which differ in structure from the granite of the rocks on which they rest. They resemble ice-borne boulders in shape. The soil is peat and decomposed granite, but on the hill-flanks are beds of sand and water-worn boulders. One bed is to be seen at the roadside high above the Dart, near Ashburton. It seems to be water-worn glacial drift, and the height is about 200 feet above the sea.

The hill-tops are capped by curious granite elevations called "tors" (heaps or mounds). These, though much weathered, often retain the characteristic shapes of ice-ground rocks.

The grinding force seems to have acted from the north-east towards the south-west.

Blakeston Tor, on the south-eastern side of the moor, is a good specimen of the class.

The cut is from a sketch made on the spot.

Heytor Rocks, about 1100 feet above Bovey Tracey, are good samples also. From the internal structure of these

Fig. 86.

granite hills as seen in a quarry near Heytor, the tors appear to be weathered remnants of an upper bed of granite, the rest of which has been ground and broken and pushed away by some power, acting chiefly from the north-east. Still lower, layers of granite have also been worn at the edges, so as to leave a smooth rounded conical hill, strewed with rounded blocks, and capped by a rounded tor. The granite breaks into angular fragments, and weathers into strange shapes.

The worn surfaces are very clearly seen for about 200 feet below the top, and a few remnants of grooves can there be traced. These last are very faint, and much weathered. Without other indications, and long practice, they would be wholly insufficient evidence, but taken with the rest, they too point to ice moving from N.E. to S.W.

If the N.E. is the weather-side, most of the loose stones ought to be found pushed over into the shelter. In fact, most of the loose boulders which are strewed about Dartmoor are to the westward of the tors, and to the westward of ridges, and of the range itself. The forms of the hills generally, when seen from a height, agree with this theory; they are all rounded. Whatever their composition may be, whether they are "granite," or "killas," or "elvan," igneous or sedimentary, upheaved or not; they are steep towards the south-west, and slope towards the north-east, like hills mentioned above.

On the hill above Wistman's Wood (see vol. i. p. 31) is a great boulder as big as a house, which seems to be a "tor" pushed bodily from its base towards the point from which the prevailing wind now blows, as shown by the trees.

From Shetland and Orkney to Devonshire, at certain elevations, there is a recurrence of the same rock-forms which are held to be old ice-marks in Scandinavia, Switzerland, and elsewhere.

Brentor, near Tavistock (see map, vol. i. p. 232), is at a lower level. The shape is like that of hills in the valley of the Forth, with similar bearings. The rock at the top has the general shape of ice-ground rocks, but it is so weathered, worn, and grass-grown, that nothing like a groove was made out. The general shape of the hill seems to point to a grinding force acting from the direction of Bristol, at a height of about 700 feet above the present sea-level. Hence this spoor runs out to sea, unless some of the boulders and loggan-stones of Cornwall prove to be erratics and perched blocks. No Cornish ice-grooves are known to the writer. According to Sir C. Lyell, the southernmost extent of "erratics" in England is to the north of Dartmoor.*

If ice-floats of former days resembled ice-floats off Labrador now, there may have been an easterly limit, beyond

* Antiquity of Man, p. 280.

which ice-floats could not pass. But that limit seems to have included Kent.

In 1860, a party of fishermen were creeping for what they might find at the bottom of the sea off *Margate*. They got hold of something heavy, and thinking that they had netted an anchor, or something better, they dragged their prize to land with much labour. It was a big rounded stone of the pattern of those which form terraces about the Torneå. It was something so foreign to the sandbanks, gravel, and chalk-cliffs of southern England, and to the experience of the fishermen who found it, that they hoisted the stone to the end of the pier, and there it was shown as a curiosity.

From Muswell Hill and the Thames' mouth, the S.W. curve leads to Southampton Water.

In many of the chalk-glens of southern England, rich alluvial flats are flooded to irrigate meadows. The bright clear sparkling wealth of water in the rivers is divided and made to spread and wind hither and thither. The green grass and the water-threads of silver and crystal weave themselves into a pattern of graceful curves, and this waving, moving, brilliant, wet carpet, is spread on a yellow floor of flint gravel, peat, and clay, laid in a white chalk-groove. At Stockbridge, in one of these glens, shoals of trout and greyling are daily tempted by the best of British flyfishers, armed with the best of London tackle. From constant practice and long acquaintance, these fish and fishermen have learned so much that great skill spills little blood; but as a good fencer is a dangerous foe, the man who kills two Test trout a day is apt to kill most elsewhere. A stranger used to wild fish finds highly-educated trout too cunning for his rough hand; but if fish will not take, it is well to take to something else.

The old spoor which was found at the North Cape is here.

This valley, which ends in Southampton Water, is terraced, and the terraces are as plain as they are in Scandinavia. From *Stockbridge* four shelves are very clearly seen on the western side of the hollow. The alluvial flat in which the Test winds is about a mile wide, and it rests in a chalk-groove. The solid chalk crops out where the plain ends. Close above the plain is the first horizontal shelf, and it is well marked at several places, and on both sides of the glen. The second shelf is about 100 feet higher; and the whole series may be thus roughly expressed. The only tool used was a pocket aneroid :—

Feet.	
200	hill-top.
180	fifth.
160	fourth.
150	third.
100	second.
10	first terrace.
0	alluvial plain.

The whole country is cultivated, and there are few hedge-rows. The colour is uniform—green in spring, yellow in autumn, brown when the fields are bare. When light is favourable, and attention directed to the terraced shape of these rounded chalk-downs, the whole landscape seems pervaded by horizontal lines. Though all the chief outlines are swelling curves ⌒ ◡, a great many of the hills have slight notches ⌋‾‾‾‾‾‾⌊ hewn out at corresponding elevations on both sides; and from these, horizontal lines of light and blue shadow mark the terrace of erosion, which surely marks an ancient water-level. All theories of lakes are vain here.

The chalk is covered with a very thin layer of soil and

rolled flints. Many of these on the watershed are water-worn pebbles, like those which are found on sea-beaches ; others are only partially rolled ; others are like flints newly broken out of the chalk. These stones look like water-work, and here it must be sea-work. A well-preserved set of terraces

Fig. 87. Terraces at Stockbridge.
Casting a small fly over heavy fish.

occurs near the hill-top to the west of Stockbridge, opposite to the peat-pits. A hedgerow shows the waving outline of the hill very distinctly. These terraces are about fifty feet apart, and might easily pass for works of human skill, "parallel roads" or fortifications. They seem to be very well preserved marine terraces of erosion, and there are ten or a dozen of various sizes. Lower down the valley they recur. On the road-side, near a place called *Hazledown Hill*, close to the watershed of the valley of the Test, three small horizontal ridges of broken and rolled flints, skinned over with fine turf, again recur at elevations at which the aneroid barometer marks the same level—namely, heights somewhere between 200 and 150 feet above the level of Stockbridge.

From Hazeldown Hill the way is clear to the glacial drift on Muswell Hill; and these terraces carry the sea-level over London along the line of this last curve. It passes from the mouth of the Thames to Southampton Water; from the last patch of British glacial drift yet described by good authority, down to the English Channel with its broken chalk-cliffs.*

To men who "live at home at ease" all this may seem to be impossible, or mere vague speculation. A man who has never seen ice upon the sea, and who thinks that rocks were created in their present form, is apt to suspect a latent joke in "sea-margins" in corn-fields. A Londoner who had not tried to construe a stone, would stare agape at the notion of ice floating over St. Paul's, or the nearest steeple, where the weathercock has whirled ever since he was born. To such men all modern geological change seems impossible, and English ice a myth. But those who will accept a rough translation of a stone record may rest assured that floes and bergs passed over the site of London, when Muswell Hill was capped with glacial drift.

The northern "glacial period" is still within easy reach.

The *Times* of August 4, 1863, gives the official report of the loss of the *Anglo-Saxon*. It narrates that on the 25th of April 1863, the vessel fell in with ice and foggy weather south of Newfoundland. The engines were slowed, and as the ice

* It is right to state that a sixteen mile walk to Muswell Hill, without a guide, and a long search about the foundations of the new building, and elsewhere, failed to discover the patch of drift in question. It is there, but it was found by chance, and it is now buried. If any one should fail to discover marks described in these pages, he may think of the old saw which says that "bad seekers are bad finders."

became thicker and the fog denser, the engines were stopped. The vessel drifted till ten on the 26th, when the ice being somewhat less compact, she was moved slowly ahead till two, when clear water was reached. Steam was then set on, and the vessel went ahead full speed towards Cape Race: she was about lat. 46° 54′ N., and soon after she ran aground, and was wrecked in a cold fog at Clam Cove in Newfoundland.

If she had been on the European coast, she would have been in the Bay of Biscay off La Rochelle, south of Brittany and the drowned land of King Grallon. The ice would have been north of the Pyrenees (whose name means "*ice-peaks*" if it be Celtic) where signs of glaciers abound, she would have been near the latitude of the place where works of human art were found associated with remains of reindeer.

If she were sailing over Europe, she might have been over the lake of Geneva, off the high coast of Switzerland, or in the Sea of Azov, under the lofty Caucasian coast, and north of the moraines of the Lebanon.

In the *Times* of June 17, 1864, another wreck in the same latitude is thus recorded:—

ICE IN THE ATLANTIC.—By the arrival of the Allan steamer *Peruvian* we hear of the loss of two vessels belonging to this port—the *Philanthropist* and *Highlander*. The former was on a voyage from Liverpool to Quebec, and was lost in the ice on the banks of Newfoundland on the 11th of May. The crew were picked off the wreck by the bark *Wolfville*, and taken to Quebec. She was a ship of 805 tons, and was built in St. John, New Brunswick, in 1852. Her present owners we have been unable to ascertain, as she very recently changed hands. The second vessel, the *Highlander*, was bound from Quebec to Fleetwood, and was, says the telegram, "lost near St George's Bay," but it is supposed through contact with ice. She was a perfectly new ship, having only been built this season at Quebec, and was, when lost, on her first voyage, coming over to England, we believe, for sale. Both vessels had valuable cargoes, and were fully covered by insurances, partially if not wholly effected in London.—*Liverpool Courier.*

If the Arctic Current came through the English Channel, the same climate would descend upon the English coast.

Drift, shells, ice-marks, and rounded terraces, record that a frozen sea, 2000 feet deep, did in fact pass over the sites of London, Edinburgh, and Dublin ; over Snowdon ; over Scotland, Ireland, and Scandinavia ; and some of the highest marks left are fresher than the sculptured pillars of the temple of Serapis, which sank in the Bay of Naples, stayed under water for a time, and rose again.

The force which lifts and lowers land is still active in Greenland, Iceland, Scandinavia, Labrador, England, Italy, Sicily.

The same paper which recorded the evil deeds of Jack Frost in summer 1863, also recorded abortive efforts to escape made by the imprisoned cyclops Fire.

Accounts from Messina of Friday last state that the volcano of Mount Etna is vomiting fire and lava. A new eruption is threatened in the direction of Bronte. The inhabitants of Catania are terrified at the formidable noise and the shower of ashes and stones falling in that direction. The population of the mountain have made preparations to quit their dwellings. Their horses are saddled, their cattle gathered together, and all their household furniture packed up to be ready for immediate removal. Prayers are being offered in the churches, and the relics of saints are to be exposed to the piety of the faithful. Terror prevails among the entire population.

The memory of an English earthquake is still fresh. There was a small volcanic eruption in Iceland in 1862. We live in a period of active geological change, though few men think about Frost and Fire.

The water-meadows at Stockbridge, like the hills, furnish occupation for unskilled anglers. Every dry watercourse gives samples of " denudation" and " deposition" by streams. Every tame stream gives a lesson which may be used to master the ways of wild streams, which are too deep to be easily seen through. In the middle of a weir, about ten yards wide, behind which was a " head" of water three feet deep, a sluice was lifted so as to make a strong rush through a still pool in a lower watercourse.

A certain latent mechanical " water-power," expressed by the broad arrow at E., was stored up behind the dam. The same force of gravitation makes rain fall, stops a wagging pendulum, and works a drop and the surface of the ocean-pool into spherical forms. By raising a sluice at E., a certain amount of this power was freed, and set to work on water at rest in the river-pool.

From one direct force, which tends to produce direct movement downwards towards the earth's centre in all latitudes and longitudes, and from the movement expressed above by the form ⊥, a series of very complicated vertical and horizontal movements resulted in the stagnant pool below the weir in the Test.

At the head of the pool, at the spot where the falling water escaped from under the sluice at E., whirling jets spouted up. In the strongest downward rush, westward towards W. waves rose highest, curled round, and broke eastwards, up-stream towards E. A complicated set of curves, jostling streams and waves, crossed and recrossed the line of direct movement from E. to W. Surface-waves rippled and broke on the shore in every direction. At the tail of the pool was a shallow, and the whole of the bottom was overgrown with fine water-plants. Each of these was a tell-tale to point out the course of the stream below, and floats on the surface showed movements there.

These seemed a movement from every direction.

Because there was a rush from east to west in the middle of the pool, two eddies whirled opposite ways about the points N. S. in the diagram. The weeds mapped out the currents. A stick thrown into the rush at E. turned back where the

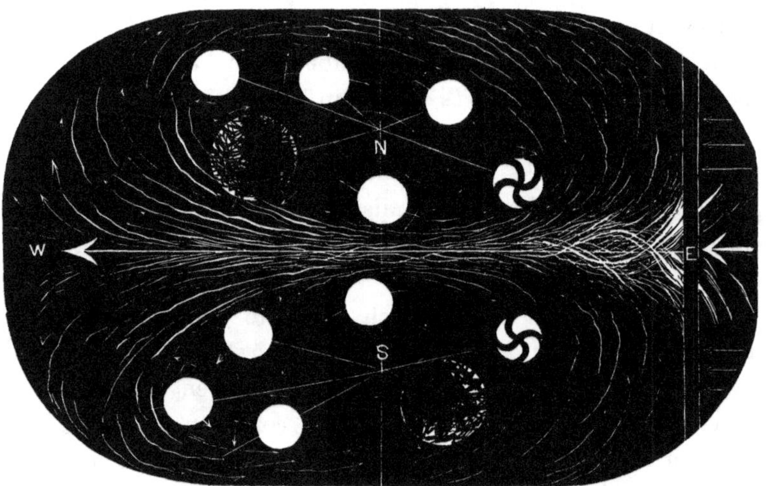

Fig. 88.

weeds turned and whirled round the point N. Two stacks of dry reeds (expressed by circles and white spots), thrown one on each side of the rush, revolved in opposite directions about their centres of revolution N. S. They described ellipses, and turned on their axes in the directions shown by arrows; and so the floats waltzed over the sunken forest of weeds, which showed like movements at the bottom of the transparent stream. Not one reed had passed over the shallow when the evening flies rose out of the water, and trout seemed disposed to dine. The experiment was simple, any child can see the result, but all the mathematicians that ever lived might have found occupation for their lives, in striving to comprehend the curves that resulted from the action of the direct force of gravitation which stretches a plumb-line.

No special talents or mental tools were used by philosophers, to discover this natural force of "gravitation;" it is something patent and manifest to all, though no human mind can account for it, or explain it, or calculate the effects of it.

From the stagnant pool the river Test leads back to the watershed, and to the rain-cloud which rose out of the sea. No special talents or mental tools need be used to discover the second force which tugs at the cable of a fire balloon, beside the force which tightens the cords of the car. The effects of this force are hard to calculate, the mode of action is wholly unexplained, but the force is manifest as daylight itself.

The Atlantic is a big pool to cover single-handed; arctic currents are heavy streams; those who venture in are apt to get out of their depth. From Lapland to Southampton is a long cast; but, nevertheless, the small fly has fallen very near the southern haunts of heavy fish. The last cast over London and the watershed of the Test may chance to rouse a shoal of geographers, geologists, and surveyors, better worth raising and harder to catch than Test trout; and this is the point of the first hook dressed to tempt such readers.

As two sets of floats and two small water-systems revolve and circulate in eddies, in a small pool, and in the largest pools that can be seen; so, according to meteorologists and bent trees; authority, maps, and observation; the atmosphere and local storms,—the largest and deepest streams in our world whirl and move; turning opposite ways, on opposite sides of the Equator in the Northern and Southern Hemispheres. The reason seems to be, that two mechanical forces, which are at rest when evenly balanced, move air opposite ways when one or the other is in excess.

So also, according to theory founded upon facts, of which some are stated above, the ocean circulates within narrower bounds for the same reasons. Because it circulates, and tends to move north and south upon a surface turning eastwards, main currents move diagonally ; and the coldest and heaviest tend westwards. For the same reason floats revolve and circulate about the Poles, as the stacks of withered reeds did in the pool, as froth does in every eddy, as clouds do in the air ; and as the coldest are also the hardest and the heaviest of floats, those which tend westwards make the deepest marks.

It is admitted that this double engine, made of air, water, and ice, has done the work of "denudation" and "deposition," which geologists study, survey, and describe. It is argued that the tool-marks of each part of the natural engine ought to be known, and that large work done by regular and constant movements in air, and water, and ice, ought to be, and is in fact, symmetrical.

It is easy to build clay-maps in shallow pools, to watch currents and eddies, study their action, and seek to apply knowledge, so gained from experiment, to larger things. The pastime is lazy, healthy, and frivolous, as any idle angler can desire.

The map (vol. i. p. 496) is intended to show that forms which are attributed to denudation coincide with general movements in air and water, some of which correspond to movements in a river-pool, and which seem to make a pattern of curves upon the rough moving surface of the globe ; that all the largest indentations about the Equator trend westwards, all the chief coasts on the eastern side of continents, and many mountain-chains, cross meridians diagonally as currents do. It is argued that hills and hollows, ruts and ridges, which are less in proportion than sand-lines on a boulder, may be tool-marks of a natural graving-engine, worked by fire and frost.

As a mayfly rises from mud, through water into air, and dies, so the mechanical forces which drive this part of the engine seem to rise and fall.

The world's heat, which is always found when sought underground, and the sun's heat which is added from without, evaporate water and expand air; the power seems to move water and air to the limit where force radiating from the earth's centre is expended, or overcome, by force converging upon the centre, whence rays of heat and force diverged.

In one word, the natural engine seems thus far to be driven by two opposing forces which bear various names—

"gravitation"	and	"levitation,"
attraction	and	repulsion,
condensation	and	evaporation,
contraction	and	expansion,
crystallization	and	dispersion,
weight	and	heat,
water-power	and	steam-power,
weights	and	springs,
freezing	and	boiling,
Frost	and	Fire.

The engine seems to be driven by converging and by radiating mechanical forces, and by the will of Him who made them, and who said "Let there be light, and there was light," in the dawn of time.

And so the pursuit of mechanical force leads round to the place from which this long journey began, and a further search requires a fresh departure.

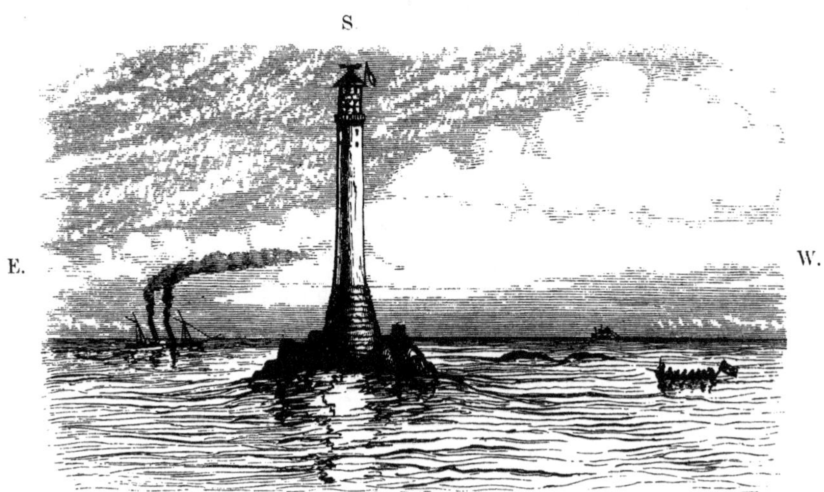

FIG. 89. "THE SCILLY BISHOPS." Lat. 49° 51′ N.

The last of the British Isles. From a sketch made 8th July 1859.

The rock above water is higher and longer than the Eddystone. The building is probably the most exposed in the world. Spray goes over the top, which is more than 150 feet above the sea-level. The rock, so far as the shape of it could be seen or felt, resembles a Devonshire tor; *e.g.*, Blakeston Tor, p. 221. For a contrast in climate in a similar latitude, see below, and p. 248

CHAPTER XLIII.

BELLEISLE CURRENT—AMERICA.

In the summer of 1864 a holiday trip to North America was so arranged as to test glacial theories above stated. The Arctic Current and Gulf Stream were twice crossed, and their climates felt at sea. Icebergs were seen in July about lat. 49° in the Atlantic. Cape Harrison in Labrador, the Straits of Belleisle ; the coasts of Newfoundland, Cape Breton, Nova Scotia, New Brunswick, and of the States, as far south as Washington, were visited. The curve (see map, vol. i. p. 496) which passes through the Straits of Belleisle was followed through Canada and the Western States to St. Louis on the Mississippi. Various cross-routes and high points on the Alleghanies were selected, traversed, visited, and examined for ice-marks; the Mammoth Cave in Kentucky was visited for its own sake ; and the following are some of the results :—

Cape Chudleigh, the most northern point in Labrador, is in lat. 60° 54′ N. ; Cape Harrison is in 55° ; Belleisle in 52°. The Shetlands correspond to Cape Chudleigh ; Londonderry, Stranraer, and Newcastle, to places near Cape Harrison ; Killarney, Cork, Gloucester, and Colchester, to places near Belleisle. There is no good chart of the Labrador coast. The interior is unexplored. There are no high mountains and no glaciers in the country, so far as it is known to trappers, Indians, fishermen, and settlers along the coast. The coast-line is low, rocky, and glaciated. All the hills, rocks, and islands, are rounded. There are few cliffs, and very few beaches; but vast numbers of rocks, reefs, and islands, and many long fjords. Hamilton Inlet, for example, is 150 miles long. The climate is very severe. In July and August 1864 many of the harbours were frozen, and patches of snow lay close to the water's edge at places which correspond to watering-places in North Wales. Heavy pack-ice reached to the horizon opposite to Hamilton Inlet on the 1st of August 1864. Between Belleisle and Cape Harrison, islands of ice were constantly in sight. The largest of these were in the offing, and resembled isolated rocks, like the Bass or Ailsa. Some were aground and stationary for a fortnight, others had moved away when the vessel returned.

It was very difficult to estimate their dimensions, but many certainly rose 200 feet above the water, and one near the shore rose 300. Smaller bergs were aground amongst the islands and in the fjords, and many of these were from 50 to 100 feet high. Smaller fragments, called "growlers," about the size of ships and boats, were drifting everywhere, and bits as big as hogsheads and barrels were rolling in the land-wash. The temperature of the water was generally about 37° and 40°. The air at sea was about 40°, but on rocks and islands the temperature of the air was far higher in clear weather. The whole of this drift-ice was working in shore, gathering in eddies behind points, and shooting off eastwards where points jutted out into the Arctic Current. The movements were analogous to those of floats in a river—sticks, leaves, froth, or ice. The coast is now rising between St. John's in Newfoundland and Cape Harrison in Labrador. Rocks have been marked, and the marks have risen; boats now ground on solid rocks where they floated twenty years ago; rocks which were seldom seen now seldom disappear at high tide; harbours are shoaling; beds of common shells are found high above the sea; raised beaches are seen on hill-sides in sheltered corners; and blocks of foreign rock are perched upon the summits of islands and on the highest hills near the coast. The rocks are much weathered, and very few striæ were found. Those which were found aimed up-stream. At Indian Island, lat. 53° 30′, near the lat. of Hull, they pointed into Davis Straits, at a height of 400 feet above the sea; at Red Bay, in the Straits of Belleisle, they aimed N. 45° E. at the sea-level. In winter the sea is frozen near the coast to a thickness of 18 inches or more; in spring the northern ice comes down in vast masses. In 1864 this spring drift was 150 miles wide, and it floated past Cape Race. From a careful examination of the water-line at many spots, it appears that bay-ice grinds rock, but does not produce striation. The tops of conical rocks have been shorn off. The shape of the country is a result of denudation. No matter what the dip and fracture of the stone may be, the coast is generally worn into the shape known as "roches moutonnées." It is impossible to get at rocks over which heavy icebergs now move; but a mass, 150 miles wide, perhaps 3000 feet thick in some parts, and moving at a rate of a mile an hour, or more, appears to be an engine amply sufficient to account for striæ on rising rocks, which were under water when sea-shells lived above them, and were buried on them. A cube of ice cut from a stranded berg, and floated in sea-water, rose one-tenth above the surface. At this rate, a cube 300 feet high is 3000 feet thick, and would ground in 2700 feet of water; one 30 feet high is 300 feet thick, and will ground in 270 feet. In winter anchor-ice forms at the bottom; it must therefore form readily about the base of stranded bergs. The mass which was 150 miles wide was therefore a floating glacier, armed, as glaciers are, with stones, gravel, sand, and mud, moving along a definite course, from N.W. to S.E., from Cape Chudleigh to Cape Race, and at a rate which no glacier equals. Work done by it ought to resemble glacier-work. At the north end of Newfoundland the stream parts. One narrow rill flows S.W. through the Straits of Belleisle, and carries small bergs even to Anticosti in the Gulf of St. Lawrence; the main broad stream is shunted westward, and moves from N.W. to S.E. It was crossed about lat. 49° on the 16th of July 1864. Numerous large bergs were seen; the temperature of air and water fell when the stream was entered, and rose again when it was left behind. The stream was crossed again in November, and the same change of climate remarked, but no ice was seen on this voyage. The tail of the stream reaches lat. 36° 10′, and it carries large bergs to these regions, which correspond to Gibraltar and North Carolina.

If such a current flowed over America, marks left by it ought to correspond to these movements. Striæ ought to run from N.E. to S.W., where the stream could flow directly; from N.W. to S.E., where it was shunted by land placed as Newfoundland is now placed.

The summers of 1863 and 1864 were remarkable in Great Britain and Canada for their unusual warmth; in Labrador and Newfoundland they were unusually cold, wet, and dark. Early in March 1864 the sealing-fleet left St. John's in the latitude of Nantes, tried to force a passage through the pack, and, failing in that perilous attempt, they worked up the coast inside to Toulinguet, about the latitude of the Scilly Isles. At this promontory a shift of wind drove the ice inshore, and the whole fleet was beset for a month. About the end of April this mass of northern ice got adrift, and broke up. It carried the fleet with it, and thirty vessels were utterly destroyed, smashed, and ground up. One was forced up on a pan of ice, drifted past St. John's, and was rescued about Cape Race by a tug-steamer sent out for the purpose.

From these facts it appears that a warm summer only increased the intensity of the cold by setting more ice adrift in the north; that a glacial period now exists in English latitudes; and that the books above quoted accurately describe the normal condition of these regions of the earth.

If America were now submerged 3000 or even 2000 feet, the Arctic Current might flow S.W. to St. Louis on the Mississippi; but it would be shunted eastwards by high grounds in Nova Scotia, New Brunswick, and the Northern States. According to theory, striæ ought to run generally from N.E. to S.W. in the central district; from N.W. to S.E. on the Atlantic shores of the Alleghanies.

Ice-marks in North America appear to coincide with this theory, so far as they were observed in 1864. They did not appear to coincide with the other theory published by Agassiz in the Atlantic Magazine of the same year, which supposes the existence of a glacier, which extended from the North Pole to Georgia; but on this point it becomes an inexperienced writer to speak with diffidence.

Newfoundland extends from 51° 40′ to 46° 38′ N. lat. The northern end corresponds to the south of Ireland, the south of Wales, the country about Bristol, Gloucester, Oxford, and London, Barnet, Epping, St. Albans, etc. The southern end corresponds to the north of Switzerland, the Jura Chalons, and the mouth of the Loire. The island corresponds to the south of England and the centre of France. Bones of large reindeer discovered in France were found in latitudes which now swarm with large reindeer in Newfoundland. The banks reach lat. 43°, the parallel which crosses Spain near Valencia and Barcelona. In Newfoundland there are no high mountains and no glaciers; the land is low, and furrowed by hollows, which run from N. 30° E., or thereby. Many of these rock-grooves extend under water, and now contain large bays and fjords. The dividing ridges form reefs and headlands, and in many cases the ridges and hollows correspond to the strike. Heavy ices of all kinds and dimensions drift along the coasts, and over the banks, at all seasons. On the 2d of June 1863 St. John's Harbour, in the latitude of Nantes in France, was filled with heavy drift-ice; while the pack extended to the horizon of the signal-station, which is 540 feet above the sea. A photograph of this strange scene was taken by a native artist.* If the land were submerged, the Arctic Current would flow through the valleys, as part of it now flows through the Straits of Belleisle. A thousand feet would sink the whole land. Watersheds between the bays ought to be striated from N. 30° E. to S. 30° W., or thereby, if drift striæ were made by ice drifting in the Arctic Current over Newfoundland. The whole country is glaciated; the shape of it has nothing to do with the dip of the rock, which is folded and bent. At places ice-marks are well preserved, but generally the rock-surface is weathered. No ice-marks were found at watersheds, because rocks in the interior of Avalon are smothered in bogs, and overgrown with an almost impassable forest; no rock was seen on the only isthmus crossed. The striæ which were found were near the coast, and seem to indicate large land-glaciers moving seawards. At St. John's, the marks run over the Signal-hill, 540 feet, from W. and N. 85° W. eastwards; at Harbour Grace, from S.

* See p. 248.

75° W. down the bay north-eastwards; at the head of Conception Bay they fill a large hollow, overrun hills, and point from S. 15° W. northwards. Vast terraces of drift stretch along the base of rounded hills at the head of Conception Bay, at Harbour Grace, and at Old Purlican, near the end of the bay, 60 miles off. At the head of the bay, most of this drift seems to have come from the hills. Opposite to granite hills are numerous blocks of granite; opposite to sandstone and slate hills sandstone and slate boulders abound; and yet large islands of ice constantly drift into this bay now, and some at least bring loads of stone. Three islands, near 100 feet high, were cruising in the bay on the 20th August 1864. As coast-ice also picks up and drops stones every year, boulders from Greenland, Labrador, and Newfoundland, are certainly dropped in Conception Bay; and probably the banks off the coast are strewed with similar mixed drift. Bergs ground on the banks every year, and some have been seen loaded with stones. Striæ and drift on shore in Newfoundland indicate large land-glaciers. The shape of the country seems due to some more powerful denuding engine, moving as the Arctic Current now moves; but no glacial striæ were found at the only isthmus crossed. The interior is unexplored, and the whole is very difficult of access. Indians who use bows and arrows, and large wild animals of northern type have the land in possession; the coast is occupied by fishermen, and by merchants who deal chiefly in fish and seal-oil.

In Nova Scotia and New Brunswick striæ seem to indicate the passage of sea-ice. A current passing south-westwards from Newfoundland would be turned aside by high grounds near Halifax. Striæ in the town of Halifax point N. 55° W., through a gap which leads to the Bay of Fundy. At a height of 550 feet above the sea, at the summit-level of the railway between Halifax and Windsor, striæ point N. 35° W. The current which flows S.W. through the Straits of Belleisle would continue its direct S.W. course through the Bay of Fundy, if the low isthmus were gone. At St. John, New Brunswick, striæ in the town and beside the suspension-bridge point N. 20° E., N., and N. 25° E. The same current flowing over the north-eastern end of the province would be turned westward by high grounds inland. On a hill near Fredericton, 100 miles inland, and 300 feet above the sea, striæ point N. 35° W., and N. 87° W. There are no high mountains in the province, and these high grooves aim at a distant horizon. Nova Scotia, Newfoundland, and Cape Breton, are glaciated throughout, and strewed with mixed drift.

On the Canadian side, striæ at Quebec point into the gulf and up the valley of the St. Lawrence; the land is terraced, boulders are perched upon the high grounds, and recent shells have been found far above the sea. These facts indicate the passage of sea-ice. The falls of Montmorenci, near Quebec, have worn a notch in a terrace of rock, above which marine shells are found. The size of the notch is a measure of the time which has elapsed since the shell-beds and the terrace of erosion were raised above the sea; for the river only began to work at this point when the land rose. This tool-mark is well seen from the town of Quebec on a clear day, when the notch is filled with dark shadow, and the terrace is a line of light.

In Maine, New Hampshire, Vermont, Massachusetts, and New York; from latitude 45° to 40° 40′; striæ found during this trip, in the latitudes in which icebergs now abound farther east, appear to coincide with the probable run of an arctic current flowing over the land 3000 feet above the present high-water mark, or less. Such a current would continue its course from N.E. to S.W. on the Canadian side, and would be turned westwards by mountains which now separate the St. Lawrence basin from the Atlantic slope. The reflected currents would flow from N.W. to S.E., as they do at the northern end of Newfoundland and off the Labrador coast. Striæ at high levels point towards the Straits of Belleisle, where the Arctic Current is turned aside. Striæ at low levels on the Atlantic slope converge upon distant mountain-passes, which would be sea-straits meeting in the Gulf of St. Lawrence, if the land were sufficiently submerged ; and the Arctic Current would then flow through these passes. Horizontal striæ on the shoulder of the highest peak in this district aim N. 25° E. and N. 20° E., at 1992 and 2307 feet above the sea. If these marks on Mount Washington, in lat. 44° 15′, were made by heavy icebergs floating through a strait like Belleisle, the nearest land on the horizon was then far away. Lines produced in the direction of these marks skirt the sources of the St. John and Penobscot rivers, which flow into the Atlantic, and of the Chaudiere, which falls into the St. Lawrence near Quebec. In this direction the land is far lower than the shoulder of Mount Washington. Produced in the other direction, these lines pass over Long Island near New York. There, glaciation is conspicuous in the latitude of Madrid, as it is in the park at Stockholm ; but the direction of movement was different at the low level of New York. Two hundred miles away from the White Mountains striæ near the top of the Catskill range, at 1935 above the sea, point N. 40° E. over low grounds, up the valley of the Hudson, into the wide pass which now contains Lake George and Lake Champlain, and which lately contained the bones of a whale buried in drift. In the other direction, this mark aims into a gap. On the watershed of the gap, at 2115 feet above the sea, a complicated system of cross marks aim N. 77° E., and S. 77° E.

In the opposite direction, all these point into a hollow, which would be a strait passing through the Catskill range westwards if the sea were 2200 feet deeper than it is now. These sets, the highest marks observed, point N. and E. At lower levels the marks aim at passes N. and W. For a distance of 12 miles, and up to a height of 1800 feet, horizontal striæ on the Catskill escarpment, and in the low country beneath it, aim at the lowest ground on the distant horizon, which is between the Adirondak and Green Mountains, and leads through the valley of the St. Lawrence back to the gulf. This certainly was a sea-strait when the whale swam in it.

Fifty-seven miles below Albany, on the Hudson, near high-water mark at Barrytown, opposite to the southern end of the Catskill range, the striæ turn and point N. 8° W. At New York, in the central park and near Broadway, about lat. 40° 40′, at six different stations, striæ aim N. 21°, 30°, 36°, 37°, 39°, 45° W. Some of the stones in this central park contain large plates of mica, and may have come from the White Mountains, or from the "azoic" regions about the Adirondaks. Others may have come from Labrador, for they match rocks in that country. Further north, on the Atlantic coast, a system of marks seems to converge upon a chain of lakes in Maine. A line produced N. 55° W. from Eastport strikes the Pemadumcook Lake. Lines produced N. 14° W., and N. 28° W. from Portland, avoid the White Mountains, which are visible at a distance of 90 miles, and strike the Mooselookmaguntic Lake near Saddleback Mountain, about lat. 45°. These converge upon a low watershed. A line produced N. 25° W. from Boston skirts the western side of the White Mountains, and enters a wide pass which leads to Canada. If the direction of the highest striæ of this series be taken as the direction of the main arctic stream, N. 25° E. to S. 25°

W., it would strike against the White Mountains, Green Mountains, Adirondaks, and Catskills, and glance westwards to Eastport, Portland, Boston, Albany, and New York. It would escape from passes in the main range, as the Arctic Current now escapes through the Spotted Islands off Labrador, and through deeps between the sunken banks off Newfoundland.

On the other side of the mountains, marks in the valley of the St. Lawrence correspond in direction. At Montreal Mountain, striæ point N.E. magnetic; at Brockville, they point N. 45° E. true; at Niagara Falls N. 20° and N. 5° E.; at Buffalo N. 20° and N. 13° E. But, while a general south-westerly direction is thus marked by strong deep lines, other lines cross in all directions. At Brockville, for instance, a deep groove three or four feet wide aims N. 45° E., and all lines in it down to hair-lines aim in the same direction; but on a neighbouring rock a cross system of smaller grooves aims N.W. almost at right angles to the general direction; and at Prescott, the only marks found aimed N. 20° W. The water-lines of the great lakes and rivers are not striated, though much worn by winter ice. These variations in a wide plain accord with the erratic movements of icebergs in summer, the strong markings seem to agree with the general combined movement of the spring drift.

So far these fixed marks agree with the probable movements of an arctic current. In order to make the marks, a polar land-glacier would have to climb more than 2000 feet out of the Gulf of St. Lawrence, over the shoulder of Mount Washington. According to other marks it also climbed over the watershed of the St. Lawrence into the Mississippi basin, and reached lat. 39°, which seems an impossible feat for land-ice to accomplish.

Though other observers have found striated rocks south

of Buffalo, in the central district none were found during this expedition. All the rock-surfaces found in the Western States were either weathered or water-worn, though many were newly uncovered. Fossils project half an inch at many spots. But glaciated boulders were found near St. Louis, at Indianapolis, Lafayette, Fort Wayne, Crestline, Upper Sandusky, and many other places near the watershed of tributaries of the Ohio and St. Lawrence. Many were found between lat. 39° and 40°, in Ohio, Indiana, and Illinois. Not one south of 39° in these states, or south of 41° in Western Pennsylvania. At St. Louis, Vincennes, Louisville, Cincinnati, and Pittsburg; along the banks of rivers, and beside railways, no single specimen could be discovered. At these places, and in Kentucky, further south, near lat. 37°, the rocks are covered by thick beds of pure clay and fine sand. South of a line drawn from lat. 41°, long. 81°, diagonally, south and west, to lat. 39°, long. 90°, near St. Louis, no glaciated boulders were found. A short distance north of the line, blocks of Laurentian gneiss as big as bullocks are scattered broadcast over the flat prairies.

The nearest fixed rocks of the kind are about Lake Superior, but stones of the very same size, pattern, and material, are on the top of the Catskill range, on the top of the Green Mountains, on the shoulder of Mount Washington, on the highest ground near Buffalo, on the high grounds near Niagara, at Brockville, on Montreal Mountain, at Quebec, on hills beside the Straits of Belleisle, on islands near Hamilton Inlet in Labrador. Similar stones are strewed over Newfoundland, Cape Breton, and Nova Scotia, at the head of the Bay of Fundy, and all down the Atlantic coast as far as New York. None were found at Philadelphia, Baltimore, Harrisburg, or Washington. Water-worn drift abounds at all these places, but no striated gneiss boulders were found there. On the banks of the Potomac and at Washington are large stones in clay, but none of those found were striated. At Harrisburg is a similar deposit. Icebergs and rafts of coast-ice are carrying northern drift stones in the Atlantic, and if America were submerged the Arctic Current might carry them as far as lat. 39°, long. 90°, for Atlantic bergs reach lat. 37° in long. 47° W. If a polar glacier carried these stones they ought to be found in great moraine heaps at the end, but nothing like a terminal moraine exists in the prairies. For hundreds of miles the plains are almost as flat as the sea, and where the country rolls, sheets of drift cover the rolling plain, as snow covers it in winter. The stones and clay were surely dropped from melting ice-rafts, as snow is shed from clouds, and as stones are now sown in the Atlantic :—broadcast. Observations made in America so far agree with observations made in Europe.

In a series of papers in the *Atlantic Monthly* for 1864, Agassiz attributes glacial phenomena to polar glaciers which reached lat. 36° at least, and were 6000 feet thick in lat. 44°. A theory espoused by Ramsay, Geikie, Sir W. Logan, Agassiz, and such men, is worthy of careful investigation. The observations above recorded seem rather to indicate the action of polar currents, like those which exist, than the existence of polar glaciers of these dimensions. The facts above stated may swell the pile on which a just opinion must be founded at last. The question turns on the denuding power of the Atlantic drift. The forms into which the land has been ground by some ice-engine closely resemble glacier-work; if the Atlantic drift is too small to account for the work, the polar glacier is the only resource. After seeing glaciers and sea-icebergs at work, and hearing the accounts of those who are familiar with the polar sea-drift, the writer holds to the opinion expressed above, and takes his stand on the iceberg for the present.

Fig. 90. MAGGOTY COVE AND HARBOUR OF ST. JOHN'S.

CHAPTER XLIV.

GLACIAL PERIODS.

ONE general conclusion arrived at is, that the mean temperature at the earth's surface may now be as cold as it has ever been, though climate has varied at particular spots.

In Britain, for instance, there has been a recent "glacial period," whose marks are perfectly fresh ; but according to theory, partly founded upon these marks, it was a period like that which now prevails on the banks of Newfoundland and the coasts of Labrador.

Mr. Hopkins (quoted by Lyell, chap. vii., *Principles of Geology*, 9th edition, 1853) calculated in 1852 that the snow-line and glaciers would reach the sea in Wales and Ireland—

1. If the Gulf Stream were diverted.
2. If land in Northern Europe were depressed 500 feet.
3. If a cold current swept over the submerged area simultaneously.

The British marks above described seem to prove that a cold current did sweep south-westwards over Great Britain, at a time when the land was submerged about 3000 feet ; and that glaciers did reach the sea in these countries till land rose to the level of 1400 feet, or thereabouts.

There has also been a recent glacial period in North America, but, according to theory, it was only the marine climate, which now exists to the east in corresponding latitudes. Sir C. Lyell has pointed out that the glacial period of the Southern Hemisphere comes still nearer to the Equator; and if similar conditions prevailed in the northern half of the world, the cold might drift as far there.

In chap. vii., *Principles of Geology*, it is pointed out that Captain Cook found snow many fathoms thick extending down to the brink of sea-cliffs in lat. 59° S., which corresponds to Northern Scotland; and that he found the perpetual snow-line coincident with the sea-level in lat. 54° S., which corresponds to Yorkshire.

In the *Illustrated London News* of 18th June 1864, is a woodcut and a description of a collision with an iceberg on the 4th of April 1864, in latitude 54° 40′ S. About midway between Melbourne and Cape Horn, the screw-steamer 'Royal Standard,' while sailing with a strong breeze, suddenly ran into a dense fog, and shortly afterwards she ran against a cliff "six hundred" feet high. After bumping and scraping along this floating island for more than half a mile, and suffering great damage, the vessel rounded the end of the cliff and so escaped. She made her way under jury-masts to Rio de Janeiro. In the earlier months of the same year, the *Himalaya* and other vessels returning from Melbourne found these seas "beset with icebergs." At the rate of 1-9th above water, this berg was 5400 feet thick, 4800 feet under water, and 600 above. In latitudes corresponding to the Mourne mountains, the Solway Firth, Cumberland, and Durham, the sea is beset with hills of ice a great deal thicker than all that is visible of the British Isles. If the sea were level with the top of Ben Nevis, a berg of this size might touch the top, scrape the bottom of Loch Linne, 500 feet below the present sea-level, and rise 600 feet above water still. Changes of climate, and glacial denudation, which such fleets might accomplish, are not easy to calculate. Sailors, familiar with bergs off Newfoundland, affirm that even these are insignificant to bergs commonly seen off Cape Horn.

There are plenty of glaciers in New Zealand, about Cape Horn, and in South America; and very large icebergs, 150, 250, and 300 feet high, and two miles in circumference, have been seen adrift off the Cape of Good Hope between lat. 36° and 39°. These last were in latitudes which correspond to Gibraltar, parts of Africa, Syria, Cyprus, Candia, Asia Minor, Persia, Cabool, Japan, and Washington.

Sir Charles Lyell long ago imagined possible distributions of land and sea which might, as he argues, produce great general changes of climate over the whole earth.*

Having climbed thus far, some well-established facts begin to wear a different aspect.

If marks in Scandinavia and Britain do in fact prove that a cold current changed the climate of Western Europe, then similar currents may have done as much elsewhere. It is not necessary to assume a general glacial period in past time, because marks of ice are found on rocks in countries where the climate is now excessively hot.

It is proved that glacial action once extended a great way from the Swiss mountains; and that fact has been used to support the argument for a period of intense cold. But if ever there was a Baltic current east of England, Switzerland was on the other side of it, and the Alps and Pyrenees must have shared the influence which chilled Scotland.

The highest Swiss mountains are about 15,000 feet above the sea; their perpetual snow-line is at about 8500, and glaciers

* In his address, Sept. 14, 1864, at Bath, he attributes a former extension of alpine glaciers to the submergence of land, now the Sahara, where marine shells have been found.

slide to within 3000 feet of the sea-level now. The mean temperature below is about 55°; but if Western Europe were sunk 3000 feet or more, to the level of boulders on Beinn Wyvis and Driom Uachdar in Scotland, and on the Dovrefjeld in Scandinavia, then the Baltic Current, which carried Scandinavian boulders into Poland, might also wash the base of the Alps. They are in the latitude of Nova Scotia, where the mean coast temperature is 41° instead of 55°. At this rate the high Alps would still be 10,000 and 12,000 feet above the sea-level, in regions where Glaisher found snow falling above England, in June 1863, when the surface temperature was 66°. Alps 12,000 feet high, with a mean temperature of 41° at the base, and a cold sea passing westwards, might well breed glaciers large enough to be launched as icebergs if Scotland and Scandinavia were chilled and frozen also. When the land rose, these alpine glaciers would dwindle if the climate warmed as the sea fell, but they might take a long time to shrink to their present size.*

Cold is not easily driven from a fortress of which it has long held possession. It takes a long time to get the winter's frost "out of the ground." If the tail of the polar glacial system passed near the Alps, existing glaciers may be remnants of a large local system, like that which once covered Scandinavia, and is now dwindling away there.

If the Mediterranean were the receptacle of an arctic current laden with icebergs launched from the Alps, and drifting over France, Italy, Austria, and low lands then under the sea, there might be a local glacier system in Syria, and icebergs in latitudes which correspond to seas off the Cape of Good Hope.

* Hitchcock, an eminent American geologist, found what he considered to be ancient sea-beaches, at about 3000 feet above the sea, in Switzerland.

Hooker found an ancient moraine beside the cedars of Lebanon, and photographs of the Holy Land show rock-forms which strongly resemble ice-work.

Still further south, in Africa, snowy mountains now exist. If the cold stream ran that way, these may have bred glaciers at the Equator itself.

As described by Captain Grant in a lecture before the Ethnological Society, in June 1863, the country about the source of the Nile has a glaciated form. Some parts of it were said to consist of "flat-topped hills, with outbursts of granite; rounded masses are lying upon each other; there are saddle-backed hills whose western faces are steep and broken; and large loose stones are scattered about." As snow was in sight, and moraines are in the Lebanon, as the climate of this raised African plain is temperate now, a glacial period is possible even about the sources of the Nile.*

In Central Asia is a large system of local glaciers in the Himalayas, which are well described by Hooker. According to that traveller these glaciers are now dwindling away, for their marks extend far beyond their present limits. Are we therefore bound to assume that the whole world is getting warmer?

The snow-line of the Himalayas is now at 15,000 feet, and the mean temperature at Delhi is 73°. On the coast of China, in the latitude of Delhi, the mean temperature is 64°, according to Dove's Isotherms. But if Behring's Straits were wider, the climate on the eastern coast of China would suffer. There is a cold current there now, it would be colder. According to Kotzebue, there is a striking contrast in the vegeta-

* This guess is left as first printed. It is not founded on any personal knowledge of the place; but as the Sahara is now proved to be a recent sea-bottom, Alpine or Scandinavian boulders may be found there.

tion on opposite coasts in Behring's Straits, where no wider than the Straits of Dover; the western American coast is well-wooded, but the eastern Asian coast is bare and barren. A current runs inwards on the American side, and a miniature arctic current is believed to run out on the Asian side.

But if Behring's Straits were as wide as the North Atlantic between Greenland and Scandinavia, so as to spill the Arctic Current south-westward along the mountains of Chinese Tartary, and over the low grounds of eastern Asia past the Himalayas, and over India; then, even though the glacier-system of the Himalayas were lowered nearer to the earth's centre out of the cold and into the heat, the cold would gain if the sea were chilled, and the mean temperature at the foot of the hills changed from 73° to 64°, or to some lower temperature.

If mountains 28,000 feet high were lowered to 18,000, and stood in chilled water, with a climate like that of England at the coast, then the snow-line would be lowered, and Indian mountains might well breed larger glaciers.

They might even launch icebergs, and send stone-fleets south-westwards to choke harbours on the African coast, and do glacial work about the sources of the Nile.

In North America a glacial period reached latitudes which icebergs now reach in the Atlantic, and it appears that the continent was submerged about 3000 feet during some part of the "glacial period." Eminent men hold that it was a period of intense cold and enormous glaciers. The writer believes that it was a period very like the present, during which the Arctic Current has changed its course, and land has risen and sunk about 3000 feet.

The changes of level required to swamp continents and

change the course of ocean-currents, are not so large as may be supposed.

500 feet would sink the source of the Volga and drown the most of Europe.

2850 feet would sink the source of the Danube; 4500 would sink the Elbe; 1250 feet would sink the lake of Constance; 800 feet Basle; 1400 feet the Clyde; and boulders are perched on higher European watersheds, in Scandinavia, Scotland, Wales, Ireland, and central Europe.

At 4575 feet, on the Dovrefjeld, granite blocks are on mica slate (Von Buch, etc.)

At 3000 feet, on Beinn Wyvis, mica-schist is upon slate.

At 3000 feet, on Driom Uachdar, gray granite is on slate. All these are at places where transport by local glaciers is out of the question. On the Jura mountains, erratics derived from the Alps are common at about 3600 feet, and they too may have floated on ice-rafts, according to this theory of a sunken land now raised in Europe.

In Asia, the Ganges runs out of a glacier at 13,000 feet above the sea. How much would sink China is not ascertained, but most of India would be drowned by a depression of 4000 feet.

In America, 630 feet would sink Lake Superior, and the bottom of Lake Ontario is below the sea-level now. If ancient fossil-shells of marine origin are sea-marks, most of the high land in the world has been under the sea at some time.

If terraces be sea-marks, there are terraces on Snowdon, and on the Alps, according to Hitchcock, at 3000 feet; high up on the Himalayas, according to Hooker; and at about 3000 feet on the White Mountains in North America. Sea-shells were found at 3000 feet on Snowdon, by Mr. Baumgarten, in 1847.

There are cold climates, glaciers, and glacial action in

spots all over the world, wherever mountains are high enough to reach the cold, so as to catch and condense the clouds. If such hills stand on the western side of an ocean stretching nearly from pole to pole, and are washed by a cold stream, as in Greenland, any quantity of glacier-work yet found may be accounted for, without assuming any great universal change of climate at the distance from the earth's centre which is now high-water mark.

Though climate has changed place, it is not proved that the snow-line has sunk and risen again everywhere.

One of the last writers who have specially studied this subject, in speaking of Scotland, says :—

"In whatever way the change was brought about, there can be little doubt that when the land began once more to rise the temperature had likewise risen."

This accords entirely with what has been said above. But the following passages from the same page do not :—

"The submergence of a large tract of land would tend to ameliorate the climate. . . . The depression seems to have been general over the north of Europe, though probably varying greatly in extent in different regions." *

According to the theory now submitted to the merciful consideration of able judges, any depression of land that lets an arctic or antarctic current flow past an eastern coast will not ameliorate but spoil a good climate ; and such depressions in Europe and elsewhere probably caused the last "glacial period" in Great Britain and Ireland ; perhaps in the Alps and Pyrenees, Italy, Greece, Syria, India, America, and it may be in Nubia also.

There is yet another theory which will account for larger

* *On the Phenomena of the Glacial Drift of Scotland*, by Archibald Geikie. Glasgow, John Gray, 99 Hutchison Street. 1863. P. 102.

glaciers if icebergs of the dimensions described are too small to account for the ice-marks.*

It may seem paradoxical, but if the general temperature of the earth's upper crust were a little warmer, and solar radiation the same, there might be more glacial action.

The southern slopes of the Himalayas ought to be warmer than the northern, and glaciers ought to abound most in the coldest side, if glaciers resulted from cold alone. It is not so in fact, because glaciers result from cold and heat. Many English sportsmen have described these regions. Hooker gives a reason for the abundance of glaciers on the warmest side of the hills; Maury tries to explain like facts, in America and elsewhere, in his "sailing directions."

There is often a clear hard sky to the north, behind the ridge, when the southern districts are shrouded in mist, and deluged with rain, below the snow-line. Warm moist equatorial winds which sweep over the hot plains of India come loaded with transparent vapour. While thus expanded, the vapour only serves to intensify the heat by refracting the sun's rays like a lens, but when these hot wet winds meet the cold air of the high mountains, they are cooled and contract, the vapour is condensed into mist, the lens is spoiled, and the clouds drop their loads while they screen the snow from the sun. These big snow-heaps spread an awning of cloud in the air, to shield them from light.

The winds which pass over the Himalayas have but a scanty remnant of their store to bestow upon the northern slopes and high plateaus of central Asia; they carry little to the polar regions, to which the cargo was first consigned. To use Maury's illustration, the wet is squeezed out by cold, as

* For a theory of this kind, see *Quarterly Journal of Science*, 1864; and a lecture delivered at the Royal Institution, by Dr. Frankland, Jan. 29, 1864.

water is wrung from a sponge. There is a clear sky on the northern side, and the snow which does fall there melts rapidly, or evaporates, because the sun's rays are but little impeded by clouds in the lens of air.

If there were more water in the air generally, there would be more clouds; and these would form most at the coldest spots, because, in the Himalayas and elsewhere, that is the result of evaporation and condensation on the largest scale.

A confirmation of this opinion is given by the weather of 1863, 1864. In Britain and Canada the summers were very warm and bright; in Labrador and Newfoundland unusually cold and very misty. There was more evaporation at one place, and more condensation elsewhere.

If the whole of the sea were frozen, there could be few clouds; but if the whole world were warmer, there would be more evaporation everywhere, swifter movements, more condensation about the Poles, and more glacial action at high levels and latitudes.

The same thing takes place in Scandinavia, apparently for the same reason.

Warm wet south-westers, loaded with moisture, picked up from the warm Gulf Stream, fly over the sea and the low islands off Scotland, but they begin to drip as soon as they get to high land. The rain-fall at Inverary and Gairloch is far greater than in the Western Isles and Shetland; but when the clouds reach the snowy land about Bergen, they pour. About the glacier districts there are floods and snow-storms when there is clear weather close at hand. When the winds get to the high grounds, about higher watersheds further to the north and east, they have still a remnant of snow for Sneehætten, but there is not enough to make snow-domes and glaciers. The summer sun clears most of Scandinavia, because the sky is generally clear to the east of the hills, and the sky is clear because Bergen and the west coast glaciers have cleared it. From Bödals Kaabe, glaciers stream down almost into the sea; but there is no glacier worthy of the name at 8000 feet above the sea further east, and still further inland, at Sneehætten and Röraas (chaps. xiv. to xviii.)

The Bergen glaciers catch the Scotch clouds when they land, and hold them till they are well-nigh drained.

Snæfell, in Iceland, is another case in point. It stands far to the west, and has a local glacier system; it often gathers clouds from a clear sky, and rivulets pour down from it while neighbouring tops are clear of mist and snow, and rivers which flow from them are all but dry. It is a cloud-condenser, distilling glaciers from the air.

Iceland itself is another example. All the large glacier-systems are on the south, and in the centre of the island; no glaciers approach the sea on the northern coast (chap. xxv.)

Every floating iceberg is surrounded by a veil of mist, which preserves the cold mass by stopping light. The wetter and warmer the air is, the thicker is the fog which results. Fogs on the banks of Newfoundland, near the borders of the hot and cold water, are peculiarly dense (chaps. xxiii. xxiv. xliii., etc.)

On a bright day after a shower of snow, the shadows of posts in Hyde Park are often marked out in lines of snow, when the rest of the ground has been cleared by sunlight. Of two vessels of water in sunlight and shade, on the opposite sides of a house, the one on which light falls most loses most weight by evaporation.

The following is the result of an experiment. *19th June* 1864.—Two glass vessels intended to hold milk in a dairy, were partially filled with garden mould and water, made equal in weight, and exposed on opposite sides of the same house—on the north side under a verandah, on the south side on a pillar. 22*d June.*—After about forty-eight hours weighed. Weather fine; strong S.W. breezes, and bright sun during the day; clear sky at night; no rain.

Shade	94½ ounces.
Light	76
Difference	18½

Sun-light is a force which lifts water, but it is turned aside by any screen which casts a dark shade.

But if the whole earth were warmer, the sea would be warmer and would evaporate faster, to form more clouds, to give more shade to the ice-condensers, which now exist, in spite of sunlight, even on the tops of volcanoes.

If Himalayan, Scandinavian, and Icelandic glaciers exist because there is a warm sea and a bright sun at the Equator, it seems to follow that they would grow larger, and that polar systems would move faster, and so get further into warm regions, if more power were applied at the boiler-end of the caloric engine.

The same result follows if more fuel is burned under a still, or if colder water is poured on the worm; in either case the liquor flows faster. If weight be added in one scale, or taken from the other, the result is the same on the balance.

Because there are large glacier systems in Iceland, close above boiling water and molten stone, there may have been glacial periods on a far warmer globe. But the present state of things appears sufficient to account for all glacial phenomena yet observed.

Yet another theory has been started to account for glacial periods. It is assumed that there are regions in space which are colder than others, and that the solar system passes through these frigid zones at stated periods. These regions are as yet beyond the reach of a mere traveller, and the ice-records which he has endeavoured to translate do not seem to reach far back or recur at intervals. If anything is to be learned about fossil climates, patient grubbing in mud and ashes may do more than soaring at once after astronomers into infinite space.

The way upwards lies downwards at first. A breaker falls headlong, but the spray rises, and the force of the fall builds up the sea-beach. We must wade through water to dry land, and grope in darkness before we can reach light.

FIG. 91. A BREAKING WAVE. From a photograph.

THE END OF PART I.—DENUDATION.

CHAPTER XLV.

DEPOSITION I.

NATURAL SCIENCE—FORCE—ENGINES—TOOLS—MARKS.

In the preceding pages an attempt has been made to show that some branches of geology may be studied experimentally.

Small engines, which are worked by the natural forces which work natural engines, imitate nature ; and if all mechanics are parts of one system, that which is learned from one engine applies to all. So in studying "dynamical geology," working-models are useful aids.

Men can neither alter the laws of nature nor oppose them with success ; they must obey ; but they can work with nature's powers by obeying nature's laws. An engineer cannot stir a boat by stuffing a furnace with ice and a condenser with embers ; but by using heat and cold in the natural order of heat below and cold above, pistons are lifted and lowered, and steamboats are moved horizontally round the world. We are too short-lived and short-sighted to see with bodily eyes large geological movements and changes, which, in long periods of time, take place in air, sea, and land, about us ; we cannot even hope to see the whole of the outside of the ball on which we dwell ; we cannot get at the inside of it at all. The comprehension of any part of this engine is out of our reach, because we cannot even see the works. But models may be worked by the aid of natural forces, and when the models are engines of manageable size, their mode of action is more easily understood. We may learn something about the large engine, by watching how small ones work.

There are many things which men know but cannot explain, many facts which we are incapable of understanding. We cannot explain why we fall in air, sink or swim in water, and stand upon earth. We know the facts, but do not explain them by calling a force "gravitation," and by talking of "gases, fluids, and solids," and their "specific gravities." But in striving to reach unattainable knowledge, some has been reached which is power when applied to small engines; and which gives some vague notion of the largest engine of all. Astronomy is learned from the fall of weights, and the flight of small projectiles. Geology may, in like manner, be learned from geological toys. Human minds cannot grasp the ideas of infinite size or smallness, space, time, or number; but those who think are driven by facts to perceive that these incomprehensible things must be. If there be a limit anywhere, what is beyond it?

Men can never understand the great engine which works in infinite space, for they cannot even comprehend an atom; but that is no reason for ceasing to strive. An old Scotch saw says, "Aim at a gown of gowd, and ye'll get the sleeve o't." In striving to understand how mountains have been made, we may set natural mechanical forces to build and demolish molehills; we can construct and watch our little engines. In seeking abstract knowledge, things of practical use—shreds of the golden gown—are found. By experiment, designedly or accidentally made, men have learned all that they know about the engine with which they travel through space; and they have used their knowledge to make small useful engines to carry them round the deck of their spherical rolling ship.

By geological experiment, human minds may gain more knowledge of the engine, under hatches, and by imitating it gain more power. Engines are worked only by using natural powers ; these were found out while searching ; the most ignorant searcher may chance to find a treasure, even on board of this our argosy which circles round the sun.

Water and steam power are treasures, but only applications of natural force to human engines.

It took a long time to "invent" a water-mill, and a clock, and other engines worked by weights. The hydraulic cranes which now wave their black iron arms like living giants, and lift and pour out cauldrons of molten iron as a man lifts a pail of water, have only appeared in modern times ; but gravitation, which works all these engines, had been pouring rivers and oceans upon the earth, and steering it amongst other stars, before there were men or millers to use that natural mechanical power. Like it, steam is no human invention, and its application to engines is nothing new. It is told that one of the many so-called inventors of steam-engines gained his first knowledge of steam-power from the clattering lid of his mother's kettle. He was but a young discoverer, an observant scholar and imitator ; and yet his mind has swayed other minds and inanimate matter, ever since he applied the knowledge which descended to him from the first inventor of kettles, and was left by him as a growing fund to benefit all engineers. The human inventor did not contrive a force ; he found one, and so gained power which he used. There is, in fact, no single mechanical principle in any human contrivance, which had not been applied to some natural engine, long before the principle was "invented" and "patented" by men.

The first savage who boiled a root unwittingly used steam-power and burst boilers, in the food which he ate. A human mind had swayed the movements of matter, and had set a caloric engine to work when a man had purposely kindled a fire. But the application of heat-power is far older. Whatever the antiquity of men, and kettles, and fires kindled by men to boil kettles, may be, boiling springs, volcanoes, the world, heat, and light, are older than men and their weak inventions. The tool-marks of the old engines record part of their history on rocks.

In striving to understand the records and the engines, the best course is to seek after the powers employed, and set them to work when found.

If the minds of men who only discovered a use for weight and heat still sway the minds of engineers, and through them and their engines sway the movements of inanimate matter, a greater Mind can at least do as much with the universe and the minds of its inhabitants. Earnest striving to solve problems in natural science leads to this belief. We can neither see all the face nor reach the works of our own little world, nor can we hope to understand even that one wheel in the great engine; we cannot by searching find out its Maker; but we cannot do better than study his works. The more we see of them, the plainer it must appear that such an engine had a contriver who governs it.

In making geological toys to imitate parts of the engine of nature, all natural mechanical forces yet discovered may be employed upon all materials within reach, and all available wits set to watch results and turn knowledge to practical use.

Millers have learned to use gravitation with water-weights, in spite of river-floods; engineers may learn to use the world's heat, in spite of volcanic eruptions.

It has been done in Italy. If Icelanders would use hot

springs which have worked for centuries, they might have winter-gardens and hothouses; they might boil their mutton for nothing and sell the soup; they might at least warm their houses and cow-byres, irrigate their hay-fields, and wash in the hot water which runs to waste at their doors. If miners would but direct the natural underground heat-power which moves air in deep mines, they might save human lives, and the cost of power expended in ventilation. If we could learn to store up and use the heat-power which lifts water above ground, and so works all rivers and water-mills, there is plenty of spare sun-power to work all the heat-machines on the earth. Magnetism has been pressed and sent to sea as pilot; that giant may, perhaps, be set to harder work. Electricity is errand-boy and link-man, gilder and doctor, and strong enough for any place. Light paints portraits, kindles fires, and tells the shape and composition of distant worlds. Light, too, may be harnessed and set to work in time.

Towards useful discovery the study of natural science tends; it can lead to no ill, for the further we go on this path the nearer we get to truth. Natural science is not taught at English schools, and so much the worse for those who studied there. Some school of philosophers taught that the world stood upon the back of an elephant, and the elephant upon a tortoise. It was lawful to learn this much, but it was impious to ask what the tortoise stood upon: no one knew that mystery, and no one ought to seek to know it. Once it was impious to assert that the earth went round the sun. But now this reign of authority has ended. According to modern views, unstable ground may be cut from under the feet of the tortoise, and the sun does not go round the world, human authority notwithstanding. We may now seek truth anywhere and everywhere without offence; but

English scholars must seek it for themselves if they choose this path.

Natural philosophy is now open to all; but hitherto it has been little taught. Any child can and may make experiments. Every successful effort to find a cause is a fresh gain to all; the search for truth can lead to no ill if each step is made upon solid facts. All paths lead two ways, and study may lead to error; but those who travel the wrong way ignore facts or misunderstand them. He who sets his cart to drag his horse, mistaking effects for causes, may travel fast; but he can never rise. All inorganic forms which have been accounted for, record movements; all movements which have been explained, have causes. Any attempt to decipher these records and discover movements, forces, and causes, ought to lead up towards the great First Cause, whose mind and will contrived and made the natural engine of the universe. Every fact and finger-post, on every path tried, aims at this central truth, as the compass aims at the Pole.

An attempt has been made thus far to rise gradually from small engines and their marks to larger ones, from draughts in a room to trade winds, from raindrops and gutters to ocean-currents and geological denudation. A further attempt will be made to show the use of working-models in learning the unwritten history of great events; of things which are too big to be seen by little men; of changes which occupy longer time than human lives. The deposition of sedimentary strata, and their upheaval, follow after the denudation which made the chips. The way upwards lies downwards at first, for all paths yet tried lead inwards, and aim at some underground central force hidden there.

CHAPTER XLVI.

DEPOSITION 2—TIME 2—TEMPERATURE—LIGHT—AIR—WATER—WINDS—WAVES—FORM.

TIME.—In chap. ix. an attempt was made to show that a rate of denudation proves the ancient date of a recent series of events in the geology of Iceland. A rate of deposition is another measure of past time. If the surface of the world has been ground down and worn away so as to produce certain sculptured forms, the chips must be somewhere, and the rubbish-heaps in proportion to the work done, and to the time spent upon it. We judge of a carpenter by his chips; and so we estimate other work. It is manifest that a vast number of trees have been sawn up at spots in Scandinavia, because of the heaps of sawdust on shore and below the mills, in the river and in the river-bed. An old mine is known by large rubbish-heaps. An old furnace is known by large hills of cinders. Ancient and long-continued human occupation of the coast of Denmark, is proved by large heaps of oyster-shells, gnawed bones, and such contents of "kitchen middens." The evidence for time is equally good if the carpenter has struck work, or the saw-mill has stopped, or the mine is "knocked," or the furnace "blown out," or the men who ate the oysters are eaten by worms.

So it is with sedimentary rocks. They are chips; and, from their thickness, it is plain that a great number of engines, of some kind, have been hewing rocks for a very long time, and shooting the rubbish into the sea, to be carried and packed. So deposition may equal denudation, but cannot exceed it.

In most cases, the only attainable measure of denudation, and the only time-keeper, for past time, is the size of these beds of rubbish. River denudation in Iceland is older than Icelandic history; so is glacial denudation. The discoverers named the land, and the 'ice' did not grow there in a day. A rate of glacial action has not been found, and it certainly varies. The machine is working full speed in Greenland; it has struck work in Britain; and it is working half speed in Scandinavia. Taking the present rate in Iceland as something like a medium rate for many ages, the measure of the work done is the quantity of mud now carried out of the groove in which ice works.

An old fisherman's test for clear water may be used when a better guage is wanting. Fish will not take a fly in muddy water, probably because they cannot see it from their haunts at the bottom; and the test for fishable water is: "Wade ye in to yer knees, and when ye can count yer ten taes she'll fush." In the sea off the west coast of Scotland, shells are visible in many fathoms. In glacier-rivers in general, and in large Icelandic rivers in particular, the fisherman's test shows water as thick as the muddiest of Scotch rivers in the wildest spate, or the water in London when Faraday dropped his card on Father Thames, and found him filthy. Wade into the Hvitá up to the ankles, and the bare feet are wholly hidden from the eyes by white mud. Most of the Icelandic rivers are like it, and wont "fush" at all. The Hvitá is a broad, deep, rapid, thick, gray stream, larger than the Thames, and all the mud is ground by glaciers from igneous rocks. The quantity of mud in a gallon, and the number of gallons which pass in a given time, would give a rough measure of the work of denudation accomplished in this basin. If the beds of sediment could be found and identified, they would equal the groove made. Beds of rock-chips cannot be referred to the several grooves whence they were taken ; but chips do not escape from the world ; and because all sedimentary rocks are chips, and denudation at the fastest known rate is slow, all history must be as nothing to the geological time which is measured by sedimentary rocks. Modern geology deals chiefly with rubbish-heaps of this kind, with their transport and packing, and with the order in which the layers are laid. Except in the case of glacial drift, no attempt is made to trace stones to parent rocks in position ; but deposition clearly results from denudation, from transport of materials, sorting and packing ; and all these operations occupy time.

FORM results from movement, and movement from Force. The forms of sedimentary beds record movements, and the forces which caused them : and they are thermometers also, for they register temperature.

If the packing of a bed of silt records water-work, it also records some temperature greater than the freezing-point of water at the earth's surface. Pebbles and grains of sand, which retain their shapes though cemented together, record that a temperature *less* than the melting-point of the stone has endured at the spot ever since the bed of silt fell through unfrozen water. The maximum limit of temperature at a particular spot is thus recorded for the whole of the time during which this particular form has lasted.

The *Forces* which pack silt, by moving air and water, are the same which work denudation, and the engines and tools are the same. Loose stones are carried, sorted, and packed by rivers and land-ice, by ocean-currents and winds, by waves, and by floats which are strong enough to carry such weights. The fall of the sediment is a result of *gravitation*, the rise of the water results from *heat* as it appears.

The forms are the tool-marks of these engines, and by learning the marks, ancient work may be assigned to the engine which did it, and to the mechanical force which drives the engine.

In order to learn the marks, the engine may be watched, or, when any part of it is out of reach, another part may be watched, and the lesson so learned indirectly. We cannot get to the surface of the air, but we can watch waves on the surface of water, and study the barometer ; we cannot get to the bottom of the sea, but we can watch the air-engine at work upon snow and sand-drifts on shore, and study the sea-beach at low tide. We can see the tools at work.

Waves.—When a fluid is moved by any force, the smooth surface takes a form which indicates the direction of movement : if solids are moved by the moving fluid, they too are packed into corresponding shapes, which may endure to record what happened at a particular time and place. In order to recognise work done by an old wave, the thing to study is an existing wave.

Waves on a stream.—A stream of water, or of any other fluid, while flowing over an uneven bed, or in a narrow channel, curls over and forms waves. The water is dragged downwards, but it is also thrown upwards and from side to side by reflection from impediments, and it moves in curves, which produce wave-forms above, and wave-marks below.

By knowing these wave-forms anglers know where to seek fish, and boatmen how to avoid stones. In deeper water similar forms betray reefs and sandbanks ; on dry ground silt-forms record the passage of currents, and of departed waves, even waves in the invisible air. In any bed of sedimentary rock, similar forms record similar movements.

FIG. 92. WAVE-FORMS AND WAVE-MARKS.

We are driven to assume that water, and other fluids, consist of particles, and that they jostle and rebound ; that the shapes of waves upon running streams result from the directions in which force and resistance act upon these particles.

When fluid and solid particles, dry dust, sand, small shot, and similar materials, are poured down a slope, wave-forms and movements resemble each other in all the streams. In sorting dust-shot, a stream is allowed to escape from under a sluice, and the shot, in rolling down a board, moves like water in a "lasher." A single ball or a big stone leaps down-hill in curves, which agree with wave-curves on water-streams. Waves which the wind drives along the surface of stagnant water, also resemble curves described by solids. A ball played on a billiard-table bounds, and rebounds ; jostles other balls, and moves on the plane as waves do in a pond, or like tidal waves reflected from continents. We may assume that fluids consist of particles which also jostle and rebound.

If a marble is driven against one end of a row of marbles, the driving force and the motion pass from ball to ball through the series ; and the last ball moves till the force which moved it is transferred elsewhere ; or, being changed, disappears. If water consists of particles, then water and loose sand make a series, and motion and force pass through it to the last particle which records the movement when it stops. Some force—sunlight, for example—moves air ; and the wind stirs the sea, which stirs sand ; the last grains of this series take the form of water-waves, on the sea-beach and in deep water. The sand-form records movement in water, air, and light, if light be the force which started this train.

Water-waves produce waves on sand. Waves in air also produce like forms in dry dust. Waves of sound are copied in dry sand spread on a sounding-board, and on water in a musical glass. Photography and photometry record movements in light, or movements caused by light, and philosophers have come to believe that light is but an effect of systems of waves moving in some unknown fluid, as sound-waves move in air. Each of these things—water, air, and the fluid whose waves are light—is capable of moving other things.

The moving force which moved the first particles in the series, of which the last retains the recording form, is the force which did this work ; if light moves the air, light makes the ripple-mark on the beach. Are we to stop there?

In the row of marbles a hand and a human will were in the series, and the will moved the last marble. In silt-beds and old stratified rocks, the chain of cause and effect may seem endless ; but the ultimate cause of the ripple-mark must be will also, unless there is movement without a cause somewhere short of the will. Unless there is a will at the end of the train of machinery, sand, or the sea, or the wind, or the light of the sun, or some other inanimate thing, moves without a cause ; which is contrary to experience, and therefore cannot be assumed in any train of reasoning. We never find marbles and billiard-balls, shot and shell, moving without a cause, and most of their movements can be traced back to human will: why should larger or smaller particles, worlds, or atoms, move without a cause, more than these?

Forms which result from denudation and from deposition are as figures on a dial-plate which record movements; from them the moving force may be sought through the works: the further men can reach the better, if they pause to think of Him who said, Let there be light, and feel that they are looking at the works of their Maker, when they study natural science, and the tool-marks of His engines.

CHAPTER XLVII.

DEPOSITION 3—WINDS 2—WAVES 2—WAVE-MARKS.

BECAUSE the works of nature are too large for human inspection, working-models of them help comprehension. Immediate causes are learned by watching the rapid growth of form. The wind is invisible, but smoke and waves are not; and through their visible forms and movements, invisible movements and forms may be seen.

When wind blows along the calm surface of still water it does not move in straight lines, horizontally; it strikes downwards, and rolls along, driving the water-surface before it. On a windy day, where a mountaineer has fired a moor, the white stream of smoke flying over the brown heath rolls as it flies. It rolls, and breaks, and surges over the plain, as the wind does. It flows down hill into a valley, and rolls up the opposite slope; and where the smoke strikes visibly, the brown heath bends before the invisible wind. When some farmer is burning weeds near a hay-field, the waves on the sea of green fit into the curves of the smoke-cloud, and the smoke betrays the immediate cause of the movement, though it is invisible. Air does not flow in flat sheets or straight streams, but rolls as water does in a river. Because the river rolls, sand is packed into the shapes of waves, on water, heath, and grass, which are driven by rolling streams of air.

When a breeze begins to stir the glassy surface of a lake, floats move slowly along, while tiny waves and floats rise and fall, advance and slide back, as they are pushed by the wind, and pulled down by weight. The surface "ripples," and moves as far as the force can drive it. The far end of a canal grows deeper when the wind blows along it. Large lakes rise to leeward; high tides coincide with strong gales at sea. Water is driven by the wind, and the shape of a wave suggests that it is moving water driven up over water at rest, and falling back when the force has done all it can to push it over and make a breaker of a roller.

The force which moved the air is transferred to the water, and from particle to particle; and thus a "curl on the water" grows; bigger waves grow, and some large ones even move faster than the wind, and so foretell approaching storms.

The force which is thus transmitted is also reflected, bent aside, accumulated, dispersed, accelerated, and retarded. So the forms of waves, and their movements, are complicated and hard to comprehend.

Horizontal movements.—Waves, moving upon the surface, are not straight continuous ridges, crossing the path of the wind; but short curved ridges, moving and spreading in many directions. Waves on any puddle are like sea-waves in this respect.

Barnespool at Eton is a sheltered pool, walled round, and spanned by a bridge. When the wind blows strongly from the west, curved systems of small waves are driven in under the bridge; they strike against the walls, and curl round the piers, and they rebound from side to side. The force which moves the wind is transferred to water, transmitted through a series of water-particles, bent aside in passing the pier, reflected from the walls, and finally recorded upon a minature beach. These small systems are very complicated, and as hard to comprehend as larger wave-systems, but they are better seen, because the whole pool can be seen at once. The waves can be watched from the bridge, bending, crossing, and re-crossing; meeting, passing, rebounding from the walls, and gradually fading away into a calm at the sheltered end of the stagnant pool. Barnespool was the sole teacher of this science at Eton.

It is easy to draw and map out these wave-systems, and to apply the knowledge to larger systems of waves. It is easy to see how invisible particles of water move, by watching the movements of solid floats. There is no general movement in the water, but there is a slow drift on the surface. Apples, orange-peel, bits of ice, and other things which float deep, advance slowly towards the calm, but they do not move steadily, or in straight paths. They move as the water does, up and down, forwards and backwards, describing curved paths, like waltzers or tumblers, who whirl and roll while they advance. The whole of these movements clearly result from the force which moved the wind, and that is sunlight, according to modern science. The beach at the end is the tool-mark of the engine driven by some mechanical force. It is a photograph.

What is true of this puddle is true of larger ponds.

The Serpentine, in London, is a larger sheet of water spanned by a larger bridge, under which waves pass. Waves at the far end cannot be seen from the bridge, but they can be followed and watched. The systems move fastest in the middle; they are retarded by the sides, and so form loops, as they do under every arch. At the end, the loops beat upon a concave dam, and the waves are reflected; they return and meet at a focus, where the force which drove them is accumulated. The waves leap highest in the focus of the wall, and there they disperse, and set off again, moving back against the wind which drove them forward. At the sides of the canal, two systems of breakers cross each other diagonally. One is the side of the loop which is moving forwards, the other is the side of the reflected loop which is moving backwards. Orange-peel and water-logged apples leap and rock to and fro, advance and retire, as water-particles must do; and ducks in search of food paddle about under the wall, and use their experience of reflected force to avoid shipwreck. Force, from which all these complicated movements result, is still the same; and the shape of the gravel beach, and piles of drifted rubbish upon it, record the movement and the force.

The same thing is to be seen wherever there is a beach.

At Weymouth, the waves of a large bay dash against a concave sea-wall, and rebound. Systems of large size may be seen advancing from the horizon, and retreating from the wall; crossing and recrossing, and meeting in the focus, as truly as invisible waves of sound and light meet in the focus of a reflector. The waves driven by an accumulation of force leap up to form cones and pyramids, and jets of spray; and the sea boils.

From *the top of Portland Island*, which makes one horn of this bay, still larger Atlantic waves are seen moving rapidly up channel. They are retarded by the ebb, are accelerated by the flood; they are turned aside in passing the Bill of Portland, curl round into the shelter, and roll into the bay. They are reflected from the beach; the force is accumulated in the focus, dispersed beyond it; ships at anchor and water-logged buoys rock in the sea; and one side of the Chesil Bank records these movements, and the amount of deflected force expended in building this beach behind Portland.

The whole is but an enlarged edition of Barnespool, more difficult to see and harder to comprehend, because larger. A whole system is seen from the bridge at Eton ; ten minutes will carry an observer from one end of the Serpentine to the other; but from Weymouth to the Bill of Portland is a day's march, and the wide Atlantic is beyond.

On *Isle de Rhe,* near Rochelle, on the coast of France, stands a tall lighthouse, called Tour de Balêne. It stands upon a sandy point, with well-marked sea-beaches. Outside the point is a long flat shoal, at the end of which stands a

FIG. 93. CROSS-ROLLERS AT ISLE DE RHE, NEAR ROCHELLE.
From a sketch made from the Tour de Balêne.

second lighthouse on a rock which is covered at high tide. Big waves rolling in from the Bay of Biscay and the Atlantic hit upon the end of this shoal. They are most retarded where the water is shallowest; and so the long curved ridges become loops, bend and curl inwards. They do no more than smaller waves do on points in Barnespool; but from their greater size these cross-rollers are very remarkable, and do very remarkable work. One moving system thus bent on a shoal beyond the limits of vision appears to be two systems moving diagonally upon opposite sides of the shoal, the point, and the lighthouse upon it. The long rollers break and form a moving network, whose knots are tall crested " white horses" advancing directly upon the end of the spit; while the meshes are green rollers, crossing each other at right angles, and breaking heavily on opposite sides of the point.

The bent sea-waves converge and meet at their focus below the lighthouse, as rays of refracted sunlight converge and meet in the focus of the lens above. The form of the sand-spit records this movement, as the Chesil Bank, and miniature banks in the Serpentine and in Barnespool, record the movements of smaller waves there. But in this case the pool is too large to be seen, and harder to understand for that reason.

Tides are but larger waves harder to comprehend, and driven by a different variety of force. If ordinary sea-waves result from the radiating force which moves the winds, these appear to result from the converging force of gravitation, which drags water towards centres, outside of the circles which bound the sea. Tide-waves rise under the sun and moon, and follow them westward; but they too rebound, and their vast and complicated movements have not been fully unravelled.

Where tides have been mapped and so brought within reach of human vision, the movements of tidal waves appear to agree with those of common waves, which are impeded in wandering over the surface of smaller pools.

It is not necessary to study uncontrollable tides or Atlantic waves; a knowledge of this part of the engine may be fished out of every puddle. The advance of the tidal wave in the Bay of Fundy, where the rise is from 40 to 75 feet, though it is one of the grandest sights in nature, is but a large copy of the flux and reflux of broken waves in any creek, or on any sandy beach.

When something of the movement of waves has been learned, marks made by waves on sand and gravel beaches are comprehensible ; and similar marks, wherever found, can be referred to their immediate cause, and their meaning so far interpreted. Till the movements of waves are studied, their marks mean nothing, because their language is a foreign speech.

At p. 340, vol. i., a lesson taught by the ebb-tide is set down as it was learned on a Highland strand ; it is good for all strands, new and old, if only they retain the tool-marks of Deposition by waves.

Old ripple-marks on the millstone grits of Yorkshire, in quarries near Pately Bridge, are still as perfect as they are on a strand from which the ebb has just retired. When a new surface in the quarry is laid bare, ripple-marks are the same in shape, size, colour, and material, as ripple-marks in the sea. Tracks of creatures which wriggled, and crawled, and hopped, and walked about on the wet sand ages ago, are as fresh upon the stone as similar tracks made within the hour. It was recorded upon one slab that water had moved first towards the north-east, and then towards the south-east, or that two systems of waves had crossed. The surface so marked by moving water was left dry, marked by moving creatures, and dimpled by falling drops of rain or by rising bubbles of some gas. This surface now is solid rock ; thousands like it lie over it and under it, like pages in a book ; many thick beds of sandstone are piled like volumes stacked in the corner of a room. The system stands low in the series of geological records, but far above the floor. The beds in these quarries have been shattered, broken, distorted, disturbed, upheaved, crumpled ; big angular rents, fissures, and fractures, are there as plainly seen as fractures made with gunpowder and sledge-hammers. Some of the rifts have been filled, and in some of these are valuable metals, which are worked. Since the veins were formed, the sides of the crack have moved, for there are slickensides in the veins ; they have moved in various directions, for marks on the smooth surface cross each other where they have rubbed. Since all these movements took place, the broken edges of the broken beds have been ground away and rounded off—" denuded" into the shape of the Yorkshire hills and dales.

But in spite of all these and many other changes, and of all the time which has elapsed, the tool-mark of a tiny water-wave, and the spoor of living creatures, record certain facts in language too plain to be gainsaid or misunderstood.

Low down in the geological pile of stone books, on a spot in a crumpled torn page of millstone grit, it is recorded that long ago there was deposit and packing of silt in fluid water, which moved as water now moves on the nearest sandbank in the Humber ; that plants grew, that living creatures crawled, and that rain fell from the air. There is no human standard measure for such denudation and deposition, or for such time as this ; but the form registers the working of the old engine, which still works.

The climate of Yorkshire is also recorded within certain limits. The water was not frozen ; it was not steam, nor was it too hot for animal and vegetable life. The coal vegetation which succeeded resembles tropical vegetation of the present day. It is probable that the climate was warm. Sometimes an inorganic shape is laid bare in the Yorkshire quarries, which has no counterpart on cold misty northern shores, and these shapes tell their story more certainly than fossils. It is only probable that a plant like a palm-tree had a similar nature; it is not certain. It was probable that an extinct elephant lived where the climate was hot; but it has been proved by the discovery of woolly hair beside mammoth's bones, and on the skin of a mammoth, which fell out of frozen ground about Behring's Straits, that the fossil elephants which lived about the "glacial period" were provided with natural coverings to resist the cold which prevailed in England when English mammoths lived.

The trees of the coal-formation may have flourished in colder climates, though they are like the tropical vegetation which now feeds elephants. No experiment can test conclusions drawn from the shape of a fossil shell, and from the habits of living things; but inorganic forms record facts which seem never to vary. Frozen mud, mud packed by waves, and sun-baked mud of the present day, must, so far as we know, be like mud baked, washed, or frozen, at the time when the first bed of silt was formed.

Beside, and mingled with ripple-marks, certain inorganic forms are occasionally laid bare in quarries near Pately Bridge, which seem to mean baking rather than freezing: a warm climate in the place where millstone grits are found. One seemed to be a form moulded in sand, partly by air. Dry-looking white sand, apparently blown by the wind, is scarcely bound together, and rests loosely where it fell upon a strange, brown, rounded form, whose section shows minute bedding. It seems as if a bank of sand and mud beside a runlet had been well baked till it cracked, that the edges were rounded off by tides or floods till a definite form, a tool-mark of deposition and denudation, was moulded in sand. Then came a sheet of brown mud or a green coat of vegetation, now reduced to a colour, and over this the dry white sand appears to have drifted. Then came a deluge of clean gray sand, which buried the whole, hid it and preserved it till it was quarried by Yorkshiremen in search of paving-stones. The whole document must be read together before the record is understood.

Ripple-marks are familiar to geologists, but other inorganic fossil forms have not been much noticed, though they are equally worthy of attention as records. Ripple-marks abound in sedimentary rocks of all ages. In the old rocks of Orkney are ripple-marked slates. In the oldest of Welsh slates, where no trace of life has yet been found, ripples are perfect. In these old, unaltered, sedimentary beds, which have been tilted, shattered, baked, and crumpled, the hard blue surface of a flag when newly bared is often rippled as plainly as the nearest mud-bank. But in older Canadian beds which have been more altered, even these marks are obliterated.

Where the form exists it tells its own tale ; it tells that the fusing point of the rock has not been reached at the place since the mark was made ; that the freezing point of the fluid which packed the sand or mud was not reached when the waves moved. But when the form has been obliterated at one part of an altered bed, though preserved elsewhere, it proves that some other force has been at work since the sediment was packed by waves.

The alphabet of form is to be learned from engines working on the surface of the globe ; but inscriptions to be read are stored below, and some of them are harder to read than ripple-marks, because they were written underground.

CHAPTER XLVIII.

DEPOSITION 4—WINDS 3—WAVES 3—BEACHES.

THE most characteristic wave-mark is a beach. It is a form like that of waves which beat upon it, one which can only be understood by watching waves. A more beautiful thing than a big wave is not to be found in nature. Many a pleasant dreamy hour has the writer of these pages spent in watching Atlantic rollers sweeping on from the blue distance to thunder in against the Scottish coast. A green glassy ridge comes rapidly on, glittering in the sunlight; heaving, growing, swelling, and mounting up, as it comes nearer and nearer; growing steeper and steeper as it reaches shallower water. The top is ever pushing on over the base; the base is constantly held by the sea-bottom, and pushed back by the undertow. The steep ridge of water becomes a wall, and the wall a hollow curve like a sea-shell, and then the moving hill rolls over its base, and tons of water fall headlong down with a crash. The broken water rushes on like a rising tide of white foam, and leaps up in sparkling fountains of spray, and the flood drives all that will move up hill till the force is spent. The falling tide of the undertow rushes back with the force of a mountain-torrent as broad as the shore is long. Every stone is moved; the beach is constantly worn by waterfalls equal to the height and weight of the wave, and by torrents equal to the depth and breadth of the undertow. Between high and low water mark the beach takes the form of a solid wave, because pebbles are packed by water-particles which transfer the force which moved them to sand and stones. The beach driven by water has a curve like the back of a wave driven by the wind, and each ridge of loose stone leans against a rock, or rests on the back of the ridge before it. The woodcut is from a portrait of a heavy rolling Cornish wave which came from the west, curled round the Land's End, and was returning

Fig. 94.

westwards, rushing furiously to land against a strong wind, in a narrow bay with a sandy bottom and a pebbly beach. The curling head was hurrying over the base to reach the English shore, and a silver plume of spray streamed back like a mermaid's hair, or a horseman's crest.

On the far side of the creek the retarded wave was seen lagging and breaking before its time on a pile of loose angular stones, the broken chips of a fallen cliff; and these, as the water burst amongst them, and roared over them, stirred and rolled, and rattled and groaned, and ground themselves to powder. When the larger tidal wave ebbed, and these Atlantic waves were driven back, a dry beach remained. It was the track of the invader who will some day sweep England from the face of the earth, unless some underground ally lifts her cliffs out of reach of the sea.

This beach was a steep bank of boulders and pebbles, with a broad slope of gravel and fine sand at the base. The larger stones were below, driven as far as the wave could drive them ; smaller stones were above, tossed up by the recoil of the blow ; the gravel was at the top of the slope, dragged there by the undertow ; the sand was lowest and furthest out, where the force of the downward stream was nearly spent, or balanced by the advancing wave ; ripple-marks, stream-marks, and the rest of the smaller tool-marks of deposition by waves, were on the sand.

A solid wave of sorted stones rested upon the rock where it broke, and the shape of it was like that of the wave which was driven by some invisible force. The force which shaped the beach was that which moved air and water, and the invisible wave of force may be like the fluid wave and the beach.

One result of this action is the formation of new land. The sea builds dams, and rain-water fills up the space behind them with silt. Behind the Chesil Beach, near Portland, a lake is formed, and rivers are filling it with mud. Near the Start Point is a similar lake divided from the sea by a broad wave of boulders. The lake is below an ancient sea-cliff, and is rapidly filling with mud and reeds ; it is full of fresh-water fish. At Borth and Traeth Mawr in Wales, are similar beaches. At the head of Breidfjorð in Iceland are larger beaches of lava boulders, behind which are pools of sea-water, and fresh-water ponds; and rivers still flow through openings in this lava-dam raised by the sea at the far end of a bay.

Near Snæfell is the most remarkable beach of all. It is a great black natural mound running across a valley, so as to dam back the drainage waters, and hold in the ebbing tide. The crest of the ridge is composed of smooth egg-shaped blocks, larger than a man's head, tossed about in the wildest confusion at the top, and more neatly packed at the base. The

FIG. 95. BOLANDS HOFVDI. August 16, 1862.

A cliff of columnar lava, interstratified with ashes, and resting on coarse hard breccia of rolled pebbles. The talus beneath the cliff is chiefly sand; it makes an angle of 32° with the horizon, and is the only pass along this shore.

seaward slope lower down is fine black sand, strewed with brilliant shells, like those which are found in boulder-clay. The back of the mound has a different steeper curve and slope. The whole is as near the shape of breaking rollers which fall upon it as the materials of which it is composed will admit.

Small stones have been thrown over the mound like spray, and rest where they fell. It is a solid roller, which has not reached the shore. The shore of the inland lake is strewed with pumice, and suchlike volcanic materials, and is haunted by flocks of birds. The whole structure rests upon a foundation of igneous rock, and is the work of fire arranged by water. If this beach were found anywhere ;—in a quarry, or on a hill-side, it would tell of waves as large as those which fall upon it : ocean-waves, which may roll without a break from the South Pole to the beach at Snæfell.

The district of *Myra Syssla* in Iceland seems to be land formed in this way. Beneath high broken precipices, which look like sea-cliffs, a wide tract of boggy flat land slopes towards the sea. It is traversed by ridges of gravel, which have the form of dilapidated beaches, and between these the whole country is a quaking bog, through which occasional rocks appear. But these old beaches are far higher above the sea than modern beaches, and they are not horizontal. They prove that the whole land has risen unevenly. They mark a late change ; and if similar changes took place in early times, they too should be recorded somewhere amongst the old beds.

At *Màlar* in the north, at the end of a deep fjord, where big rollers cannot now come, are similiar raised beaches, with small moors and bogs resting in hollows amongst the boulders. In Scandinavia are many similar marks ; and they are found high up on the Himàlayas.

At the head of the large Newfoundland bays, which face the Atlantic—Conception, Trinity, Bonavista, etc.—beaches of this pattern form ramparts along the whole shore. Some are bars under water, others run from point to point like moles or breakwaters ; fishing craft anchor behind them, rivers form brackish pools on the land side, and silt-beds gather in the still pools. Icebergs drift about in deeper water outside, and there drop stones carried from Greenland. Higher up are terraces of larger glaciated Newfoundland and foreign stones, confusedly mixed with sand, rolled pebbles, and beach-stones. These in their turn rest upon glaciated rocks, which have risen, and are still rising. In winter, bay-ice packs old chips along the shore. In spring, rivers dig materials from old terraces to build new deltas behind new sea-beaches. The bays are like Myra Syssla, the Miry Shire of Iceland, but in Newfoundland the sea-bottom has not emerged, though it is rising; and the low ridges are now parallel to the sea.

Surely these beaches may help to explain the osar and kames of the glacial period.

In North America raised beaches abound. They were first described by Hitchcock, and they are conspicuous on the White Mountains, Green Mountains, and elsewhere, at great elevations. They appear to be sea-beaches, formed like those now forming in the bays of Newfoundland, and ebbed dry in glens which were bays in the glacial period. Those which were most exposed (the highest) are, like the beach at Snæfell in Iceland—confused stone-heaps tossed about and irregular in shape. Those which were sheltered by rising points are like those now forming in the bays of Newfoundland. At the head of one glen, at Gorham in the White Mountains, a laminated terrace of fine sand and mud, disposed horizontally, appears to be a delta formed in still water at the end of a bay. The formation is about fifty feet thick, and from its position may be a fresh-water deposit formed in a lake which burst outwards through a distant terrace, and left the glen for the railway to occupy. Upon this delta, if such it proves to be, large glaciated boulders are piled.

The translation of the whole record made on the spot in 1864 is, that ocean-currents, icebergs, and bay-ice, drifted along the course now followed by the Grand Trunk Railway, and dropped foreign boulders in still bays and straits, which are now glens and passes amongst the highest of the Alleghanies. The American author who followed Chambers thought he saw raised beaches in Wales; and sea-shells have been found there at 3000 feet. He also thought that he saw the spoor of the sea in Switzerland at similar heights. Till sea-shells are found there, and in the White Mountains, there is room for argument; but there is little doubt that these so-called raised sea-beaches are marks of waves in water, in air, and, it may be, in light.

A ripple-mark is then a copy of a ripple; a beach copies a larger wave, and both are marks of deposition, and tool marks.

This mark is a thermometer like the rest, and it is also a water-gauge.

The beach is formed at the water-margin. If land rises, or water sinks, the beach is left high and dry. If land rises "straight away" from the earth's centre, if one spoke of the wheel grows longer, the old beach-mark is level there. It is like a storm-beach; a higher mark parallel to the lower beach, and to the sea; a curve on a higher sphere, further from the centre. If land rises unequally; if it bends upwards like a bubble, or tilts up like the lid of a box, the beach-mark records that change; for it was made horizontal.

If the whole sea has sunk down; if the sea-level is nearer to the earth's centre and the land where it was, the old beach-mark must record that fact also: it must be found at the same level in all parts of the world if the whole sea-level went down at once.

If the land has risen at one place and sunk at another; if it has grown up like a dome, and sunk like a bowl elsewhere, the beach-mark records the fact by its distance above the sea which has a regular curve everywhere.

In reading this larger record, the denuding action of waves must be considered. On coasts above mentioned no beaches are formed at exposed points. The rocks are bare; but they are broken or sawn, or otherwise worn and marked between wind and water. Some are drilled, pierced, or blown up, so as to form pot-holes, caves, and arches; others are cliffs, and under some of these are fallen talus-heaps.

It is a question of temperature and tides, rise and fall of level, whether waves demolish rock, or pack chips at the water-margin. So if the real beach is found anywhere, a worn shelf at the same level is not far distant. One is at the point if the other is in the bay. Woodcuts at page 357, vol. i., are meant to illustrate this fact; which the author of "Ancient Sea-Margins in the British Isles" pointed out long ago. Applied by him to Scandinavian records, the rule proved that Finmark rose like a bow. If sea-margins were traced round the world, they might perhaps prove that the land has waves like the sea.

The changing form of a breaking wave is hard to copy, its movements puzzle mathematicians; but these facts appear to be right so far as they go. The form of a wave drawn by light may be compared with other fixed forms; and photographs of breaking waves, made and bought for the purpose, have led to these conclusions. The woodcut, p. 261, is from a photograph. With it compare the portraits of snow-waves (pp. 293, 298), the cuts on pp. 272, 286, 299. Compare these with the portraits of clouds, vol. i. p. 33; of trees bent by the wind, pp. 31, 59. Compare the cuts in chaps. v., vi., vii., viii., which illustrate movements in air and water, with real waves, snow-drifts, and beaches; and these forms will seem to be copies of movements and records of force, the spoor of the sea and the wind, and natural photographs. The force which makes a collection of fluid particles move, and take a certain form for a time, when transferred to solid particles makes them move in the same way, and take like forms, which endure. In fluids the form cannot last; in solids it may. In the photograph reflected light so acted as to pack solids in certain forms; the water-wave was copied in silver by light-waves, and it has the shape of the beach.

The wave and the beach, like the photograph, may result from waves in light.

Thus form appears to record that light acted as force, directly or indirectly, through other materials,—as RAY-FORCE, which is only perceived where it has accumulated at the end of a long train.

The cut below is a copy of a natural snow-photograph of an air-wave made this year.

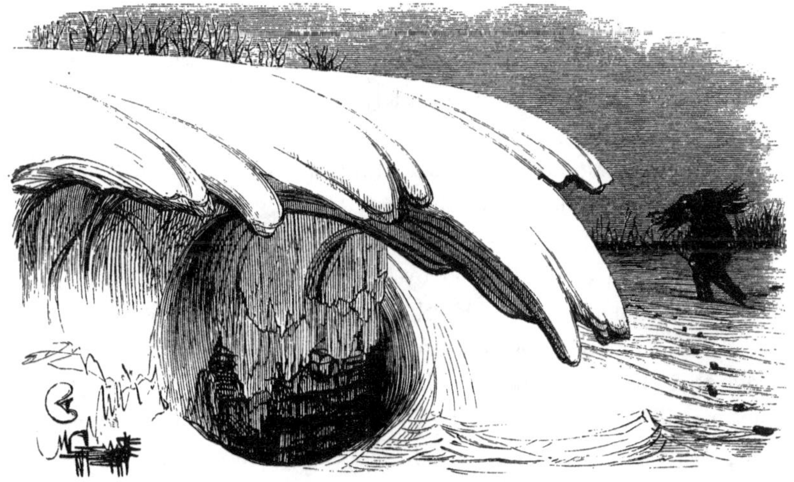

FIG. 96. A SNOW-WAVE IN CHESHIRE.
Sketched from nature, January 28, 1865, after a strong breeze of wind. Horizontal distance from the edge of the snow-breaker to the wall on which the hedge grew, two feet eight inches.

CHAPTER XLIX.

DEPOSITION 5—WINDS 4—WAVES 4—STREAM-MARKS.

TAKING form to be a record of force, and the force which makes a ripple-mark and a beach to be Rays, acting through a chain in which air and water are links only, then similar marks ought to be found at all links ; for instance, where water has played no part in packing the chips of denudation.

If water-waves are moved by light acting through air, then there must be waves in the air, and they too must leave their mark, if they move solid particles. Moving currents of air do in fact produce well-marked forms directly in solid materials, and these may be compared with fluid wave-forms and their work ; with ripple-marks and sea-beaches, new and old.

Ripple-marks and wave-marks upon a beach only show the last direction in which some force acted ; and marks of the very same pattern are formed upon snow, dust, dry sand, clouds, etc., by air. They are also formed by boiling water in hot springs, and in steam-boilers. Old ripple-marks and wave-marks need not be the work of a sea like the sea of our times. They only prove that the marks were made upon beds of solid particles by some liquid or gas ; and that the temperature then was somewhere between two extremes—the melting point of the packed solid, and the freezing point of the fluid which packed it. These marks do not record that they were made upon *sea-margins,* for they are made by currents of air moving at the *bottom* of the air-ocean, and they are made at the sea-bottom as far down as we can see, or feel with a plummet. On the very top of Eyriks Jökull in Iceland (see vol. i. p. 429), where the temperature can rarely exceed the freezing point of water, the snow was found to be beautifully ripple-marked by the wind at a height of 6000 feet or more. The marks proved that the temperature had not exceeded the melting point of snow since the particles of snow were arranged, so water was not the fluid which made the mark; but the temperature may have fallen to any point between 32° and the freezing point of air (if it has one), and if air made the mark; or it might have been made by any other fluid or gas, if there were a doubt about the composition of the atmosphere at the top of the hill.

On a lower hill-top in the Farö Islands, in July 1862, at places where snow had lately melted, bare gravel was arranged in regular ridges and furrows; sometimes running up and down hills, but always running nearly north and south, and always at places fully exposed to the west wind.

The largest stones were in the hollows, the finest upon the top of the ridges, which is also the case on sea-beaches. The stones were about the size of apples, walnuts, hazel-nuts, peas, and small shot. The ridges were about a foot apart, and at one place the hill-side looked like a ploughed field some forty yards square. The apparent cause was the flowing of small streams from melting snowdrifts. But the same form recurred where that explanation would not suffice—for example, on level places; and it never occurred at places sheltered from the west wind, even where melting snow-drifts were on slopes above beds of gravel.

These were tracks of the invisible wind, large ripple-marks made by air-waves in deep air, on beds of gravel loosened by frosts, and driven by currents moving eastwards at the bottom of the atmosphere.

Similar forms occur in similar materials, in many parts of Iceland at lower levels, at Helgafell and elsewhere. So the air has waves for a depth equal to the height of the tallest hill in Iceland, and the sea may have them at the greatest depth in the ocean. Such marks are common on Scotch hills, and further south ; and any one who has walked over a bare hill-top or on the sea-shore in a heavy gale, may have seen and felt gravel rolling and flying before the wind.

This is a mark which a geologist would be apt to attribute to water, if he found it in an old rock ; yet water has nothing to do with it. It simply means that some force moved gravel from west to east, and that the temperature has not been hot enough to melt gravel since it was so packed. The form is but a copy of a wave, and in this case it is a copy of an air-wave at the bottom of the air.

At the Geyser, where water flows from the spring at a heat of 212° or thereabouts, the stone which it deposits as it cools is beautifully ripple-marked in tiny waves, which cross the direction of the moving stream.

In steam-boilers the earthy material which is deposited from boiling water has a ripple-marked surface, which shows the direction of the prevailing movement within the boiler.

A ripple-mark upon a bed of silt, old or new, only proves that some force caused motion in some fluid, and in a particular direction, and that the material moved has not been greatly altered since that time.

The engine set to do the work may have been made of any gas or fluid, at any temperature above its freezing point ; it may have been air far below zero, or high-pressure steam ; but the maximum temperature, within certain limits of time, at any spot is fixed by a ripple-mark on any material, at some point below fusion in the substance marked.

The lowest ripple-mark in the geological series proves that the rock upon which it is found is a rubbish-heap, and that the fusing point of that rock has never been passed at that place since the rubbish was chipped off and packed. It does not prove that climate was the same as now at the surface, or under the sea, which rippled over Laurentian sand.

Air, the last link in the shorter chain, makes other marks in packing solids. In England, where snow is the exception, great snow-waves, solid white rollers, and stationary breakers, may often be seen after a strong gale. Entangled half-melted snow-crystals driven by the wind may be likened to silt moved by water-streams, and the surface of the snow-bed to a sandbank below the sea. But snow-crystals stick together more than sand; and drifted snow-heaps resemble water waves more closely than sea-beaches. Snowdrifts are air-marks on solid water, dust-copies of air-waves. When a strong gale blows, drifting snow takes the shape of the currents which move it. Drifts gather to windward and to leeward of anything which rises above the surface, and so drifts change the direction of the wind. The wind splits upon a post; so a point of snow of a particular shape forms to windward of the post, and another heap of a different form gathers to leeward in the shelter.

A heap of snow changes the direction of the wind and affords shelter; so waves and ridges of snow cross the direction of the gale, and these roll slowly on piecemeal, taking the form of rolling waves of air. When a wall or a hedge stops a drift, the wind whirls the snow over the top, and into the shelter, and makes a snow model of the curved path.

It is a copy of a breaker, a snow-beach arranged by a sea of air.

In high mountains, these snow-waves are often of gigantic size. They are snow-beaches, the drifts of many winters, and the work of prevailing winds, which have blown for ages at odd times, so they are not regular in form; but in the High Alps, and in Iceland, snow-beds may be seen curling over high cliffs, like the crest of a vast roller in act to fall upon a beach. When snow is drifting, the whirling movement of the air which models the curved form of the drift is apparent in the movements of snow-flakes driven over the hills. Of such

FIG. 97. SECTION OF A SNOW-BEACH. Copied from a drift in the south of England.

drifts excellent copies are commonly made by the help of light. But an English snow-drift is as good an illustration of the principle as the largest snow-heap in the world.

What is true of snow is true of dry sand. The material will not retain form so well as snow, but the movements are the same, and dry sand records them imperfectly.

Sand in water retains form worse than it does in air, for it is easier moved in the fluid which partially floats it; but the arrangement of sand by wind upon dry ground explains the packing of silt in water where it cannot be reached. It is ocean-work, but work done by waves in the deep air.

On the sandy plains of Iceland these sand-drifts are

well seen. Long points and ridges form to windward and to leeward of every stone post and plant. Large ripple-marked sand-waves roll over the plain, and stop in every shelter. The air is filled with clouds of moving sand, which fly from drift to drift, and from hill to hill, like spin-drift from the waves of the sea. Clouds of fine ashes rise up, and float along hill-sides like mist, and dust gets everywhere. In the shelter, drifts assume the angle at which dry sand can rest in still air. To windward is a sloping hill, to leeward a sand-talus, whose angle is about 32°. But when sand is wetted, and acquires more cohesion, it copies the form of the breaking sea-wave more nearly.

Near a pool of water, damp sand forms a perpendicular or overhanging wall on the sheltered side, and a slope where the bank is exposed.

FIG. 98. DAMP SAND BEACHES PACKED BY AIR-WAVES NEAR A RIVULET IN ICELAND.

All these sand-forms are but modifications of wave-forms, and copies of air-waves ; and they may be seen wherever there is drifting sand.

Near the Findhorn in Moray is a curious tract covered with moving sand-hills.* The sea throws up wet sand, which dries, and the prevailing south-west wind drives it eastwards along the coast.

Great yellow hills, 100 feet high, are the sand-waves of

* This district is well described in *Wild Sports of the Highlands* (chap. xx.), Journals of Charles St. John ; Murray, 1846. See also *Natural History and Sport in Moray*, by the same author ; Edin. 1863.

this sandy sea, and though they move with extreme slowness, they have covered up whole farms within historic times. In the trough of these waves, old wheel-tracks and ploughed land, the stone implements of a forgotten race of savage Scotchmen, even golden ornaments, are occasionally laid bare by the wind; and the old surface of the "land under the waves" reappears for a time. It is like the rest of that part of Moray—a mass of boulders.

When the wind blows, the movement may be watched. Close to the ground yellow streams of fine sand may be seen waving from side to side, and bounding from point to point, in curved paths, like the wind which moves them. Wherever there is a hollow, sand rests in the shelter. The trough of every ripple-mark fills gradually, but the back of the miniature wave is constantly wearing away. A grain of sand does not fly or roll straight on and continuously; it moves in curves, and travels by fits and starts. It is turn about—the lowest grain beneath the crest of a ripple, then the highest in the trough, and exposed to the wind. It rolls up the back of the wave, shoots over, and falls like the crest of a breaker; and then more grains fall on it, and shelter it for a time. But while the upper surface is thus moving to a certain depth, a lower stratum of damp sand takes time to dry and move. Sand in motion is rolling over sand at rest, as sea-waves roll over still water.

The larger hills advance on the same principle. The slope to windward turns the wind upwards, and loose sand rolls and flies up-hill before it, rippling like waves upon an ocean-roller, till it takes a final leap over the hill-top, and falls into the shelter. There it may be watched falling and sliding down, and forming a perfectly regular slope of sand—a talus in the calm. The base is continually advancing in the same direction as the wind, and a succession of strata are being deposited there at an angle of 32°. Amongst these hills, chiefly in the hollows, bent, whin, and other plants occasionally, take root and flourish. They stop the movement where they grow, but only for a time. The sand-waves march steadily on. The crest follows the trough ; the whin-bush is buried in the middle of a hill 50 or 100 feet high ; and by the time the buried plant comes up behind the wave, it has long ceased to live. When the wind blows from the east, or from any other point, the movement changes. The shape of the sand-hills is irregular, but the prevailing wind is from the west, and form shows it.

This sand-flood, in its eastward course, meets the Findhorn river flowing north. The water is too wide to be crossed at a bound, except in very high gales ; so the sand falls into the water. The river washes it out to sea, and the sea washes it up the firth ; treats it according to the fashion of sea-waves, and throws it up again for the wind to deal with.

When the tide ebbs, the sea-bottom is exposed, and there is no single form upon dry sand that is not to be found upon a wet sandbank, when the tide ebbs far enough for the banks to be seen. Stream-marks on shore explain old sea-marks.

There is, however, this notable difference between land-drifts and sea-drifts :—the sea-forms are all flatter and lower, and the reason is plain. If a conical pile of dry sand is made in air by pouring sand upon a flat base through a funnel, the sides will make a certain angle with the horizon, about 32°.

But when dry sand is poured through the funnel into water till the cone reaches the same height, the sides make a very different angle :—the slope is far greater, the base broader, the sides of the hill less steep. It is still a conical mound, but it is a flatter cone. So sand-drifts and sand-waves, made by currents of water in water, are generally less steep than the same form, made by currents of air in air. But both result from the force which moves air.

The bottom of the sea cannot be reached directly, but by feeling with the lead its shape is pretty well known in many places. It is nowhere flat, but is ripple-marked everywhere—varied by hill and dale, by sandbank, shoal, and hollow channel. Where currents move, sand-forms which result are alike on shore, in air, on beaches, and in soundings. Snowdrifts and sandhills show what is taking place at the bottom of the ocean, and why there are drifting hills and dales even there.

Sedimentary rocks are chiefly old rubbish-heaps packed in the sea. In the coal-formation beds are worked out, so as to leave casts of their surface. Beds of ironstone, for example, are worked in Lanarkshire, and the roof of the mine gives a sandstone cast of the bed below it, after the bed has been worked out. In some of these mines the form of the roof is that of mud-banks now visible at low water in the Firth of Clyde. There are domes which covered mounds ; and wedges which filled hollows like watercourses. The roof and floor approach each other where the mud was washed away, where the trough of the mud-wave was.

Similar forms recur in every sedimentary bed. These are old sea-marks ; they may also be old photographs. According to the evidence of sand-drifts, snow-drifts, and old rocks, that which is now going on above water goes on under it, and has been going on since sand and dust, water and air, were moved by sunlight, heat, and gravitation. The surface-forms of old silt-beds do but record that forces which now work, have worked air and water engines, and that sunlight, which is a force, may have worked the tools. The guide to the force is still form. The tool-mark points out the tool, and that leads to the engine, and to the power which works it, and to Him who set the task, and created a power when He made light.

CHAPTER L.

DEPOSITION 6—BEDDING—RAIN-MARKS.

According to an old saw, "Because the mountain would not go to Mahomet, Mahomet went to the mountain." He did the best he could under the circumstances, and men who study nature can do no more. The frog who tried to grow too fast, burst ignominiously ; if he had been content to look at his world with tadpole's eyes at first, he might have lived to grow and learn modestly from little things around him. If both ends of a chain of cause and effect are out of reach, it is best to study the links which surround us, and "creep before we gang."

It is impossible to watch the packing of silt in the deep sea, it is possible to watch a similar process in shallow water and on shore ; on the sea-beach ; amongst the sand-hills of Moray ; amongst snow-showers and snow-drifts. It is impossible to watch the progress of a tidal wave from Cape Horn to England ; smaller Atlantic waves are apt to sicken those who swing over them ; but waves in a puddle may be watched at ease throughout their course, and from these small things a large lesson may be learned. Because moving water-mountains go their own way, and will not be controlled by little men, little waves have been summoned from little ponds to act the part of their giant kin, and work denudation and deposition on a small scale. This much may suffice to explain what was meant by learning to translate old geological records, by watching geological engines now at work, and by making miniature engines in imitation of them.

Air and water are engines which work deposition, and the chief mechanical power employed about the work is the gravitation which sinks the silt in water, or makes the sand or snow fall in air. Therefore experiments made with water, silt, and weight, are but natural operations on a scale suited to small observers. It is easy to make ripple-marks, and beaches, and all surface-marks of their class, by stirring a muddy puddle : it is equally easy to make small geological for mations grow rapidly, and watch the whole process at home. One heavy clog on geological study is the impossibility of watching the progress of work ; but if this difficulty cannot be overcome it may be circumvented. Gravitation may be set to work in a glass tank. As an illustration the following arrangement was made :—

February 12*th*, 1863.—A glass tank with flat sides was half filled with Thames water as supplied in London. A glass funnel was placed in a retort-stand, so that the end of the funnel touched the water near one end of the tank. Through this channel finely divided materials of various colours and specific gravities were poured in the following order :—1. "Silver sand ;" 2. Coarse granite sand from the Scilly Isles ; 3. Fine pipeclay mud, squeezed in with a sponge ; 4. Coarse yellow sand ; 5. Silver sand ; 6. Yellow sand ; 7. Very fine dark river mud, part of a ball in which a mud-fish was brought home from the river Zambesi in Africa ; 8. Silver sand ; 9. Zambesi mud ; 10. Silver sand ; 11. Zambesi mud ; 12. Silver sand ; 13. Pipeclay to make a white surface. In spreading from the channel through which they fell, these materials formed themselves into a conical mound (vol.

i. pp. 378, 380); but the base of the heap could not spread beyond the glass walls, and the edges of the forming layers were seen through them. Four vertical sections of a stratified mound were seen forming at different distances from the channel by which the materials entered, and they varied in shape, colour, and material. No one of them presented thirteen flat layers arranged in the order in which the materials were poured; instead of thirteen beds there were nearly thirty. A large river brings down mud, sand, gravel, and larger stones of varying size and weight at various seasons. An ocean-current may carry various substances at different geological periods; it may carry the shells of tropical infusoria, or floating moraines; but whatever the materials may be, the same gravitation which packs it in the sea worked in the glass tank, and there the operation could be watched. At first the water was thick with small suspended particles of all the materials poured in. To imitate nature, bits of ice were floated at one end, N., and sun-light was allowed to shine on the other, S. (vol. i. p. 68). This arrangement of temperature moved the miniature engine, and it worked accordingly. The water about the ice cleared, and a thin layer of clean cold water floated, because that water was about 33° (vol. i. p. 75); but columns of cold water (about 37°) sank down from the ice (p. 78), and the falling streams carried suspended mud rapidly downwards. Wherever an iceberg is melting, the same thing must happen on a larger scale. On the outer surface of the glass the downward curve of movement was shown by vapour condensed on the glass. Wherever a cold mass stands in warmer air, like movements and condensation of vapour result. The curves of temperature were shown within by clouds of mud, as curves of temperature are shown by clouds in the air (chap. v.) As these mud-clouds fell, layers began to form on the uneven surface below, and these followed every curve which had resulted from the method of pouring in the heavier and coarser materials. Horizontal layers of falling silt formed in the water, and sank gradually, settling upon each other, but varying in shape as the currents of cold water moved them from N. to S. below while warmer currents moved them from S. to N. above.

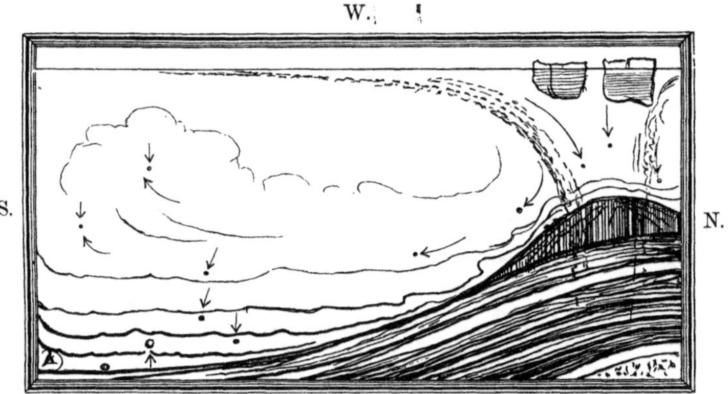

FIG. 99. A WORKING MODEL OF A MARINE FORMATION.

Wave-marks and ripple-marks were formed on the surface of the mud, and fresh layers were seen to form against the glass. The heavier particles forced their way through the falling shower, and these beds, in forming slowly, assumed a very complicated structure.

White clay and brown mud separated and mingled, and took strange branching tree-like shapes, like those which occur in mottled sandstones. These are called "dendritic concretions," and have been attributed to electrical action; in the tank they resulted from mechanical action alone. The bed of silt, in gathering weight, squeezed out the water, and the water in rising displaced and pushed up the lightest particles of mud. Through a lens the operation was seen; some grains were falling slowly, as snow falls in still air, others were rising in jets and fountains of water squeezed out by the growing weight above; others again were drifting before the currents, as snow and clouds drift before the wind. When the water cleared, the surface of the mud was a white surface of deposition with current-marks, the sides of the mound a section of a small geological formation; and the whole operation had been seen from beginning to end. Temperature and gravitation had been set to work a small engine, and it packed silt as the sea does.

By March 24 the surface of the mud was covered with minute water-plants, green and brown, which grew from their invisible seeds and spread from centres. About these plants minute bubbles of gas formed, and more formed beneath the mud, amongst the sand, and under the plants. In expanding, these gas-baloons lifted plants, sand, and mud. When the raising power of the gas had gathered sufficiently, a net of green, studded with shining balls of gas, and with sand and mud entangled in the meshes, rose to the surface, and there hung suspended till the gas escaped. Then the system fell slowly down again at a new place. As there were currents in the tank whenever the sun shone, upward, lateral, and downward movements and transport of inorganic materials resulted from this minute water vegetation, and from the arrangement of temperature which worked the engine.

Similar action must result from the chemistry of vegetation and sun-light wherever water-plants grow upon beds of silt; and old sedimentary rocks must record movements like those which were seen in the glass tank.

The tank was kept as a microscopic vivarium, in the hope of finding some African monster. It was covered with a sheet of glass, but exposed to air and light; and by July the water was peopled with living creatures hatched in the mud. They could be seen with the naked eye, and better still with a lens or microscope. They played and fought and gambolled in their forest of tiny plants; they died and were buried in the stratified beds of their little world. They were chiefly home-bred Thames-water monsters; if any were of African descent, they were eaten up by hungry English crustaceans, or overlooked. While these lived, they too helped to shape the silt-beds above which they swam; they left their tracks on the surface, and their dead bodies fell amongst the withered plants which formed the upper layer in this bedded sedimentary deposit.

By December 22 a layer of water six inches deep had been lifted up and carried away by the sun; evaporation was rapid while the weather was hot, and no condensation—no rain, had made up the waste. Meantime the vegetation had become a thick mat on glass and mud, and the water-fleas were numerous, active, and ravenous. The top of the sand-heap had risen above water, and had become a circular island, similar in shape to islands of boulders in the Baltic, along the Swedish coast. By stirring the puddle, the island was worn by miniature waves; and beaches and terraces were worn and built, "eroded and deposited" near high-water mark. As the water fell lower a repetition of the disturbance made a series like those shown above (vol. i. p. 334).

Lastly, a stream of water poured through the old funnel cut water courses in the island, and built deltas in the water about it (chap. x.)

So within the compass of a glass tank many natural phenomena may be imitated and watched:—denudation by water-streams, the habits of crustaceans, the growth of plants, the formation of surfaces of deposition, and the deposition of beds of silt : geology, natural history, and botany.

It is needless to enlarge upon this toy. It is obvious that a working section of river-mud may be got anywhere by planting a glass under water ; a glass tumbler and a handful of mire will show the process of geological deposition at home, to any one who will condescend to learn from common little dirty things. It is impossible to get at the bottom of the sea ; but if sunken mountains be out of reach, it is very easy to make mole-hills like them in a glass tank, by imitating nature, and by setting natural forces to work natural engines of small size.

Having thus taken one small step under water, the next stride is upwards on land. We cannot get at the bottom of the sea, but we live at the bottom of a sea of air, and deposition of strata goes on about us.

The rocks with which geologists now chiefly deal are stratified sedimentary beds, in which plants and animals were buried ; most of these are made of chips which were ground off solid rocks, and fell through water. The formation of beds by the falling of heavy solid particles of frozen water through air is a similar process, for it is an effect of gravitation, and it can be watched ; snow-drifts are formed by streams as sand-banks are. The snow-formation only endures so long as the temperature is less than 32°, but while it lasts it is a fusible geological formation of sedimentary beds.

Like these, Icelandic strands, deltas, and plains, are made of fragments of fusible frozen lava, which would certainly melt again at some high temperature. While they last these also are parts of a " fusible sedimentary geological formation." The snow-formation is but the last of a series, fusible at a lower temperature than those upon which it falls. Sandstone beds are like the rest; beds of a silicious sediment which is melted in making glass. Lava and silica, like water, become vapour in a sufficient heat, for they colour flame. Geyser water holds silica in solution, silicious shells extract it from sea-water. Snow is but a sediment easier to melt and harder to freeze than the rest: all sedimentary rocks are fusible: all their materials sink when cold, solid, and heavy; flow when fused; rise when hot and light. One sedimentary bed packed by gravitation and a circulating fluid is as good as another for studying the process of mechanical arrangement, and a snow-bed is the easiest to get at in the series.

In lofty mountains these sedimentary water-beds may be seen resting upon sedimentary beds of like form. Avalanches and landslips fall from lofty cliffs, and their fallen debris takes the same talus-slope. The mechanical action is the same, though snow and grit melt and freeze at different temperatures. In Iceland snow-beds occasionally alternate with beds of ashes, which fall during eruptions, and drift at all times; the packing process is the very same, it must still be the same, at the bottom of the sea. It must have been the same ever since gravitation worked deposition there, or anywhere.

In some geological books it seems to be assumed that all strata are deposited flat. It was not so in the model, it is not so on shore, and it cannot be so on the uneven sea-bottom.

When snow falls on rough ground, it is unequally deposited even in a calm, and silt must be unevenly spread for the same reason.

Let the dark line represent the outline of a sea-bottom, or of a hilly country, and it is evident that beds of snow or silt must be deposited irregularly; at various angles, at different levels, and in different quantities at different places.

Every snow-bed undulates with the ground beneath it, and many beds slope because deposited upon a slope. The snow-shower which forms a bed on the top of a cliff, makes another at the foot, and a third in the ditch. For that reason,

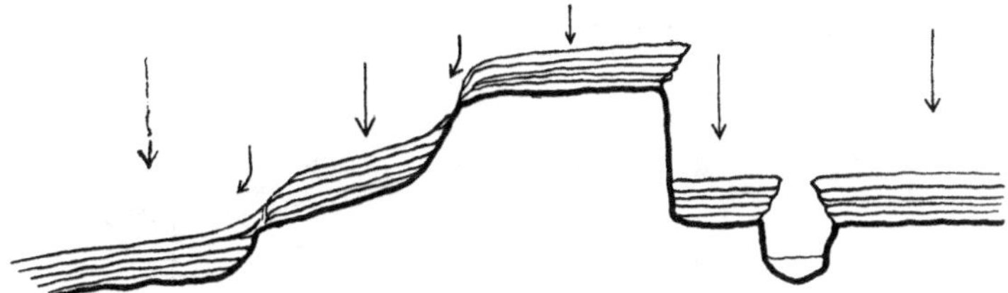

Fig. 100. Stratified Snow-beds forming.

sloping or separated beds of rock do not necessarily imply disturbance, for they too may have been deposited upon a slope, or simultaneously at different elevations. This evident truth is proved by every streamlet, and on every strand —where road-dust has been swept into a gutter and left, where a rivulet flows over sand into a sea, where the tide ebbs and flows now, and in geological sections of old rocks.

When snow drifts, beds dip down-wind as they form; when sand is moved by a river, the beds dip down-stream. In the upper reaches of the Tana, in Norway, the river meanders amongst beds of sand, which it covers in floods, and through which it cuts sections at other times; the beds dip at all manner of angles, but they all dip one way. The same is true of Icelandic river-plains, where travellers may ride for many miles over deltas of ashes and mud, alternately fording rivers and riding over dried sand-heaps packed by the winter floods. On the wide strand about Mont St. Michel, in France, where the tide ebbs and flows over sands for six or eight miles, sections made by streams show that stratified beds are not deposited as flat layers in the sea, but may be deposited in layers sloping opposite ways, where the stream which packs them ebbs and flows.

At Goat Island, in North America, beds of gravel, etc., are packed upon glaciated rock, and the form of the packing shows that water formerly moved towards Buffalo, instead of flowing

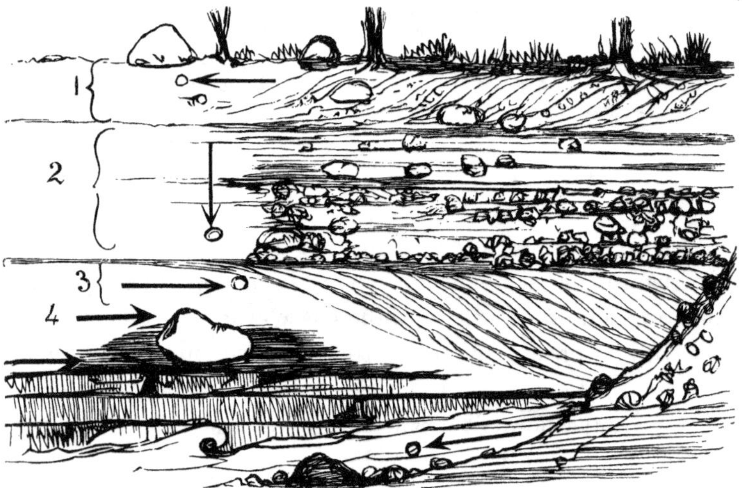

Fig. 101.—Drift-beds on Goat Island, Niagara.

from Buffalo to Niagara, as it now does. No. 1, the highest of the series, dips down-stream, and was probably packed by a river. It contains fresh-water shells, and consists chiefly of gravel and sand.

No. 2 is a bed of stiff clay, containing scratched stones, many of which are foreign to the district. Because this bed is horizontal, it is probable that it was formed in still water, upon a flat base. The lower part of No. 2 is a series of horizontal beds of gravel, coarse sand, clay, and scratched stones, the lowest of which rest upon a flat surface of reddish sand.

No. 3, the sand, contains no stones, but is disposed in thin sweeping beds, which have a general dip of 15° S.E. This bed was packed by water, moving south-eastwards ; but till the packing of silt had been watched in models, in snow-drifts, on strands, or elsewhere, the record could have no meaning. No. 4 is a bed of clay containing large blocks of a kind of rock which is not found to the south-east, but abounds to the north of the spot. The rock below this bed is marked with glacial striæ, which indicate the passage of heavy ice towards the south and west. The river Niagara flows the other way at the foot of the bank, and it has cut a channel through all these beds of drift, and through some of the upper beds of glaciated rock. Reading this old document by the help of snow-drifts, the meaning seems to be, that during the time of 4 and 3, water and ice poured as the arrows point ; that during the period of 2, water was at rest, and things fell through it ; that during the packing of 1, it flowed as it now does, from the watershed towards the sea.

At the watershed, near Fort Wayne, some hundreds of miles away, a similar record confirms the first. A section of a gravel-pit shows—

1. Gravel and rolled stones ; no stratification visible.
2. Numerous beds of fine sand, *horizontal.*
3. Ditto, with occasional small rolled stones, *horizontal.*
4. A series of beds of sand and gravel, all dipping *towards the south-west*, in all twenty-four feet thick. These indicate a stream flowing south-westward over this watershed of North America.
5. A bed of clay, about three feet thick, containing large, polished, and striated boulders of rocks, which are found *in situ* to the north, beyond the great lakes.
6. A bed of fine white sand.

The translation made on the spot is given above (pp. 245, 246); the language was learned on the strand described chap. xxii.

This land in North America seems to be an ancient sea-bottom. Atlantic currents are sorting tropical infusoria and glacial debris off Newfoundland; it is not possible to get at the bottom of the sea there: but the gravel-pit at Fort Wayne may explain what is now going on in the Atlantic, if the strand, the snow-drift, and the glass-tank, have been understood so far.

If sedimentary rocks were formed in old oceans, this lesson applies to them all. At Kreuznach, near the Rhine, is a sandstone quarry, where beds are of different colours, and their arrangement is very well seen. The section is like No. 3 in the woodcut, p. 312. But beds which rest on each other dip opposite ways, and record that water ebbed and flowed, or changed its course, while the stone was silt falling through the sea. This so-called "false bedding" is true deposition, and great currents may have packed large beds on the same plan. These forms abound in old rocks.

The mechanics of deposition may be learned from models. The outward form and internal structure of sedimentary rocks record movements in fluids, and they are registering thermometers within a certain range.

Eyriks Jökull (vol. i. p. 429), and other large mountains of bedded igneous rock in Iceland, appear to rest upon a thin bed of sand and cinders. Because of "false bedding" in this thin layer, it was packed by water which ebbed and flowed; if so, Iceland probably rose from the sea. Four or five thousand feet of igneous rock are spread above the bed of tuff, which is near the level of the lower plain in the woodcut, and the crust has been broken and ground into mountains and deep glens. Lava-floods have poured over the surface out of rifts. But the fusing point of a frozen lava clinker has never been reached at the bed of tuff since the clinkers froze and fell there, because the false bedding is preserved, and because the black glossy cinders retain the shape which they had when the white ashes were packed about them. The form of a sedimentary bed proves that the fusing point of the material has not been reached since the bed was packed; and the rule holds whether the bed is made of mud, snow, gravel, or Laurentian gneiss; whether it was packed in a toy on shore or in the deep sea. A great deal may be learned from little things; much may be fished out of dirty puddles; but every student who will condescend to make scientific dirt-pies on the plan here indicated, must set his wits to work out contrivances to illustrate his own special study. There is room enough and to spare in the field, though many are working at geological deposition and bedded rocks. Let one more familiar example of learning from little things suffice.

Ripple-marks, wave-marks, beaches, and bedding, are marks made when loose materials were under water or awash. Other marks can only be made upon plastic surfaces in air. These, like the rest, record facts, but the language must be learned before a record can be read, and the easiest way of learning a language is to try to speak it or write it.

A rain-mark was made upon a plastic surface in air, because half an inch of water would shelter the surface from the rain. But in order to learn the meaning of ancient rain-marks it was necessary to see marks newly made—Sir C. Lyell saw them in the mud of the Bay of Fundy. It is very easy to imitate nature in this case also.

Every shower of rain makes its mark on still water. Each drop makes a dimple and starts a radiating system of circular waves, which, like other waves, may be refracted, reflected, and focussed, accelerated or retarded. They meet, and cross, and jostle, so that the water-mirror becomes a rippling pool. But when the shower is over the waves cease their gambols, and the lake is a mirror again. A shower may fall on a plastic surface—on mud, clay, dry dust, snow, or any other such material—and there the dimples may retain the shape given by the falling drop. The mark is a tool-mark, the dint is made by a drop lifted, carried, and dropped by the engine which works denudation and deposition; and the tool-mark may be so placed as to record very ancient work done by the same machine. Rain-marks endure when the plastic surface is baked, frozen, or otherwise hardened.

It is not necessary to travel far in order to learn this language. The scrapings of the streets of London are chiefly powdered igneous rocks, ground up to a tough mud by carriage wheels, and scraped into heaps by scavengers. The wet sludge forms a surface almost as smooth as that of a lake, and it sets gradually as the water evaporates. After a summer shower this smooth mud is often dimpled with regular cups, and each of these is a cast of a drop of rain, which fell there. Each is a tool-mark, and a record. Road-scrapings bake in the sun, and freeze hard in winter, and the mud when dry may be further hardened by baking, so as to resemble some of the old rocks upon which fossil rain-marks have been found. It is so far a record. But if the material is fused by greater heat, the record is spoiled and lost. In the summer of 1862 a thunder-plump made a very beautiful set of dimples on smooth mud faces, and filled the pockets of cabmen with the silver of pedestrians, who feared the rain, and thought the mud a nuisance. One, however, who came from a rainy land and wears old clothes, watched the shower and the mud, and went home to try whether the shower could not be set to work for him.

An old cigar-box was filled with wet plaster-of-paris, and when the plaster was beginning to set, the box and its smooth white-faced contents were turned out of doors and watched, to see what the rain would do to the plaster, and what rain-marks really meant. They meant that the surface was smooth, plastic, and above water; the shape, size, depth, and direction of each cup recorded the shape and size of a drop, the force with which it fell, the direction from which it came; the slab recorded the number of drops that fell within the area of a cigar-box during a certain time. When the plaster set it became a register, and it will last till it is destroyed. It is easy to see how the drops fell, to determine the quarter whence the wind blew, and the force of it; and similar marks found upon old rocks of any age record similar facts. But rain-marks do not record climate, as some have argued. Hailstones bury themselves in snow and cold mud, so the climate may have been cold. Drops which fall from clouds of steam escaping from a boiler; scalding drops which fall upon hot sticky mud, beside the boiling springs at Krabla in Iceland; summer rain or winter sleet; all make the same marks. The climate in old geological times may have been very different from existing climates, though rains fell and winds blew slanting showers down upon smooth plastic mud.

Like other marks, this class register temperature. The mud was not frozen, for it was soft when the mark was made: it was warmer than 32°; it was colder than 212°; it did not boil, for the surface was smooth and wet. The fusing point of the material which retains the mark has not been reached since the mark was made. Within these limits a rain-mark registers temperature, and it proves that the whole earth was not covered with water at some unknown date. Like the island in the glass tank, some part of a bed of silt was above water when the rain fell.

To get at past climates other marks are used, and they form a separate branch of study.

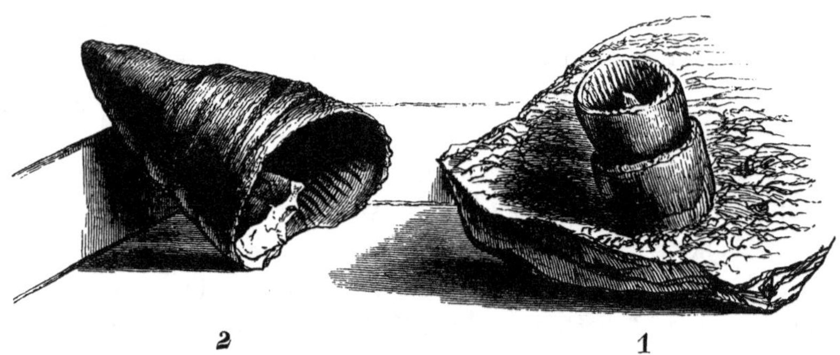

FIG. 102. FOSSILS.

2. Broken from the limestone wall of the Mammoth Cave, Kentucky, near the "River Styx."

1. From a "weathered" limestone surface preserved under a bed of yellow clay on a hill near St. Louis, on the Mississippi.

These specimens illustrate one denuding action of rain-water. It holds carbonic acid in solution, and it dissolves insoluble carbonate of lime by transforming it into soluble bicarbonate. When a limestone rock-surface has thus been dissolved, and worn, and washed away, insoluble silicious fossils project. These, by their preservation, prove that a rock-form was not sculptured by mechanical force alone. The hills about St. Louis were not sculptured by ice, though limestone-hills near Buffalo were. The Mammoth Cave, and the shapes of hills about it, are chiefly chemical work, because fossils project from the sculptured surface of the stone.

CHAPTER LI.

DEPOSITION 7—FOSSILS—ALTERED ROCKS.

LIKE other shapes, the forms of plants and animals are thermometers.

Because an organism lived, the average temperature where it lived was, during its life, somewhere between 32° and 212°, freezing and boiling; that is, if the extinct thing was made like most of those which exist. Even lichens will not grow in extreme cold, and vegetable cells burst in boiling water; an animal made partly of albumen and water is frozen in ice, and is coagulated and cooked when a submarine volcano makes the sea boil. Living things can resist extreme temperatures for a time; but nothing now living can long survive boiling and freezing. Because a sea-plant grew, and a fish swam, their average climate was probably somewhere between these limits; and their shapes are registering thermometers so far. If species is known, climate may be guessed from the haunts and habits of living things of the same or like form. An arctic shell means cold water, a palm-tree warm air, and things like them similar climates. But organic forms, which are unlike living things, do not so closely record temperatures. Sedimentary beds, with water-marks, rain-marks, and fossils, together record the former existence of land under and above water; with an atmosphere and a climate fit to support life. Because the fossil form has been preserved, a stone, or bed of stone, has not been fused since the materials took their shape.

Fossils are time-keepers also.

The water-formation exists as solid, fluid, and gas ; solid snow and ice, fluid water, gaseous steam and vapour.

When temperature falls to a certain point, a crust of ice forms and floats upon fluid water, while vapour rises, is condensed, crystallizes, and falls as snow. If it falls upon plants and animals it smothers and preserves them, as silt does, and far better. If wetted and frozen again, the snow becomes ice, and the buried plant or animal freezes. Till this formation is melted, it is an altered crystalline sedimentary formation containing fossils. The famous frozen Siberian mammoth was so well preserved in frozen gravel, that dogs fed upon the flesh when the ice which contained it thawed. In any other sedimentary bed the skeleton, or a cast of some part of the creature, might have remained, but the flesh would have yielded to natural chemistry. That fossil proved that temperatures less than 32° had prevailed at the place from the date of the mammoth's burial in ice. It was an old formation, because mammoths have long ceased to live. English ice now melts every summer; Arctic ice does not. A perch preserved in English ice records the date of his death within a few months, because of the known climate, and implies a late formation, because his race exists. We know that the Arctic ice which contains an extinct mammoth, is older ice than English ice which contains a perch. One is less than six months old, the other far older, but how much older is not recorded. We know these facts from observation. If we did not, the fossils alone would lead to the conclusion that the perch ice was the newest water formation, because perch exist and mammoths do not. But if a perch were found in ice under a mammoth, buried in snow, these relative positions would prove that some perch lived before the mammoth died, and that the lowest bed was the oldest in that series, though it contained fossils of existing species. Like slates on a roof, these two portions of past time overlap, and their extent is only known in one direction.

Fossils.

Living Mammoth	———— .	
Living Perch	———— . ————	Perch now alive.

In the first place, relative position proves the relative age of the fossil; and when that has been ascertained, the form of the fossil is like an index-number on a page. The uppermost layer is the newest, unless the series capsized: because snow and sediment both fall. When two human graves were found above each other under the foundation of an old church, history gave a date and position older relative dates.

Christian church wall	——— .	date known—A.D. ?
Human grave	——— .	probably near the time.
Ancient do.	——— .	older, but uncertain.

If in this case the bones of buried men differed, the lowest had the type of the oldest race, and such bones thenceforth mark ancient graves. The buried form became a time-keeper, for such forms lived before the year A.D. ?, when the church was founded.

Thus out of form, species, and superposition, vague geological dates are constructed with fossils, and slowly built up into a skeleton-history of part of the world's crust. The study is like turning over the leaves of an old saga, in which events were recorded year by year. Those which are mentioned in the uppermost page happened after those which were first written down; and when the place of an event has been learned, it marks the place of others which happened before or after it. Fossils in upper beds died after those which are buried under them, and the lowest human grave was first filled.

Position gives the age of a fossil, and then the fossil alone gives position. A stone is like a torn page which records a known event. If written by a man who was at the battle of Blenheim, the page must be placed below the Waterloo page —for Blenheim soldiers had become extinct before Waterloo; and above the page written by the Icelander who described the battle of Clontarf as a recent event. But the fossil record is not a history, it is but an index, and by no means complete.

Position even without fossils gives a relative date for beds of rock.

A bed of snow resting on ice on a pond gives three dates. The water was there before the ice formed, and the snow fell upon the ice—snow is the latest formation, water the oldest of these three. In Iceland, beds of silt are on lava in lakes, ice grows on the lakes, and snow falls on the ice. Of these five the lava-crust is the oldest, and still older fluid lava once flowed under the frozen lava-crust.

There is a regular series whose position depends on temperature and specific gravity; a series liable to disturbance, and frequently disturbed.

COLD AND WEIGHT.

1. Water as vapour in the air—condensing and falling.
2. Lava and ashes in the air—falling.

3. Water as snow, a bed of sediment—at rest.
4. Water as ice, a frozen solid crust—at rest.
5. Water as cold fluid in the lake—at rest.
6. Lava as silt, a bed of sediment—at rest.
7. Lava as a solid frozen crust—at rest.
8. Lava as a hot fluid, which escapes at times.
9. Water as steam, which is always escaping, and struggling to escape, and has blown up the lava-crust in many places.

HEAT.

The stone book of sedimentary rocks, with fossil pictures engraved amongst the leaves, has been rumpled and torn, pages are missing, leaves were of different sizes at first, whole volumes are yet unread. It is hard to read the record, and harder still to understand it. But wherever an organic form can be traced, it records a climate fit to support organic life, and proves that the page, though it may be torn and charred, has not been destroyed by fire. The fossil form is like a footprint in snow, which disappears when the snow melts, though the melted snow may freeze again. It is like a wrinkle upon the lava-crust, which ceases to exist when the lava is fused. But these organic shapes tell more than tool-marks of engines, however great. They tell of air and water, and their movements; of heat which kept them from freezing; of cold which kept them from boiling; of gravitation which bound them to earth. But they also tell of life, which made each shape a separate reproductive system, "whose seed is in itself"—a system wherein heat and weight play their parts, but are guided and governed by subtle powers, of which those who live by them here on earth know absolutely nothing at all.

As a bed of snow is altered by a sufficient heat; as loose

grains of lava-dust may be consolidated by fusing and freezing —so all sediments may change into solids.

At the Sèvres china factory, and at Minton's works in Staffordshire, and elsewhere, certain clays are mixed with water till the mixture is like a glacier-river; the sediment is washed, allowed to settle, and after a time sludge becomes mud, and a tough paste. It is then moulded and patted, twisted and worked into all manner of forms, dried, baked, and finally burned. When all is done, the sludge has become a hard flinty brittle substance, with a form which tells part of its history. One bit was made on a wheel, another pressed in a mould; one was baked hard, another burned; a third too much fired, half-melted, and so distorted by its own weight. If the miniature geological formation above described were made with coloured clays and sands, dried, baked, and burned, the sludge would become stone, and any forms impressed upon the surface, casts of small plants, or creatures that lived in the tank, or their tracks, or stream-marks made by currents, would be preserved (chap. l.)

The forms of sedimentary rocks indicate certain temperatures, an order of succession, and vague dates, for they were deposited one upon the other long ago, at times when plants and animals could live, and they have not been fused since. But there is a wide range of temperature between 212° and the fusing points of various stones, and many rocks have been baked and burned, and partially fused, as china and bricks and glass are. The lower the rocks are in the geological series, the more they bear marks of heat. Therefore, according to position, a brick-kiln or furnace heat is below, or was an ancient condition of the upper world.

Beds of slush do not turn to stone without some active cause; and the deeper men go in mines, the greater is the heat of the earth. When a volcano bursts the crust, earth-light shines out, and rocks melt like wax in the fire. It is only by watching human works that we can hope to estimate the effects of heat upon sedimentary rocks; but these effects may be watched at furnaces.

Snow becomes glacier-ice by a combination of heat and pressure; by softening, kneading, and hardening; by fusing and freezing again. ———————— Below 32°.

Clay becomes brick by kneading and baking. Finer clay becomes china. ———————— About 1100°.

Sand becomes glass by fusion with various other substances. ———————— About 1000°.

Whinstone was made into a black glassy mineral at Birmingham by fusing it. The difference in the structure of the mineral was attributed to the rate of cooling, which was too rapid for crystallization. ———————— About 1000°.

Obsidian is a natural black glass, formed in volcanic mountains. It seems to line passages in lava through which hot gases have escaped. The stone is something like a lump of sugar which has been partially fused in a candle; and cavities in lava are commonly varnished with a coat of some glassy substance of like kind. ———————— ? 1000°.

Jasper, bloodstone, and similar glassy minerals, abound in volcanic countries, and in old igneous rocks. ? ——

All these are effects of heat.

Limestone of the coal-formation, containing fossils, and other limestones, are used as fluxes in smelting iron. The stone melts and runs as lava does. It is often run into moulds, and when it has time to cool it freezes into an earthy mineral, with a glassy wrinkled surface, and a crystalline structure. No trace of a fossil remains after the fusion, and there would be little sign of fusion if the surface were gone, and the slag a large bed of stone in a geological series. 3300°.

Lavas are like slags ; *whinstones* are like lavas.

All these are products of heat, of fusion and freezing.*

The whinstone may have been sedimentary rock because it is like lava, and lava like slag, which was limestone, and was perhaps a coral reef, or a bed of shells and silt at the bottom of an ancient sea.

Fire-clay will not readily yield to heat ; it is easily baked, but very hard to fuse. One of the Lanarkshire iron-furnaces was lined with fire-clay as usual, and the first fire was lit with faggots, amongst which were branches of hazel, and furze. The furnace worked for many years with the hot-blast ; thousands of tons of iron and slag were melted in it ; but at last the walls grew shaky, and it was "blown out" to be mended and re-made. In breaking out the hearth the workmen found the shape of a forked branch, and the overseer sent the curiosity to be examined in Glasgow. The learned could make nothing of it. It looked like a bit of forked stick, but it was heavy ; it was not wood, but some mineral, so the chemist wrote back to say that he could make nothing of the specimen sent. If geologists would take a hint from this story, and repeat such experiments, they might explain the mystery of fossils altered in old sedimentary rocks. It would cost nothing to line a furnace with bricks, in which plants and shells, fish and leaves, had been packed ; the heat of the furnace is 3300°, or more, and the stones would be touchstones for temperature recorded in altered rocks.

If there has been a constant succession of life, from the earliest known fossil species down to the present day, the heat which baked rocks has never been the general climate of the upper world, since Laurentian times at least. But many

* These and many other temperatures are quoted from a Thermometrical Table compiled by Dr. A. S. Taylor. London, 1845 : T. and R. Willats, 98 Cheapside.

sedimentary beds have been baked since then, and the lowest are most altered. The heat certainly was internal heat, and the condition of beds which were buried and have been raised to the surface again would give the temperature of the lower regions, if a pyrometer scale were made with which to sound the earth's sedimentary crust.

Beds low in the series indicate internal heat, wherever these beds are found. Snow indicates external cold at all latitudes and longtitudes. Temperature, as recorded by sedimentary rocks, appears to be arranged in shells about a centre —heat within and cold without.

It has been argued that "metamorphism" is not necessarily a result of heat, because in some cases the central bed of a series of three has been altered, while the other two retain their characteristics.

To use a homely illustration, the same amount of heat would toast bread and metamorphose the ham of a sandwich into lard. When a hot sun shines on the delta of an Icelandic river in spring it warms a series of beds, which alternate, and are variously altered by the same temperature. The foundation is some igneous rock, which was fused at some time ; on that solid is a pile of loose ashes and dust, and lava-mud, sorted by the river. In winter this series is covered with ice, on which rests a layer of ripple-marked stratified mud. Over this, water has flowed, and frozen, and packed more silt ; and so the upper beds alternate.

☼ Solar heat.	1. Snow.
50°. July Isotherm.	2. Ice.
36°. Annual ditto.	3. Water.
32°. January ditto.	4. Mud.
32°. Lava under ice.	5. Ice.
3300°. Lava melts.	6. Mud.
☼ Volcanic heat.	7. Ice.
	8. Mud.

When this series is melted by the sun in spring, the ice fuses and the mud remains. It is abominable ground to ride over, for hollows cave in where the fused ice has left a roof of sand. In the mountains it is common to find the series—

Snow. Ashes. Snow. Ashes. Snow.	} Below freezing, 32°.

When the heat is sufficient, the snow is altered and "metamorphosed" into glacier-ice, but the ashes remain unaltered. If a series be made of

Fireclay ; Limestone, Ironstone, and Fluxes ; Fireclay,	} 3300°, white heat.

and heated till the slag runs, all traces of life will be obliterated in the central bed, while the other two may contain altered fossils, like the mysterious forked curiosity found in the furnace in Lanarkshire. So a bed of impure limestone between two beds of slate may be metamorphosed into crystalline marble, by a heat sufficient to fuse limestone and slag, but only sufficient to bake ripple-marked clay into hard slate.

So also a bed of sandstone, with alkaline plants, rust, and lime imbedded in it, might be partially fused into coloured quartz ; while neighbouring sandbeds, without the alkali, resisted the heat and hardened without fusing. At 1000° flint-glass melts.

In running iron and slag from furnaces, bits of wood, fireclay, brick, sandstone, and other such materials, often get entangled in the burning stream. They are variously altered by a heat of about 3300° F., but all of them can be identified, though enclosed in iron, which flowed over and round about them. In all these cases the structure depends upon temperature; and it seems to follow that a bed of silt may dry up, and so remain; or it may be sun-dried; or baked, or burned, or fused, by the heat of the earth.

The way to do a thing may be learned by seeing it done; the way in which a thing was made may be surmised by comparing finished works. An altered rock may be compared with a brick, or slag, and if they agree in form and composition, it is evidence that the rock, like the other substance, was altered by heat.

If sedimentary rocks have sunk past the brick-kiln to the smelting-house region below, crusts of lava which welled up and froze in Iceland, and which now furnish materials for silt-beds in deltas and in the sea, may once have been sedimentary fossiliferous beds, which, like some ironstones and fluxes, were silt, and now are metal and slag, because of heat.

So far, theory and models, and the effects of heat in manufacturing processes agree. Geological facts confirm their evidence.

Scandinavia, Iceland, Greenland, Labrador, and Newfoundland, are slowly rising or falling—that is to say, in these regions the solid crust of the earth is swelling or sinking; receding from the centre, or approaching it. But beds of snow and fields of ice, which form the upper layers of the solid crust, are not split, torn, dislocated, or smelted, by this movement. Parts of Europe and North America have risen from the sea, and yet the layers of soil and sediment next below the winter snow continue to be soil, sand, clay, gravel, boulder-clays, and loose materials, packed as they were at first. These beds have not been much disturbed, or altered from below. The work of geologists who have learned the alphabet of fossils has been mapped; and a traveller can now identify the uppermost layer of the country on which he stands by turning to a book. He may find out new facts for himself, but the document has been made out so far that the outline of the story told by sedimentary fossiliferous beds can be learned from a translation. There has been a succession of formations which rest upon each other, each a ruin of older rocks; and during that period the outer world was inhabited. There has been a succession of life; but when it began, and whether it was continuous or interrupted, remains to be proved.

In passing from formation to formation, the most superficial observer must remark a striking difference in the shape and structure of the rocks themselves. In North America newer rocks are to the south, the older to the north; and the contrast is very striking. In regions where the uppermost beds are of late age, the country is flat, and beds are laid horizontally, or dip very little. They are like beds of snow and drift which cover them, little disturbed. The same thing is true of beds of like age elsewhere. There are many cases of disturbance recorded in such rocks; the soil itself has been disturbed by earthquakes in Italy and in Iceland, and the ground is there riven and disturbed. Even snow-beds and ice have been shaken and melted from below in Iceland and Sicily; but, generally speaking, beds lately deposited have been little disturbed and altered. But as the American traveller works northwards, or the English geologist works westward, the case alters. In old strata every form tells of violent disturbance, every stone of great heat. There are many sedimentary rocks in which no fossils have yet been found, many beds in old fossil-bearing strata which contain no trace of life. One question left for argument is, whether these were deposited in cold water or in water too hot to support life?

It is plain, that generally the oldest known fossiliferous rocks have been much shattered and altered, and that no convulsion within human experience has equalled the amount of force to which these altered beds have yielded.

The geological sections of Wales are masterpieces of art; they show a series of folds and curves upon a vast scale. The rocks themselves record this part of their history in characters which a child can read, now that this alphabet is taught in schools. They retain their sedimentary structure, but many of them are crumpled, as snow-beds are when they slide from a house.

In the Isle of Man, at Brada Head, a cliff 300 feet high is marked by coloured bands, which sweep and bend, curve and wave, like round text with the flourishing of a writing-master's pen. The shapes of the hills have nothing to do with this internal structure; their forms are tool-marks of denudation. No possible combination of cold streams ever packed silt into such a form; no loose silt or hard rock could possibly bend into these curves without scattering or breaking at the bends. The rock must have been packed in flat or sloping layers at first; it is now hard and brittle; but between whiles it has been plastic, and then it was kneaded and welded like scrap-iron in a press. No twisted gun-barrel could record the fact with more clearness. Were these plates so welded when they were wet or when they were hot? The structure answers the question. In this cliff are dykes of igneous rock, which fill rifts, and the pattern on opposite sides does not fit. Even in beach stones and pebbles this structure is seen, and the rock looks like stone which has been burned at a furnace.

Waving white lines of quartz meander about in many a

tall cliff on the west of Scotland ; they are followed in all their windings by lines and bands of other colours, and these are now edges of crumpled sheets of hard brittle stone. They, too, must have been soft when they were folded like coloured glass in the workshop. In Ross-shire, in the forest of Gairloch, some beds of quartz rock of similar structure contain fossils, which only appear when the rock is weathered. So quartz rock in all probability was a sandbank, though it is now like half-fused impure distorted flint-glass, which melts at 1000°.

Districts where these old crystalline beds occur show other signs of great disturbance and great heat. Large dykes and upthrows of granite, trap, basalt, and other igneous rocks; veins, faults, and fissures; traverse whole districts. Measured along their edges, beds which were deposited upon each other "conformably and unconformably" are of great thickness; and yet, from "Fundamental gneiss to oolite," from "the Minch to Brora," from "Skye to the Cheviot Hills," the whole patch of the earth's crust which denuding engines hewed into the shape of Scotland, was long ago moulded and kneaded like plates of clay in the potter's hand.* No recurrence of earthquakes like those which have been observed by men, could so crush and alter such thick beds of sediment over such areas.

In Dana's *Geology* the Appalachian chain is well and clearly described. The range includes a series of long wrinkles and folds, which include rocks of the coal-formation. In travelling from Pittsburg to Harrisburg, these folds are seen in cuttings by the wayside. Beds dipping in one direction are passed by the train ; sandstones, grits, and coloured beds

* Geological Map of Scotland. By Sir R. I. Murchison and Archibald Geikie. 1861.

succeed each other in rapid succession, till the anteclinal or synclinal axis is passed.

The train runs through one side of the bend, ∩ or ∪, and thence the beds dip the other way. Coloured bands, grits, sandstones, succeed each other in the reverse order, till the next fold in the old earth's wrinkled face brings back the old series of sandstones, grits, coloured bands. The roadside is like a picture made by the Geological Survey ; the journey is a day's lesson in contorted rocks. Yet the shape even of this great mountain-chain is not wholly due to this wrinkling process. Valleys are not in the hollow curves of the strata ∪ ; neither are the hills on the top of the folds ∩. One great fault, according to American geologists, left a wall as high as the Hindoo Koosh, 20,000 feet at least; for on one side of a crack, over which a man can stride, the highest of upper Silurian beds faces the lowest of lower Silurian. But the upper Silurian wall of the raised side of this vast crack was "denuded," hewn away, and the place where it rose has been planed smooth, so that masses of grit, caught in the chink while it was open, are cut through by the surface.

Such changes mean some great force, and the lowest rocks mean great heat, according to the evidence of burnt stones.

The rocks of Newfoundland are greatly folded and fractured. An able geologist is now engaged upon a survey there. When his labours are published, we shall know something of their relative age. They include sandstones, grits, slates, and numerous beds of granite, but all these are not metamorphosed.

The Laurentian rocks of Labrador were supposed to be "azoic;" they are low in the series, if not the lowest beds known, and they resemble the old rocks on the Scotch coast. From Belleisle to Cape Harrison, the land appears to be a maze of granite dykes and altered rocks. The country looks as if a sedimentary crust had been smashed up, half-fused in hot stone, and frozen again.

The only modern natural formation which bears any resemblance to this old Laurentian gneiss, is the water-crust on the sea. Part of it is snow, part flat ice; but where a pressure sufficient to smash the crust has been exerted, the fluid water has risen through the faults, and the whole is cemented together by frozen water. It is a crust of sedimentary snow and altered snow, now forming; it is broken up and disturbed; it has faults, upthrows and downthrows, ground edges and slickensides, angular conglomerates of cemented chips, veins and dykes of ice. But underneath this old ice-crust is a fluid sea, and above it are new-fallen beds of snow, which rise and fall with the bending crust, when the tide flows and ebbs. The problem is—Did the shell of temperature which makes water boil coincide with the formation of any layer of sediment at the bottom of the sea? and if so, at what temperature did life begin in Laurentian or lower beds? Since they were first made, these old rocks have been altered by a heat incompatible with the life of anything which now lives on this world.

We have now reached the period of a water-formation. A solid crust is formed about the poles, and is forming everywhere; and if the earth is cooling, the ice-crust will reach the equator, and descend from the air to the bottom of the sea. The solid is forming upon a fluid base, and now is the period of rapid action and violent disturbance in the water-formation, which hardens at 32° or some degrees lower, at a certain distance from the earth's centre. Under the ice-formation water still boils in Iceland at some point nearer to the centre. If the whole earth is cooling, the point of ebullition may have been further from the centre and nearer to the surface in Laurentian times.

When temperature falls, movements in the water-crust diminish. There are fewer ice-quakings and sea-eruptions when the arctic winter sets in. When the ice sets the crust rests, and the slow deposition of snow is the only apparent work in progress. But there is fluid beneath, and the crust sways, and cracks, and groans, to prove that water may still break the prison which holds it. The water-formation is like the rock-formation even in this; it has a fauna and flora of its own. Minute vegetation reddens snow, birds and beasts walk on floes, fish and sea-weeds flourish under them. Esquimaux hunt and fish on the crust of the sea, and seldom tread on real earth or stone. If the world is cooling, and cools a little more, the whole sea will be like the arctic regions. If some glacialists are right, the whole earth was in a like condition during a glacial period. Snow and vegetation already begun may spread; animals may change, and adapt themselves to new conditions; Esquimaux geologists, if any survive, may be driven to speculate on the comparative age of snow-drifts and altered glacier-ice. They may recognise certain ancient drifts by works of art contained in them: the new white snow-stone, by frozen seals and extinct brown bears; the old blue snow-stone deposits, by fossil whales, sharks, lobsters, fish, and other strange marine monsters; the lowest altered solid blue ice-beds, by mammoths, seaweeds, shells; the lowest beds of all, by conglomerates of different chemical composition from any water-bed known. Questions may yet arise as to altered sedimentary highly-crystallized snow-beds, passing into compact blue ice near ice-dykes:—whether the beds were altered by pressure, or by a heat almost sufficient to fuse snow-crystals and fossil flesh, or by some other unexplained natural power, like the northern lights? The ignorant may hold, with the Esquimaux highlanders found by Ross, that the whole world is snow and ice, and that it was so created. Keen arguments may arise amongst the better informed as to the origin of upthrows of igneous ice—whether the matter rose plastic or fluid, through a crack, or a hole; and if it rose at all, why, and whence—from large or small lakes of fluid ice in the ice-crust, or from a fluid water-core which reached to the earth's centre? It may be argued that, because the coldest air is also the heaviest, there can be no fluid water under colder ice, for the coldest water would be sure to sink and freeze first at the earth's centre. The argument could be settled by experiment; but there will be a double crust under the disputants—an upper crust frozen at 32°, or below zero, resting upon a rock-crust which froze at 3000°, or some other temperature, when the world was younger and warmer, before old age had cooled its hot blood. And under these two crusts there may still be fluid water, and fluid lava at deeper depths, if there be such a thing as internal central heat diminishing by radiation into space. We, who tread upon the upper sedimentary beds of the rock-crust, wade amongst the snow of the water-formation, and skate on the winter's ice, find more heat when we burrow downwards. We see that melted rocks well up from below in all latitudes and longitudes; and when they cool sufficiently, they too form a surface-crust. Surely it is reasonable to believe that we, and the beds beneath our feet, rest upon a crust which froze upon a fluid, and which grows inwards, as ice does on a pond.

There may be many such crusts, many fluids, and many imprisoned gases underneath; but the greatest heat must be in the centre, and the last fluid drop there, if there be any truth in experiment. In every material which is melted and cooled, fused and frozen, in arts and manufactures, the crust forms outside about the warmer fluid. Water so freezes in a spherical bottle. A drop of tallow sets on the outside, and the fluid interior can be squeezed through the crust when it is formed. Wax so freezes in a mould, the outside crystallizes first, and the inside is often poured out to show the crystals. Slag cools on the same plan. So do metals—solder tin, bismuth, lead, silver, copper, iron, gold, platinum, irridium. So does lava. Because all these, and many more, cool on this same plan, it is probable that the world, whose shattered crust contains materials which are fused and frozen in the arts, cooled outside at first, if it ever was fused, and so prepared a foundation on which denuding engines built up chips and sedimentary rocks, to be the tombs of plants, animals, and men.

At a certain comfortable club, where travellers and their guests dine, a luxurious contrivance is placed on a table at 6 P. M. A large double dish of block tin, filled with hot water, is the base prepared for good things which appear and disappear later in the evening. While quietly reading the bill of fare, this engine is apt to startle strangers, for it stirs the silence of the half-lighted room, like a gong, with a bang. The upper crust of the hollow dish may be seen to undergo sudden convulsions. It jerks up at one spot, and when that jerk is expended, down goes the tin plain for another spring. Loose crumbs jump, and gravy is agitated by earthquakes, while hot springs hiss and sputter through safety valves. A traveller in search of causes finds red-hot iron under the double dish, and if he seeks further, he finds that the store of heat was taken from the kitchen fire. But where did that heat come from? A book in the library tells how an engineer and a philosopher, whirling along a railway, settled the question. They held that the heat of burning coals was solar heat stored up in plants during the coal-formation : mayhap it was taken from another store. As the heater cools the action decreases. There are frequent earthquakes before dinner; only a few bangs after it, to rouse the sleepers. Mayhap the February eruption of Etna, the English earthquakes of 1864, the sea-waves off Newfoundland, and such-like disturbances and upheavals all over the world, are caused by an old store of terrestrial heat and light now hidden beneath all sedimentary rocks in the world.

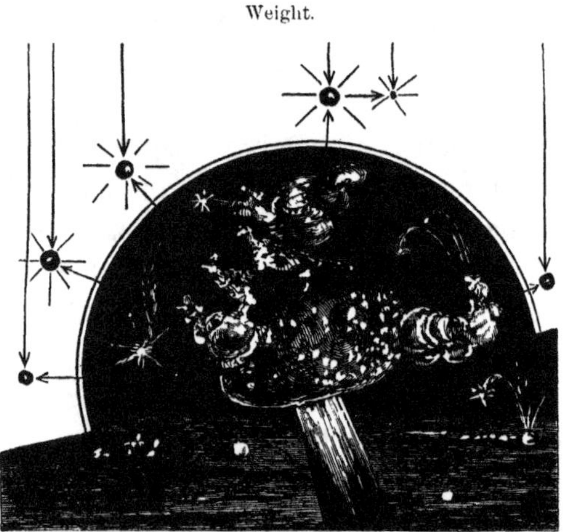

Fig. 103. An ounce of silver, prepared at Newcastle, December 16, 1863. The crust red-hot, and newly frozen; the interior, fluid, partly gaseous, and white-hot; the mass cooling rapidly in cold air. Real size. See pp. 350, 352, 358.

The arrows are intended to show the directions in which two forces acted on shining hot projectiles and luminous sparks thrown off by the metal, while thus cooling by radiation from within outwards.

CHAPTER LII.

UPHEAVAL I.

DYKES—VEINS—SUBLIMATION.

In the last chapters sedimentary geology, palæontology, and a whole series of rocks, were bored through in search of light. It would ill become one who knows so little of these sciences to say more about them. Whether Laurentian gneiss be the lowest in the series or not, it is low enough to prove that great heat has worked with great force beneath sedimentary beds which underlie great tracts of the earth's surface. If it were possible to get lower, nobody could live in the temperatures which fused these rocks. But thoughts may go there safely, if they can find conveyance; and the first step in such a journey is to seek a vehicle for thought.

When snow has fallen on a glass roof it is possible to stand under it and watch the snow melt. Warm breath does not melt glass, but it warms the roof, and the lowest bed of snow is fused. It is possible to feel heat flowing away from the hand up through the glass, and to see the effect of it on the snow above. A higher temperature would do as much for sedimentary rocks. A lamp placed under the glass cracks it, and melted snow or rain drips through: a greater heat would do as much for an igneous crust, if there be one beneath the Laurentian gneiss. In travelling from London to Cornwall, the edges of a geological series are passed downwards. Arrived at the lowest attainable bed at the surface in that direction, rocks are found to be broken as the glass was. In mines, some cracks are seen to be filled with various metals. According to one school, lodes were deposited from solution, and experiments made with solutions have proved that various metals may be deposited in chinks by passing currents of electricity through a model. Currents of electricity do pass through the earth's crust, and the bearings of metallic veins seem to correspond to magnetic currents. So far experiment confirms a theory which savours of the old battle between Neptunists and Plutonists. But in volcanic countries sublimed metals are deposited in chinks; electricity may act on metals in the state of vapour as it does on solutions. Experiments are wanting in this direction; but metals are found only in small quantities in solution at the surface now. Other materials—dykes and upthrows of igneous rock—fill larger rifts and holes in Cornish rocks: these rose hot from below, but Neptunists once believed them to be precipitates. In Scotland and in Labrador such igneous rocks form a very large proportion of the whole visible crust. Heavy metals, which fuse and sublime at very high temperatures, may exist in larger quantities in deeper layers, because they sink deep in fluid slag; and because these low rocks were melted.

In Lapland, at Gellivari, a vein of crystalline magnetic ironstone is seven miles long, and about a mile broad at the outcrop. At Rutivari, also in Lapland, is another large mass in a wide glen; a considerable hill is there made of magnetic ironstone. At Danemora, in Sweden, a similar mass of iron is quarried. At Fahlun, the copper-mine is a vast pit, like the crater of a volcano. About Lake Superior, in North America, deposits of iron and copper are on a like scale. In Nova

Scotia, hollows in veins of red hematite are hung with pendants like icicles. In many of the specimens of iron and other ores exhibited in 1851 and 1862, in London, the structure of the ore suggests fusion. Gold nuggets seem to have been suddenly cooled while in a state of fusion; and gold-bearing quartz looks like burnt stone. If ores were fused and thrown up like dykes at some places, metallic vapours may have risen elsewhere, as steam rises through chinks in igneous rocks in Iceland, and as iron has risen in Elba.

In Yorkshire the smelting of lead-ores caused so much damage to vegetation in the dales, that smelters were forced to use their wits and cure the evil. On the tops of the Yorkshire hills they built chimneys, and from these they made passages along the hill-sides, down to old furnaces in the dales. Some of these passages are three miles long. The smoke from the hearths was passed up to the barren moors, and there it now escapes harmlessly. The sweepings of these chimneys were found to contain valuable metals, which only did harm when out of place. These were sublimed at the smelting-house, and they were carried upwards by the draught. Forty tons of lead were taken out of one chimney in one year, and arsenic and other metals were also swept out of the vent. At a distance of three miles, the proportion of condensed metal in the sweepings nearly equalled the proportion lower down, and the black smoke which escapes still carries sublimed metal into the air. In this process the heat of a small smelting-hearth drove lead a distance of three miles, and it will drive it much further when the vents are made longer. It is not possible to get at the roots of lodes, but it is easy to walk down from the chimney-top to this smelting-house, and to look in at the fluid metal without being consumed.

Lead-ores contain a great deal of silver, and smelted lead is sent to Newcastle to be refined. There it is possible to see a working-model of an engine strong enough to work geological upheaval, and the mechanical power which works it is a dazzling white heat. The little engine may throw light into the darkness of the earth's past history, and down upon strata, which cannot be reached, beneath Laurentian gneiss.

In separating lead and silver many tons of impure metal are fused in a row of large iron caldrons. At one stage in the process, the temperature has to be reduced to about 550°, and it is done by putting out the fires, by stirring the metal, and by throwing cold water upon the fluid amalgam.* Though the boiling point of water is 212°, and the metal is hotter than 550°, the water does not all fly off in steam at once. Spherical masses roll upon the pool of molten lead, and these whirl and oscillate, striking and rebounding like elastic marbles, and apparently dancing on nothing. Their weight, or their resistance to the force which supports them, reacts upon the crust which forms under them, for the surface bends where they rest; but they do not touch the lead. Many of these are hollow shells of water, supported on a core of steam, which is constantly forming below, and condensing above (see p. 353).

Every now and then a water-ball as big as a musket-bullet bursts like a molten shell or breaks. Fragments large as shot of various sizes then disperse, radiating from centres, and each fragment becomes a separate rolling sphere. Some are hollow, some are not, and the steam-chambers vary in size. These roll hither and thither on the hot pool for many

* December 16, 1863. For full scientific descriptions of this process, see *A Manual of Metallurgy*, by John A. Phillips, London, 1852, p. 496. See also Reid's *Elements of Chemistry*, 1839, p. 416.

minutes, but slowly and gradually the water-spheres diminish in size and number; and they all turn to steam and vanish when they have done their work by taking heat from the metal to give it to the air. The heat which does this work is a luminous red heat which acts on photographic plates like any other light. It seems to be a mechanical force also.

If a white-hot bar of iron is plunged into water, something of the same kind happens. Little steam rises unless the bar is plunged so deep that pressure overcomes resistance; then steam explodes and scatters the water. A wet finger may be dipped into a caldron of lead or fluid iron with perfect impunity; there is scarcely a sensation of warmth, though the metal is hot enough to char a stick, or fry a beefsteak. When a mass of hot iron is under the steam-hammer, water is commonly sprinkled on it to clear it of scales; it rolls on the iron like shot on a board. But when the water-spheres are crushed flat by the heavy blow of the hammer they explode with a loud report. If a wet stick is thrust beneath the surface of fluid lead, or if air is buried by splashing the pool, rapid expansion of gas follows, and drops of metal are thrown upwards and scattered by an explosion. If water is thrown on metal so far cooled as to admit of contact between the two surfaces, then water takes up heat and turns to steam, while the metal darkens. In a short time more light from within supplies the loss of "steam-power," and the metal brightens. As a hot poker and a wet finger are protected by gloves of steam from contact with cold water and hot metal, so water-spheres are guarded and supported and shaped by the steam which forms between cold water and hot metal. Hollow spheres float on steam atmospheres, and both are repelled by strong heat. So heat-rays are force, and the brightest are the strongest.

But when this ray-power does not equal the opposing weight-power, as in the case of the blow struck by the steam-hammer, the fluid sinks through its vapour, takes in a full charge of heat from the metal, and bursts into steam. Strong heat, light, or ray-force, may keep two heavy bodies apart in spite of the whole force of the earth's attraction at its surface; repulsion and attraction do, in fact, shape fluids into hollow spheres.

While under these special conditions, the order of the water series was—

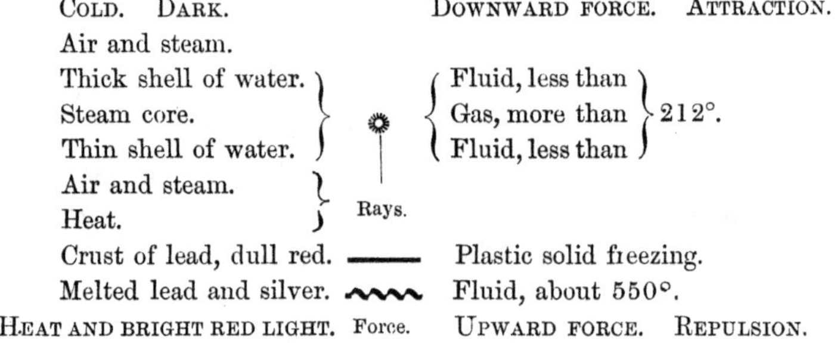

The arrangement is unstable, and can only endure for a time; but while it lasts the earth's ATTRACTION is overcome by REPULSION. A central sphere of hot gas in a shell of colder fluid is possible; to make it last, the centre of gravity and the centre of heat must nearly coincide, and continue so to coincide. If it so coincides while the mass cools, a drop of water may become a shell of ice, or a hailstone, or a snow-crystal, with a structure radiating like rays of force; but a drop resting upon a plane is squeezed out of shape by weight and resistance.

The temperature of 550°, which thus changes the form

and condition of water, is only the freezing-point of pure lead. At 550° the metal crystallizes like water at 32°. Small crystals form in the mass, and float up like ice forming in a freezing-pail, others sink like salt. If left to themselves these crystals form a crust; if stirred they melt, and disperse and crystallize again.

More crystals form as the temperature falls, and many sink, for lead is heavier than silver. Some form and stick on the cooling sides of the vessel; some unite; lead and water-ice alike freeze on iron spoons which are used to stir a freezing mess, for iron is a good conductor. In one case a measured scale marks 550°, in the other 32°, or 28°, or 14°, as the case may be; the shapes of the crystals differ, but cooling obeys the same law in this metal amalgam and in salt water. When crystals form rapidly in the lead, a great iron strainer is plunged into the pot, and it strains and gathers out a spoonful of dry granular lead-ice, from which the wet drains and trickles away. The lead-sludge is thrown into a caldron to be separately cooked, and passed along the row of caldrons; the fluid is left to be enriched, for in that fluid is the silver.

The freezing point of silver is far higher than that of lead;

Lead melts . .	612° *
Silver . . .	1873°

it takes longer to part with the heat which keeps it fluid. As water and brine are separated by crystallization at or about freezing, so lead and silver are parted at or about 550°. As brine is strengthened by adding brine, and by taking fresh-water ice away, so a pot of metal is enriched by adding a mixture of lead and silver, and by taking out crystals of pure

* There is no certain measure for high temperatures. These and other figures are quoted from works of authority, or from statements made by practical men.

lead, wetted with fluid amalgam. During the cooling of these metals the upper series is—

Cold.	Weight-force.	
Solid .	———	. thin Lead crust.
Fluid .	∿∿∿∿	. Lead and silver.
Heat .	☼	. about 550°.
	Ray-force.	

As cooling goes on a crust forms all around, above, and below, and against the sides of the iron vessel; wherever rays of heat escape; most where they escape most; and a fluid core is left at last. A large round drop, composed of these metals, and cooling in space as they cool in a cup, would have a crust of frozen lead and a hot core of lead and silver, partly fluid, and crystallizing while cooling by radiation.

When this solution of silver in lead is strong enough, more heat-power is brought to bear on the mixture, and the metals work on a different plan. They boil.

Melted amalgam is ladled from a pot into a large cup, made of bone-dust, and hot air and a strong flame are made to play on the metal surface. The mess seethes. Thick fumes of leaden steam are driven off, and fly away, with hot air and coal-smoke, through the chimney. In Yorkshire such fumes fly three miles and more. Lead and oxygen combine, and when combined, they stream through the bone filter as melted litharge; or they float on the silver, and flow over the edge of the cup. But the boiling point of silver, like its freezing point, is higher than that of lead, and fluid silver is denser and heavier than fluid litharge; so, while lead evaporates, and litharge floats and flows away like slag, silver sinks through the lighter fluid and floats on the strainer, and the rich broth grows richer still. As the lead boils off, more and more of the stock is ladled in, till the " dainty dish is fit to set before a king;" and then, with an extra force of heat, the last of the lead is driven away, and the silver-plate is cooked. The bright metal clears up like the sun breaking through mist; and it shines.

In water, lead, and silver, like effects are produced by various temperatures. The heat which evaporates water freezes lead; the heat which evaporates lead only melts silver; the force of electric light drives them all away in fumes. At the highest of these temperatures, and at the pressure of the atmosphere at the earth's surface now, water, lead, and silver are gases; all three are solids at 32°.

Mingled together, and cooling, these fumes or gases would condense in order, or combine and condense in some new order. Silver would sink in a fluid oxide of lead. Litharge would flow on the top of red-hot silver, and form a crust of oxide when it cooled, and water would become ice upon the heavier solids only after they had both fallen and frozen, and cooled to 32°. Till that point was reached there could be no rest for water, for heat would move it in escaping from the hottest, lowest, and heaviest, through the highest, lightest, and coldest of this series of three fusible solids.

The "working" of this engine is a thing to be seen. It was seen in Edinburgh class-rooms, in Spain, and elsewhere, in 1839 and 1842; at Newcastle it was seen again with a purpose, after seeing Vesuvius, Hecla, and the Geysers. Seventeen thousand ounces have been refined in one cake by Pattinson's process, first invented in 1827; 9000 ounces make an ordinary charge. It is a pool four inches deep, two feet and a half wide, and charged with from 1700 to 1800 degrees of temperature, and it is a powerful little engine to work upheaval. The pool is perfectly fluid; it shines with a brilliant white light of its own, and reflects other light like a polished mirror. When the hot breath of the furnace plays on the surface it ripples like water; when the cup is shaken the shining mirror is broken up into waves; when a white-hot cinder falls on it, rings spread as they do when a stone is thrown into water; when the temperature varies within the cooling mass, gentle currents move hither and thither, and glowing embers drift on them like fire-ships on a calm tide. The fluid surface is smooth as glass, and still when undisturbed, for silver, quicksilver, and water, when melted, all obey the laws which govern the movements of fluids; but of these three only the hottest shines. A constant play of colours and a maze of curves play on the surface with every movement and breath of air. Like a soap-bubble, or oily hot water, the fluid shining silver has a thin varnish in rapid movement, which refracts and distorts the rising light.

There is a great store of latent force in the quiet silver pool; it shines, and there is hot oxygen locked up in it. There is gas ready to expand, and ray-force only waits for resistance to show its power.

With cold the resistance comes, and the battle rages. When the silver is pure the fire is extinguished, and freezing speedily begins. First a few crystals form on the surface, then a network, then a thin skin. If a bit of cold silver is tossed in about this stage, it floats like a small iceberg, and gathers a thin raft about it. The silver-ice may be pushed about, for it is a floating body; and if pushed down, it rises again high above the fluid. It stands higher than ice in water; far higher than solid lead in fluid lead. Every point seems to act as a way for heat to escape; the floats soon take root by spreading below; and so they grow and spread, as icebergs do, in freezing water. At this stage the lustre of raised points far exceeds that of smooth plains; the rough solid hills are white, hot, and 'tell' light against smooth thin crusts in the lower regions. These tell dark in this general blaze of light. When the cooling has advanced to a certain point, and a pellicle forms all over, a stream of cold air is blown in to hasten the cooling. Then the lustre changes from dazzling white to red, the upper crust thickens, and the action becomes rapid. Molten silver is within; it is compressed by the forming shell, and hot oxygen is squeezed out of the mass. The surface at this stage begins to break up and bubble; it is upheaved; silver escapes where resistance is least, generally near the edge, where the heat of the cup keeps the crust thin and soft.

At this stage the light of the surface changes colour rapidly. Where the hot interior finds a vent, it is still brilliantly white; where the crust has set, light is bright red; where the crust is thick, it is a dark cherry red. Hills now tell dark against lighter coloured lower grounds, and the brightest spots are hollows in hill-tops and boiling holes in the plains. There is great variety in light which shines out of hot silver while it is freezing, and the same is true of all other materials which have been watched. This light, like sun-light or any other light, may be refracted and reflected: a lens forms an image of the silver on a screen; the image formed on the palm of the hand is sensibly hot. The metal is giving off light and heat, which produce their usual effects at a distance. Similar rays made water-spheres revolve above dull red molten lead, and white-hot solid iron. The silver plate is a self-luminous body, like the sun, for the time.

To prevent loss from boiling over at the edge, the workmen commonly prick the silver plate in the middle; they break holes in the ice, and the silver pool wells up like water in a pond. Then comes the time of rapid upheaval and disturbance. Bits of broken crust rise and fall like the lid of a box, and hot springs of boiling silver gush out in shining fountains of glittering light. They freeze as they overflow, and hollow pillars rise up, growing like the trees of Aladdin. They rise and grow and branch, and shed a crop of silver fruit, till they reach the point where the pressure from without equals the force within, and then, when the weight equals the heat, when the column of fluid is balanced by the gas, the tube is sealed by a silver dome, and that well in the ice is frozen.

All these quaint forms are casts of ray-force. Motion is arrested suddenly, and fountains are caught flying.

Larger holes give rise to larger tubes, through which boiling silver splashes out. Tubes grow into truncated cones, and these as they rise gradually narrow, till their limit is reached. Then they too cool and close, and a silver volcano is plugged with frosted silver. When the cone is finished, and the vent stopped, smaller vents open in the plain; and from these a crop of tubes and cones grow, till a range of hills forms on a frozen silver sea. There is scarcely a mountain form or fantastic lava-shape in Iceland, a branching shape in a metal vein, or an ice-form off Labrador, that may not be thus copied in freezing silver.

Throughout this period, the explosive force within casts showers of spherical drops whirling into the air, and each of these for the time becomes a separate system, moving in obedience to the laws which govern projectiles, and working itself into shape, because it is moulded by two opposite forces in obedience to the laws which govern force. These sparks work in the air, as they fly, while the parent plate works in its cup; and many of them cool as hollow shells about chambered interiors.

For a full hour a plate of 9000 ounces continued these displays of volcanic action; the charge of heat raised mounds of silver more than six inches above the surface, and threw silver drops to a distance of more than two feet. At last the whole mass froze, and then the rapid action ceased.

But though violent boiling ended then, so far as silver was concerned, there was still a great store of light, heat, and force in the solid. The light was cherry red in the hollows, dark red on the hills, and the light which the crust reflected was pure. The heat was still felt at a distance, the lustre was seen in hollows and cracks; and water thrown on boiled furiously, or danced as it did on hot lead.

The frozen plate was dragged from the furnace at last and weighed, and then it was cut into junks with steel chisels, and heavy sledge-hammers wielded with a will by brawny arms. It took a great amount of physical force to quarry this work of heat and cold. The internal structure was shown in the section. The mass was hollow, chambered and crystallized like slag, or Icelandic lava, or glacier-ice.

If one of the numerous spheres which were thrown off by this plate were the subject of inquiry and out of reach; if its path were known, its surface seen, its size measured, its density calculated from its movements, its light analysed, and its composition unknown; the data would not give pure silver, because of the spongy structure of the mass. If planets are made on the same plan, philosophers may have to revise some of their conclusions as to other worlds.

When remelted and run into bars and ingots, the silver takes less room, and has greater density, though many ingots are chambered still. When stamped and hammered, the metal has still greater specific gravity, greater density. It is the same substance, differently packed by natural mechanical force and by men.

Like the water and the lead, the cooling mass, during part of the process, was a solid shell with a fluid core, and during that time force worked most upheaval. The free projectiles were spherical, with crusts roughened by radiating projections, and with spongy cores.

A world arranged as a core of hot gas in a shell of fluid, with a solid crust, is possible; because that arrangement always recurs in making this experiment. It always results in certain outward forms, and these endure when the action has ceased, to show what the nature of the action was. But till the engine was seen to work, the forms had little meaning. A portrait of a "specimen of pure silver" is on page 338, and it was thus prepared :—a bent iron point was dipped into the silver and came out red-hot, with a frozen crust of white-hot silver-ice upon it. By dipping, this grew to be a smooth shining hemispherical half-frozen button, and then it was set to freeze in a draught. It cooled as the large plate cooled afterwards, but suddenly; and the fluid interior burst violently through the crust : the fountains froze as they flew ; and strange shapes resulted from their movements, and these from forces. Gravitation acted downwards towards the earth's centre : radiation from within the silver outwards in all directions : expansion acted from within, contraction from without : the radiating forms were casts of distorted rays.

The duration of the rapid action was in proportion to the size of the mass. A spark cooled as it flew. An ounce cooled in a few moments. Nine thousand ounces worked for an hour after the fires were drawn. Seventeen thousand ounces worked for a much longer time ; the mountains were far larger and higher, and the eruptions threatened to blow off the brick roof of the arched furnace, through which a window was opened to let spectators see this silver light do the work of ray-force.

The violence of the action was in proportion to its rapidity. A charge of force had to be expended, and it escaped quietly and slowly, or suddenly and with explosive violence. A small mass suddenly cooled, burst, or threw up high projections in proportion to its bulk. A like mass more slowly cooled, worked for a longer time, but did not work explosively.

Of three masses of unequal size—a drop, an ounce, and the parent mass cooling together in the same temperature—the smallest cooled first, and had the highest projections; the larger cooled next, and the largest last.

The first was cold and only reflected light when the second was still working, and shining through cracks and holes in its crust; the third was working and shining, shedding light and heat on the other two, when both were cold and dark.

That shining silver plate is an engine on which thoughts may travel a long way, in as many directions as there are rays in spheres of light and gravitation.

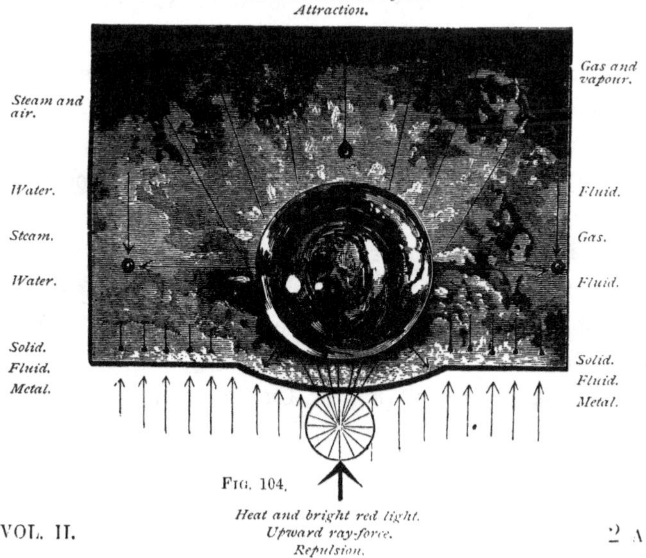

Fig. 104.

CHAPTER LIII.

UPHEAVAL 2—RAYS AND WEIGHT 2—FUSION AND FREEZING—METAL AND SLAG.

WHEN so many roads are open it is hard to choose a path. If light be visible force, the diameter of the sphere within which it works is twice two of the greatest distances yet measured from this world to another star; for light, if it shines thus far from a point in space, must shine as far in other directions.

***

Space and distance on this scale must be left to astronomers. A shorter path will lead a student to the nearest furnace where metals are fused, and there he will find ample room for him. Stars though visible are out of reach; our own little world is too big to be seen; but at a furnace it is easy to see and to think;—to watch small shining bits of our world fusing, boiling, whirling through the air, freezing and falling; to see small work done during minutes, hours, or days, and to think of material things obeying the same laws during all time. The scholar may learn one more alphabet of form by watching solids, fluids, and gases, which are parts of a great whole, fusing, and freezing, and taking shapes from forces and their fixed laws.

If any laws govern all matter, they apply to all quantities, times, and distances alike; to the least as to the greatest, to sparks and to worlds. Gravitation seems to be a law which applies to all visible material things; if visible light be an opposing force of like general application, these two may have shaped worlds in obedience to the laws of the great Lawgiver who made this round world like a little drop. Modern astronomy rests upon gravitation, which is a law discovered from the movements of projectiles large and small. Whirling worlds and still larger systems of worlds all seem to obey that one force. If they obey two, and if light is one of these, a knowledge of a second law may grow from little things. If natural philosophers will deign to study rubbish by furnace-light, and make experiments, they may learn to follow ray-power as far as gravitation in time. That is a way which lies open beyond the short path which leads to the furnace.

Let a few familiar examples suffice to explain what is meant by "ray-power." The subject is too large for unskilful hands and minds to grasp. It is dangerous even to step on such untried ground.

Gases, fumes, steam; fluids, hot water, lavas, and suchlike hot materials are now escaping through sedimentary and igneous crusts. Since this part of these volumes was first written, two volcanic eruptions have taken place in Sicily, one, at least, in Iceland; the sea was disturbed off Newfoundland in 1864, and England has several times been shaken by earthquakes. If lavas make large hills above, they must leave large hollows below the crust; it is impossible to get at these halls, but perhaps they may be seen through small holes, made on the same plan, by the same working giants, with the same materials, in small igneous crusts. Chambers abound in all frozen crusts, and frozen slags are made of fused rocks; if geology cannot quarry through the earth's crust, let her study wherever she can, and begin with slag.

Chambers in a solid are well seen in Wenham Lake ice, in impure glass, and in frozen soda-water within a bottle, which is a transparent crust of impure glass: hollows like these may be found by breaking or cutting through bread, biscuit, pie-crust, plates of sulphur, sealing-wax, tallow, ingots of various metals, and plates of slag which are opaque. Larger hollows, of like shape, abound in lavas which were fused, and whose history is known; in ores and in rocks, whose history is not so well known; but many of these rocks certainly were fused like the slag. Similar but far larger chambers also abound in the crust, from which lavas rise and stones are quarried. By watching at a furnace, the growth of a chamber and some of the resulting phenomena may be seen, and the lesson seems good for geological application. Many chambers were formed during the freezing of the Newcastle silver plate (chap. lii.); one large steam chamber, and many small ones, formed in the hot-water sphere (p. 353). Because outward forms in volcanic countries resemble those which always result from the boiling and freezing of water, slag, metals, and other materials, the inward structure of any frozen crust throws light into dark chambers under ground. Of several volcanic mountains of like shape the smallest may be seen to grow, and may be broken up to see the structure; or a transparent glass mountain may be watched while growing, and can be seen through when it has grown. The best teacher of natural science is experiment; so the growth of forms on the earth's crust may perhaps be learned in rubbish heaps by furnace-light.

Silver, cast-iron, mercury, metals, slag, and glass are smooth and 'flat' as water and other fluids while fused. The surface for the time is like the surface of the sea, part of a sphere at the end of a ray: it is like a bit of a wheel at the end of a spoke, and it takes its shape from gravitation. A freezing fluid takes many shapes. If slowly cooled, it is flat and smooth like ice on a pond. Furnace-refuse left to cool in the air sets in layers, which would be arched crusts if they reached to the horizon or covered a sea. But many of these concentric layers are bent and shattered ; projections of various shapes are on the upper surface, chambers, passages, and holes are within. Cold slag is like the silver-plate which was seen to work, and silver hollows were tracked to the surface where a mound was seen to grow ; to a student who knows this silver alphabet, the outside of a plate of cold slag tells a history : like the cover of this book, it gives some notion of the contents. The furnace gives a ready answer to any one who seeks the meaning of a new form—a new letter in the slag alphabet. The small heat-engine is at work, and the tool-marks of ray-force may be learned in that small source of light, a blast-furnace.

It very soon appears that outward forms record movements in freezing fluids : movements caused by opposing forces, whatever the freezing or boiling points of the fluids may be. In chap. viii. an attempt was made to show how transparent water moves, and why. Transparent glass moves like boiling water, and for the same reason opaque slag, while fluid, is moved by the same forces acting more powerfully. The fluid obeys the law of gravitation like any other fluid ; it falls and flows ; and, like other fluids, it boils and rises when the other force gets the mastery. The outward form of the frozen solid is a record of the struggle, and such forms are built about rays. The axis of a mound in slag is perpendicular to the plane of the horizon ; so are the axes of volcanic mountains set upon the tire of a wheel drawn in any direction round the sphere of the world. In a late book which gives sound information from behind a comic mask,* it is stated that the edge of a crater in Mexico was crowned with icicles pointing upwards. They were forms built about rays, and probably grew from vapour caught and frozen while rising out of the bowl. Hoar-frost so forms on posts, gates, rails, and trees, near wet grounds in England ; it so forms on the edge of a bowl with water in it ; and in colder regions, as on the White Mountains in North America, larger "frost-work" grows about rays which meet within the substance on which the crystals form.† If water condenses, the form grows by deposition about rays. If water, silver, glass, metal, or slag freezes slowly and gradually, the crust is flat and even ; if it cools rapidly or suddenly, the crust is uneven, and the forms either aim at the earth's centre, or at some other point or line about which they grew during a struggle. In order to catch the meaning of outward forms thus produced, they must be seen to grow ; they must be watched, as the silver-plate was at Newcastle.

The cut p. 338 is a portrait of a specimen of pure silver, which cooled as described above (p. 353). The arrows show directions in which two forces acted :—Rays from points and lines within bright hot masses of freezing silver, *outwards ;* weight, attraction, gravitation, or some other opposing force, *downwards,* towards the earth's centre, like a plumb-line ; *inwards,* towards points and lines whence the ray-force diverged. Such forms alone suffice to explain their growth when that alphabet is learned ; and slag-forms are like them in this respect.

Certain glass vessels are frosted by plunging tough red-

* *Travels,* by Umbra. 1864.

† For copies of these forms, see photographs published at New York, which may be purchased in London.

hot glass into cold water. Steam carries heat rapidly from the outer layer, and a hard shell forms suddenly. It shrinks rapidly and unevenly, breaks, and the bits curl up, while the hot layer within sticks to the shivered crust, and rises through a network of cracks. The rest of the cooling goes on slowly in heated air, and the finished work is smooth within, but rough, like broken ice, outside. The inside was shaped by air blown in through a tube; and if a glass-blower wants to make a large chamber, he blows in drops of water, which turn to steam and expand within a tough glass boiler. It expands like India-rubber, but does not shrink, for it freezes hard. The structure of transparent glass, and the shapes of chambers thus formed in it, can be seen through the solid walls. A soda-water bottle will serve for illustration, and the lesson may be learned at any glass-house.

In the case of the silver plate, a gas (supposed to be oxygen) was imprisoned in a fluid, and it acted like the breath of a glass-blower. Some of the gas escaped, but part of it was caught and imprisoned within solid walls of silver, when the metal had set. In all cases and in all dimensions like action ought to produce like results. Steam bursts hollow spheres of water, which dance above hot metal (see p. 353). The gas either bursts a prison or the prison-walls take the shape of the imprisoned gas. In the latter case, the chamber is a cast of the forces which expanded the gas and compressed it. When a stream of iron flows over wet sand, steam forms and expands beneath; the fluid iron upon the sand bubbles like the boiling water beneath it, and part of the steam bursts through; but air and steam are often caught in the freezing plastic iron while rising through the tough mass. Each hollow prison then takes the shape of the struggling prisoner. It is a hollow iron mould of the force which expanded steam and the force which contracted iron ; the shape of it records the struggle for mastery between attraction and repulsion, which ends only when the two are balanced. But when the prison has cooled, and steam has condensed, the weight of the whole atmosphere tends to crush the walls through which imprisoned heat finds a way. Domes thus formed on blistered steel, hollows in cast-iron and in other metals, are often crushed and bent inwards by weight. So ray-force and weight-force together shape this crust. Things which cannot escape—air, and water, and other substances—often line such hollows with crystals, and so leave open spaces. Other chambers have porous walls, and the hollows are filled from without long after they are made ; as caves and mines are partially filled with ice in cold regions. A slag crust is like the rest : when suddenly cooled, it is shattered or distorted. Thick plates, which have long ceased to shine, often burst asunder on the cold floor of a smelting-house ; and when they do, red light, or the brighter light of fusion, shines out from the centre of the mass.

Though metals and slag are opaque, they may be seen through by the help of air, water, ice, and glass, and by the forms which they assume while freezing. Perhaps the crust of the earth may be seen through in like manner, by learning the meaning of outward forms in slag and lava. Luminous heat expands steam, which moves the lid of a kettle, or moves the largest engine ; the same force blows a glass bottle, makes a bubble in metal, and bursts the chambered slag crust, which is made of fused rocks. The same heat melts lava, and the same forces which shape crusts on lava and slag may have shattered the earth's crust, as a workman shatters the crust of a glass jug with cold water.

The writer spent much of his childhood amongst rocks and furnaces, and there gleaned ideas which are now packed in these volumes. One great ploy was to clamber amongst sea-cliffs, another was to see iron "run." That is a sight which bears frequent repetition, though many visitors only see the dirt and feel the heat. Turner thought the colour worthy of his brush, and failed to copy it. Guthrie saw it, and preached a sermon about it—and even Guthrie failed to describe the scene. Till brushes are dipped in light, and words are real fire, the scene cannot be thus brought home. But any one who chooses to take the trouble may see a smelting-house for himself, and a student of natural philosophy will find occupation there. In Lanarkshire, the sky glows at night with the flaring red light of great fires. They glow in hollows, and shine from distant hills like stars or beacons, and the red flames which glow on the clouds leap up and sink down, panting with regular pulsations, like living things. Each of these lights may be reached by following a ray; and each is a centre of active work, in every sense of the term. There steam-engines clank, and whistle, and yell, while men rush hither and thither with iron carts, rattling over iron-plates, with loads of fuel and iron-ore. These tilt their loads of stones dug out of the earth's crust into conical, tall furnaces, whence the light shone upon distant hills and clouds.

A roaring blast of hot air is blowing furiously at the base of a heap which grows from above, and the heap burns and melts. A snow-heap melts below when it rests upon warm earth; but here the heap is made of the crust itself. At Woolwich a heap is made of old iron. The workmen heave in shot and shell, clanking chain-cables, anchors, old rails, nails, hoops, clippings, and filings; with a "one—two—three—heave;" in goes an old rusty gun which has fought and gone out of fashion, and down it goes with a crash; and so the iron heap grows to be a pile on a hot base. Snow, iron, and stone, down they all sink alike when they melt ; and when a charge is fused the base of a furnace is filled with a fluid, which takes the shape of the cup which holds it, as snow-water takes the shape of a lake-basin, or the sea takes the shape of its bed. But here two separate fluids float on each other, like oil on water ; one is heavy iron, the other lighter stone.

The lighter fluid is constantly drawn off, so a river of slag is pouring all day long from the base of each furnace. It is a miniature lava-stream, and it teaches a lesson which may be used elsewhere. Morning and evening the heavier iron is "run." With long bars and heavy sledge hammers, brawny half-naked men attack the base of the hearth. They strike, and push, and heave with might and main ; and break, and drill, and quarry through an outer crust of fire-brick burned hard as altered rock in a single day. The hand may rest on one side of the brick ; but as the quarrying goes on, a red heat, then a white heat, and lastly the bright light of fusion is reached. Then out bursts the flood, glowing and shining, flowing like a river of golden light, scattering a spray of shooting stars, which hiss and fly and vanish like fireworks at a festival, or meteors in the sky.

It is a period of rapid action in iron, but it is a period of short duration at a furnace. Moulds, called the "sow and pigs," are prepared in sand ; they are shaped like great combs, and down these trenches the golden river pours, boiling as it flows. The light changes at every moment, and the movements change like it. Stars soon cease to fly and shine, but darker drops are thrown up when the metal boils, because air and steam are escaping through it from the sand. As each comb is filled, a clay plug turns the stream, and when the whole charge is poured out, the sand floor glows with red iron-ice formed in ditches of sand. Within a few hours, this ice is "pig-iron," and by next day it is cold. Cold iron floats on fluid iron, as ice floats on water.

The forms below are casts of the mould, the upper forms are casts of the forces which made the iron boil and freeze, and a broken "pig" shows the inner structure of such a mass. The case of the silver plate is repeated, and like forms recur in iron thus manufactured in Lanarkshire and elsewhere.

At many furnaces, the operation is carried a step further. The pigs are melted again to make malleable iron, and the fluid is run into large moulds.

When the furnace is tapped, iron and slag pour out together; a bright, shining, double river of metal and stone. It curls round corners, falls over shelves, forms pools below the falls, and eddies like any other stream. The fisherman's instinct knows the very spot where a salamander might find good resting-ground, if there were such fish in that glowing pool; there are the very eddies and whirlpools which a wading fisherman sees meandering past his legs when he wades out for a long cast (p. 225), the eddies which curl behind every post in a stream of water or air (see vol. i.) But this is a double stream about to freeze, and form a double crust. When the mould is filled, bright colours play about the surface; then it darkens and curdles, and winds sluggishly as the slag begins to freeze. Floating stone bergs form and move about as froth floats on a river; as icebergs float on the sea; a crust begins to form on slag floating on iron, as crusts begin to freeze on water, on glass, lead, silver, and iron; and in a few minutes the slag-crust sets as ice did on the St. Lawrence when it set this winter. This is the slag period of violent eruption, the crust breaks, and the fluid core bursts, or wells slowly up through chinks and round holes, which glow and shine brightly in the red-hot ice. The main stream flows on below, and pours over from pool to pool as before, but the upper crust continues to grow on the surface. Flaring sparks fly through open chinks, and when caught and cooled they are cast-iron spheres, with uneven surfaces, and a crust of oxide. The iron stream below, hotter and heavier than the upper stream, gradually cools and stagnates as pig-iron did alone. The stone islands of the upper crust grow together, and join and form a red-hot solid plain, and though the iron is hid in this case, the lower crust certainly forms as it formed in sand when it was the upper crust.

When the iron freezes the slag contracts, darkens, breaks, and rises into miniature mountain-chains. The first surface, with all its cones, curves, and wrinkles, and the whole series of crusts which formed under each other, rise and fall together slowly; and all the phenomena of geological upheaval result from this stage of rapid cooling, in slag resting on cooling iron. When the iron stream has frozen solid, the upper crust remains shattered, distorted, and angular; but also bent, folded, twisted, and chambered; it bears the marks of fusion and of freezing on the surface and in every section, and all this small work was seen in progress so far. In these two crusts the time of rapid action ends when the fluid becomes solid, but there is still a great charge of mechanical force in the hot mass.

The next step in the manufacture is to turn on a stream of water, and violent action is renewed at once. The water sinks into the chinks, and rises with all the borrowed power of that tamed giant steam. Motion which had almost ceased begins again more violently than before, because this third fusible layer is more easily boiled, and harder to freeze than the other two below it. A red heat scarce sufficient to raise iron and slag by expanding the solid, throws a broken crust hither and thither by the help of steam and boiling water. The solid layers which heat the water, cool, contract upon hotter layers within, break, and let water sink deeper to hotter regions below. Steam rushes up, exploding, hissing, sputtering, scattering broken fragments, tossing heavy plates into the air, bursting chambers, grinding edges, rounding corners, driving jets of boiling water high into the air, and filling it with rolling clouds and whirling drops. At this stage it is hard to see what is going on, but there is a violent commotion; and the igneous crusts are broken up, and partly ground by steam-power, which gradually wanes, while the iron parts with the charge of ray-power, which came with it out of the furnace, out of the coals, out of the sun, if George Stephenson guessed right, or out of the cooling earth. One very common occurrence about this stage is the sinking in of the roofs of chambers. The iron contracts, and the slag roofs fall down. The decreasing action is not regular; it diminishes quickly at first, very slowly and gradually at last, in proportion to the 'energy' expended. The amount of ray-force spent on clouds of steam, in heaps of sediment, or in hot fountains, is deducted from the store in the mass of hot iron. Boiling springs sink lower and lower, those which spouted two feet rise only one, and after a time only rise a few inches; next they well up slowly amongst the ashes; and at last the water circulates quietly as warm water does in any vessel, as air does in any room. This hot-spring period lasts for many hours. There is no visible light, no violent action, but the power is not all spent, and it was bright heat at first. At this dull heat ether boils furiously, and the iron below still has work in hand.

If the water gets to the lower side of a large ingot, so as to cool that side first, the whole mass bends upwards like a bow; and all the upper formations rise upon the arch, steam-jets, hot springs, and all. Sometimes an ingot a foot thick breaks short off like a carrot from this uneven contraction and expansion, and so makes a 'fault.' It is the case of the frosted glass over again, but on a larger scale. When both sides are at one heat, the bow unbends, and the mound sinks down slowly. When the upper surface cools, the ends curl up like a shaving of whalebone laid in a warm hand, or like a flat fish laid in a frying-pan. No matter what the substance may be, expansion and contraction work the engine, and the same forces must work that larger engine—the earth's igneous crust—if there be one under sedimentary rocks. Thus at the end of a short time a bright stream, flowing like a river, and scattering drops like a spray of light, is changed into rigid, solid crusts, of metal fit for human use, and of slag only fit for the cinder-heap. The mass stands in water thick with sediment, which falls in time—a small geological formation of fusible sedimentary beds under water. In frosty weather the water freezes in turn, and in very cold weather that crust splits like the other two. A stranger who had not seen these changes take place, might find it hard to believe in the wild vagaries played by hard, cold, ugly, wrinkled, dark-gray solids, resting in their cinder-heaps now, but lissom and active, strong and bright, in their vigorous hot youth, when their bright faces were smooth and soft, before they froze.

When iron ingots and plates of slag thus cooled are broken up, the shape inside is explained by the movements observed, and shapes outside can be referred to them. The silver plate was a costly toy, and can only be seen to work at a few places; slag-plates are piled in hills and cost nothing.

Lanarkshire roads are made of broken slag. In such a path, at a hall door, the writer gathered the first-fruits of this branch of education, and there he made his first collection of igneous rock-forms. Any other child may do as much, and the wisest of philosophers may pick up knowledge in the path which leads to the nearest furnace whence light shines.

CHAPTER LIV.

SPARKS—VOLCANIC BOMBS—METEORITES.

If a reader who has followed thus far, or who happens upon this page by chance, will look back to the "contents," he will find that this hunt has run a ring. Those who have followed all the way—if such there be—have been to Spain, Italy, Greece, Switzerland, Scandinavia, Spitzbergen, Iceland, Greenland, and America; all round the British Isles; high up in the air, and down through water into the earth, with miners and geologists for guides. The quarry was viewed in the last chapter, and it went to ground in the cinder-heap whence it was started. The quarry was terrestrial light, and it is impossible to follow it deeper by any direct road.

If a geologist could crack this little round world on which he lives, and study first the whole outside of the shell, and then the kernel and the core, within and without; if he could cut it in two, like a roll or an orange, a stick or a bone, and study a whole section at once; if he could first watch the growth of it, and then crack it like a pebble, he would understand the structure better than he does. A geologist can do nothing of the sort; but every geologist wants to know what the inside of the world is like, in order that he may the better understand the outside of it. A great many able men have tried to crack that nut. In November (5th and 6th) 1863, the

Newcastle Daily Journal published a clever summary of scientific speculations on this subject, and a woodcut of a section of the globe, according to the view taken by T. P. Barkas, the writer whose signature is attached to the paper in question. The cut represents a hollow shell. The list of the famous men who have tried to solve the problem is very imposing, and it includes teachers and masters of many branches of knowledge; but their opinions differ as much as the several ways by which they sought to reach their point. In this mocking age nothing is complete without a ludicrous element; so, to relieve the darkness of the earth's interior, and lighten a heavy subject, Captain Symmes is introduced to play merry-man amongst grave and reverend actors on the world's gravest stage.

"He believed that the interior of the earth was peopled, and he invited Baron Humboldt and Sir Humphrey Davy to descend with him into the subterranean recess by an immense hole which he fancied existed in latitude 82° north, from which the polar light was supposed to emanate."

Baron Humboldt did not go; but he says, "According to conclusions based upon mere analogies, heat probably increases gradually towards the centre."

No theory ought to be accepted because of the author's authority; no man's theory ought to be ridiculed till it has been tested and found absurd; but Humboldt is a better guide than Symmes along underground footways, which lead step by step from experiment to conclusion, like ladders which reach from point to point in a deep dark mine. One leaps in the dark, the other feels his way cautiously. Parry, Scoresby, Kane, and others, have been far enough north to prove that Symmes was wrong; all experiments yet tried confirm the view taken by Humboldt. A student who will not leap to conclusions, and cannot keep pace with philosophers whose thoughts are mounted on well-built scientific cars, must take his own way, and do the best he can to reach his point. The quarry pursued was Light, and it was run to ground where it cannot be followed; but a student in search of knowledge may watch a spark flying out of a caldron of fluid iron: he may study that to begin with, and strive to advance indirectly, step by step. One who does not mind dust and ashes, and the risk of burned fingers, may fill his pockets with luminous drops of metal and slag at any furnace, and crack these like nuts at home.

Some years ago a great number of sparks were caught flying, and others were sifted out of the dust on the floor of a smelting-house in Greenock, to the great wonder of the workmen, who could not make out "what the gentleman wanted wi' that dirt." The "gentleman" had just returned from Iceland, where he had been with the purpose of studying forms which result from the mechanical action of terrestrial heat and light, and he wanted to compare certain round stones with frozen sparks; he had come to fill his pockets with dust, in order to gain light amongst his old friends—intelligent Scotch workmen—and at his old haunts, beside furnace fires. The round stones were gathered with the notion that the inside of a round world, which is hot within and hard without, and travelling through cold space, might be like the inside of luminous sparks of iron and slag, and larger drops of lava, which shone like stars while they flew through the air at first, and only ceased to shine when they froze. The student meant to compare all these with meteorites, to test his theory as far as he was able, and to say nothing about it till it was licked into some tangible shape. It has now taken the shape which it wears in these volumes, and readers who have had the patience to follow thus far—if such there be—may now judge this spark, which was sifted out of dust and ashes, at home and abroad.

The first step in the comparison was to make the frozen sparks seem equal in size to the lava-drops; and with that end in view, they were placed under a microscope, and drawings made from them.

Like forms have been found upon all such drops. The surface always appears to be dimpled with cups, and roughened with projections of various shapes: these resemble forms which abound upon every plate of slag; they are miniature copies of mounds and hollows in cast-iron, from which sparks and drops were thrown while the iron was hot; they are like hills and hollows which may be seen to grow on freezing iron and slag at any smelting-house; they are like those which were seen to grow upon silver at Newcastle and elsewhere. In one case cones and craters are on the shell of a small spherical mass; in others they are on a plane, but the plane is in reality a portion of a sphere whose centre is the centre of the earth. The round lava-stones are like the frozen sparks. They were shot out of cones and craters, and their surfaces are often pitted and dimpled and roughened with miniature craters and cones, which, in their turn, resemble shapes which abound in the lavas, and in the large mountains of Iceland, and other volcanic regions. The outer forms bear reference to the interior of the frozen sparks and "volcanic bombs;" the outer shape of the volcano to the interior of the earth. They are all shapes built about rays.

The history of "volcanic bombs" may be learned from passing events. In February 1865 an eruption broke out in Sicily, and numerous writers have described what they saw there. The following are extracts from a letter published in the *Scotsman* of the 20th February 1865:—

HOTEL DELLA CORONA, CATANIA,
February 7, 1865.

Having just witnessed an eruption of Mount Etna, I think a short account of it may be interesting to your readers. The morning of the 2d was ushered in by a terrific thunderstorm accompanied with torrents of rain and hail. But intelligence is brought us that Etna is in full eruption ; that the lava has already run so fast and so far that the road to Catania is blocked up ; that thousands of peasants have fled from their home in terror of destruction ; and that a war-vessel has left Messina, carrying the Préfet and a staff of engineers to the scene, with the view of saving life and property.

It is almost dark before we reach the steep zigzags leading up from the main road to Taormina, where we intend to sleep. On reaching a sudden turn, we see in the clouds a long undulating line of red light. It is the lava-stream—Etna outlined with a pencil of living fire. And now the low rumbling of the still distant volcano breaks on the ear, mixed up with the peals of thunder, which continues to reverberate among the mountains. As the night deepens, the clouds begin to clear away, the stream of lava becomes brighter, and the light emitted from the crater, which was at first but faintly reflected from the clouds above, becomes more and more brilliant, until the whole sky over the mountain glows with a lurid light. Here and there at different points bright jets of flame appear for a few minutes and then vanish. These, we suppose, arise from the burning of trees set on fire by the lava or the falling scoriæ. There appear to be six craters quite distinct, but situated near each other. From all these, in irregular succession, sometimes from several at a time, there are incessant discharges—huge masses of red-hot stones and scoriæ thrown to an immense height, with volumes of steam and smoke which reflect the fires from the red-hot cauldron below. The glowing smoke flickers in the breeze as if it were flame, and through it and far above it, with the naked eye, we can see the red-hot stones mount and then fall slowly back into the abyss.

I regret having omitted to note the time which these stones took to rise and fall, as that might have given an approximate idea of their size, and the height to which they were ejected. But Taormina is from twelve to fourteen miles distant in a direct line from the crater, so that the stones, to be seen at all, must have been enormous. Comparing the height to which they seemed to rise with the appearance which such a building as St. Paul's when so far removed might present, it could not be less than 1000 feet.

Leaving Taormina at nine, we drive to Mascali. The weather is a complete contrast to that of yesterday—bright, clear, and calm. As we pass along among almond trees in full blossom, through orange and lemon groves glowing with their golden fruit, the ground carpeted with young flax of the brightest green, and see the labourers following their peaceful occupations in the fields, it is difficult to realise the idea that within a few miles a volcano is breaking up the crust of the earth and spreading a deluge of liquid fire over its surface. A walk of three hours over a used but not a difficult road brings us to the lava. As we approach, the rumbling sound from the eruption becomes louder and louder ; but as the sun gains power and brilliancy, the volcano becomes invisible to the eye. A faint line of smoke along the current of lava, and a dark cloud hanging over the crater, are the only visible signs which he gives of his existence—signs which, if met with on a Scotch mountain, might be passed by as arising from moor burning. The stream of lava which we visited is said to have flowed from six to eight miles. The lava, under the influence of the bright sunshine, appears to consist of blackened scoriæ or cinders. It is only through the chinks, or where the surface is displaced by a rolling block, that the fire is visible. The current, where confined in a narrow gorge, flows rapidly—that is to say, at the rate of from two to eight feet in the minute, according to the steepness of the descent. On the flatter ground, where there is more obstruction, and where the stream spreads out to a great breadth, the progress is invisible to the eye. As in a glacier, there is a more rapid flow in the middle than at the sides, for these sometimes seemed to be quite fast, while the motion in the centre is distinctly perceptible. The portion of the current which is flowing towards Mascali, has a breadth of some two or three hundred yards, and a depth on its sloping front of from twenty to twenty-five feet. It may be approached without much inconvenience, and with perfect safety ; for although large masses are constantly rolling down, there is always time enough to escape before they reach the bottom. Men were busy carrying off the beams of the roof, with the other timber work, and filling up the cisterns with stones. When the lava comes in contact with a large body of water, dangerous explosions take place through its rapid conversion into steam. The point which the lava has reached I calculate to be about 2400 feet above the level of the sea, and the crater some 1500 feet higher, or one-third of the way up the mountain. We followed the stream towards its source, until we were driven off by the heat, the blinding dust, and the sulphureous smoke. Of the three, the dust was the most troublesome. Below us we could see the course of the current filling up the hollows and spreading over the flatter surfaces like a huge black glacier, while above, confined in a narrow gorge, it came tumbling over a precipice in a dark mass, relieved by streaks of fire. We waited until night set in, when the lava began to glow again, and soon assumed the appearance it presented from Taormina of a river or cascade of fire. On what seems now to be a glowing mass of living fire men were walking not two hours ago, for the purpose of getting some trees which had been swept down by the torrent. One tree we saw carried on shore by two men who had stood on the lava while they cut it in two. A small prize for running such a risk! They returned for a second, but were driven off by the heat and suffocating fumes. An Italian engineer who was on the mountain took some rough measurements, and calculates that the crater has already discharged eighty million cubic metres of solid matter, that the progress of the different branches added together would amount to seven metres per minute, and the length of the whole to forty-five English miles. I consider the estimate of the distance too high; and as the eruption began only four days ago, it does not seem to tally with the other calculations.

The following are extracts from the *Times* of February 24, 1865:—

Letters from Sicily, in the Malta papers, give some further particulars of the eruption, and the progress it has made. A letter from Catania, on the 12th inst., thus speaks of it:—

"The mountain indulges in a constant roaring, to which we are gradually becoming accustomed, but which at first kept me awake at night, and this at a distance of some thirty miles; so you can imagine what it must have been on the spot which I went to (Monte Crisimo), situated at about two miles N.E. of the new crater."

Another letter of the same date from the same place says:—

"Two nights ago we could not sleep for the noise, the wind blowing from the north. An eye-witness tells me there were eleven streams of lava, mostly small."

The following are extracts of other letters from Sicily relating to the eruption :—

"Aci, Feb. 7.

"The lava issues from four mouths on the south side, and varies every day in the direction it takes. If the eruption continues it will do more damage than that of 1859."

"Giarre, Feb. 10.

"Yesterday I visited Piedimonte, out of curiosity, and observed that the right branch of liquid lava was advancing with the extraordinary velocity of about a mile and a half an hour. Great damage has already been effected by the lava. At the present moment, while I am writing, all the windows of the house I am living in have been broken by concussion, which was accompanied by earthquake. The noise is like a continued cannonading, with a discharge from time to time of 100 guns all at once."

Another letter says :—

"All the world is busy talking and speculating on the effects of an eruption of Etna which broke out on the north side of the mountain, about ten days ago, at a place called Monte Frumenti. It is very violent and threatens to do much damage, as the streams of lava run east and north, and are progressing with great rapidity. I went up with a party to see it, and certainly it is one of the grandest spectacles I ever beheld. There is an incessant rumbling noise, with, every now and then, loud explosions resembling the discharge of heavy artillery, when showers of red-hot stones are thrown to a great height into the air, and either fall back into one of the craters (for there are three of them in activity), or are carried away by the streams of molten rock which are constantly flowing. It is certainly one of the finest sights I ever witnessed ; all other things appear tame and commonplace when compared with it. Shortly after the party I was with arrived at the summit near the craters a dense fog came on, and we were compelled to bivouac for the night, as the guides refused to undertake the responsibility of conducting us down until daylight in the morning ; and when we did descend we were convinced of the propriety of their decision, as the road, which we had passed over in the dark without apprehension, appeared appalling when seen by daylight the following morning. From our bivouac, 6000 feet above the level of the sea, the scene was magnificent in the highest degree. The constant thunder of explosions every two or three minutes, and the streams of lava running down, and, every now and then, setting fire to trees that stood in their way, was a sight well worth the hardship of a night's exposure on the hill-side. Some of the streams of lava are a mile wide, and have extended seven or eight miles already; as yet the mischief has not been much, as the progress of the devastating flood has been confined to the mountainous regions; but if it once descends to the cultivated parts, the damage will be incalculable. Government is doing all it can, by sending troops to assist the people in removing their goods, pumping out the water from the wells and cisterns to prevent explosions, etc.; but it is a sad sight to see the country devastated and overwhelmed by this fiery torrent, and left desolate for ages. Happy are the countries that are free from such calamities."

With these fresh descriptions, and an ordinary power of comparing great things with small, let any one visit the nearest glass-house on a day when the metal is melting and boiling. All that is so well described in Sicily may be seen in miniature through the opening in the retort—the liquid fire, the bubbling craters, the hot whirling projectiles. Let any one watch the sights and sounds about a blast-furnace, to which attention was called in the last chapter, and the action of furnace heat and of terrestrial light will seem to be identical in character, if different in degree. The lava, freezing as it flows from the base of the mountain, throws off a spray of liquid projectiles—"sparks," which rise 1000 feet, and freeze as they whirl and fly. Like them, and like any other freezing fluid, the lava-stream freezes on the surface, and the lava-ice records the rate of cooling by its shape. In Sicily it is irregular; in Iceland, where old lava-floods were larger, the crust is more compact—more like a crust on slag, which cooled slowly. The sparks are alike, though various in size and in shape. They shine as they fly; some burst like rockets, and scatter a shower of golden fire, others shoot and shine and fall, freeze and glow, and darken on the floor; and when they are found, these sparks are shaped like little worlds. They are frozen drops.

At Hraundal, in Iceland, a crater is at the upper end of a glen. It is at the source of an old lava-stream, which flowed down a hollow for some miles, and froze into clinkers. The hill may be about 100 feet high, and it is a perfect "cone of eruption,"—a truncated cone, with a funnel-shaped hollow in the top. The colour is a dusty brick red, and it stands in a broken-down crater of larger size, and of a different make and colour. The central mound is a pile of round stones, dust, and fragments. Some of the stones are as big as a man's head; others about the size of oranges, potatoes, and nuts; and most of them are distorted spheroids, egg-shaped or discoidal. They are exceedingly hard and tough, and very heavy. It took hard blows with a heavy hammer to crack these nuts; but many were broken on the spot, and a pocketful of specimens were carried during a long day's ride, and brought home. A black specimen was brought home from Myvatn the year before, and these are the stones which had to be compared with furnace sparks.

Because these stones were drops of lava, which cooled by radiation while revolving in free air, it is certain that the outside cooled first. The first crust froze, and shrank about a fluid or viscous hot core. The Myvatn specimen was somewhat like a split truffle, for the outer crusts tore, as freezing slag-crusts commonly do under like conditions. A second crust formed within the first, and a third under it, and then all three were torn, and the hot core bulged out. The "faults" remain, and their sides show the edges of three crusts, which seem to have been soft, for they bulged sideways into the rent. These three crusts differ in colour, though they are alike in structure; and in this they resemble thicker lava-crusts, and shattered cliffs, amongst which this lava-ball was found. A tap with a hammer broke this specimen, shell and kernel, and so revealed the inner structure of it. It was shot out by the earth's artillery—by a radiating force, which projected it from a tube with a chamber; it was shaped by heat and cold, by expansion and contraction, by forces acting in opposite directions, from within and from without, while it was whirling and flying through the air; it is a work made in obedience to the code of laws which seem to apply to all known objects in nature; it may be shaped like larger works. The seedling may be like the old plant; the structure of this frozen drop may be like that of the world from which it sprang. Iron sparks are like it; cups and cones, faults and fissures, dykes and craters, like those of Iceland, are on the outside of it. Point a common telescope at the moon, and the same forms reappear upon the surface of a star which shines by reflected light, and seems to be no larger than one of the iron sparks under a microscope.

Sparks and bombs resemble each other in their structure. They all have crusts and cores, and the whole mass is pervaded by tubes and open chambers, of which many communicate with each other, and some with openings in the outer shell. The outer crusts of broken specimens are built upon lines which radiate from within; joints and vertical fractures in the crusts all bear reference to points within the mass. Produced in one direction these lines converge, in the other direction they diverge. The crusts surround a core as a nut shell surrounds the kernel, and the outer layers shell off.

They are like the earth's igneous crust, as seen in cliffs; they break vertically and also horizontally. The kernel of the stone is shaped like a sponge, with tubular branching, irregular passages, and spherical hollows, built about lines which radiate as heat did, from points within the mass outwards. But all the rays are bent in one direction; like the arrows in the cut, p. 28, vol. i., or the curves at p. 473, vol. i., and in the map at the end of that volume. All the specimens from Hraundal have crusts with irregular spongy cores, built about centres of radiation and motion. After trying to copy sections by various unsatisfactory devices, the stone itself was tried as a type. Slices were made equal in thickness to a printer's block; they were inked and pressed, and here is the result.

These shapes tell of expansion within and pressure without, and of rotation; the mass shone while it was forming, and ceased to shine when the crust had formed and cooled, and such masses whirl as they fly. The first frozen shell was filled with fluid or viscous lava, and with vapours which shaped hollows in the plastic mass and escaped through them to holes in the outer crust. The last of the imprisoned vapour was caught on its way out, the prisons took the shape of the prisoners, and some of them now are crystals, which forced the prison-walls to take angular shapes. Surely this miniature geology may grow. When furnace sparks and volcanic bombs agree so well, a student may venture one more step on the ladder which has led, step by step, to knowledge and to light.

As a very eloquent, able speaker is apt to say, "Three courses are open" to every student. One is to follow some beaten path, and never to venture out of it; to choose a leader and follow him, pacing gravely over the same old ground every day, and learning every inch of it. That school of peripatetics is numerous, for the ways of these scholars are

SECTIONS OF VOLCANIC BOMBS, FROM HRAUNDAL IN ICELAND.—Printed from the Stones.

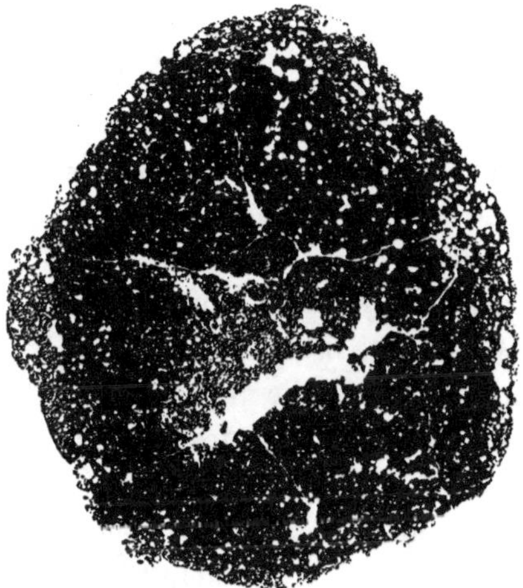

No. 1.

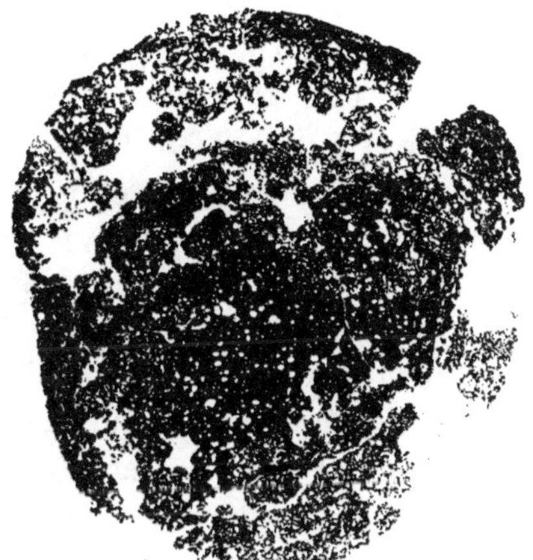

No. 2.

No. 3.

No. 1 is from a section made at the supposed equator of a flattened spheroidal bomb. The whole rough surface of it is pitted with smooth cups:—miniature craters, of which many end in tubes. As shown in the section, many of these ducts communicate with chambers in the crust. Of these some are associated with rows of small chambers, and with long irregular passages in the core, which aim at or meet in a large irregular chamber near the centre. The ends of numerous radiating and branching passages are seen in all the sides of this central cavern. The inner surfaces are smooth, and it is evident that the walls of the chamber, and of its radiating systems of ducts and passages, were plastic when they were shaped by imprisoned vapours struggling to escape from the centre to the surface. Pits, cups, tubes, craters, and cones, record the escape of miniature eruptions through the crust. If any one system of chambers is followed from the outside, the line traced is not a straight line, but a curved spoke bending backwards. That form records the direction in which the stone revolved about its axis.

No. 2 is from a similar section made with the intention of cutting an axis of rotation at right angles. The surface of this stone is not so much pitted, and one side appears to have been flattened, as by a blow. The section shows a crust with fewer chambers near the outside, and a spongy core. The same arrangement of the materials about curved rays is apparent. From their structure these two stones revolved in the same direction, right side down the page. Part of the crust of No. 2 split off in the process of cutting.

No. 3 is like the other two in structure, but revolved the other way.

easy and safe. A second course is to avoid roads—to scorn open gates, gaps, and bridges, guides and leaders, and strive to be original. That is a brilliant, dashing, dangerous course, which may lead to honour or to failure. Captain Symmes got a heavy fall and failed when he made a guess, scorned experiment, and took a header into the earth. The middle course, in this as in other cases, appears to be safest and best. It is to follow the best attainable paths quietly and steadily as far as possible, with the best guides and the best aids, and with the best comrades, who will travel towards the point aimed at; and when the wilderness is reached at last, to choose a line and take it, and go, best pace, along the best ground cautiously, like a traveller making his way through a new country, where all must do the best they can to help themselves, for lack of guides, and roads, and cars. Humboldt got to his point and gained honour, by venturing cautiously on new ground when he had followed guides and roads as far as they would lead. In illustration of these three methods of study a writer may tell a story against himself without offence.

Some years ago, after a trip to Iceland, it was agreed that a joint book should be written, and one section of it was to be written by this hand. With a head full of the subject, the owner of head and hand set out from Edinburgh for the Carron Ironworks, intending to watch the pranks of molten stone, as a key to the forms of old lavas and volcanoes in Iceland. A heavy cloud had passed over a clear sky the day before, and a loud clap of thunder had been heard. On getting into the guard's van to smoke in quiet, it somehow transpired that a "thunder-bolt had fallen in a field about half-way to Glasgow." It so happened that the guard, as he said, was cognizant of the fall of a meteorite somewhere in England. It fell through the roof of a barn, and buried itself in the clay floor; it was dug out, and it was so hot that the workmen pitched it into a pond, where, so far as the guard knew, it remained. This guard had spoken to the guard of another train, who had seen this new "thunder-bolt" fall while he was passing, and it was still blazing when the morning train passed. Of all things in the world, or out of it, a meteorite was the one thing wanted to compare with volcanic bombs and furnace sparks, and complete the chapter; and here, as it appeared, was an authentic hot aerolite blazing within a few miles. Of course, it must be got at any cost. The friendly guard made the ticket all right, and from his box we saw a tall flame, ten feet high at least, blazing in the field where the lightning had been seen to fall. It rose from a hole in the earth, about which fresh turf was scattered, and a great deal of water was flowing out of the ground. The writer is perfectly well aware that he will never "set the Thames on fire" himself, and he has little hope of seeing that feat accomplished. To raise such a blaze out of water did seem beyond the power even of a thunder-bolt; but water decomposed and recomposed makes the oxy-hydrogen blow-pipe and one of the strongest of fires. There was the flame—a fact to be accounted for somehow. "Three courses were open:" to rest content with the information and leave the facts unexplained; to leap to a conclusion and hire a lot of men to dig out the meteorite; or to go to the place and investigate. It seemed best to get out at the next station and walk back along the known road to the field; then to clamber through a gap which was seen in the hedge, and see what was to be seen at the spot. The point was reached at the cost of a wet walk of some miles and a few scratches. There was the blaze sure enough; a tall fact ten feet high, roaring; and at the base of it water was welling furiously out of a clay-pit, for all the world like a boiling spring in Iceland. A very simple experiment extinguished the aerolite theory: the water was quite cold to the touch. "A blower of coal-gas had been fired by the lightning." That was a jump, and a fall was the result: the steady school stayed at home; the middle course found out the truth. Leaving fire and water to fight their battle, the wet traveller went to the nearest house and asked an old woman when the lightning lit the gas. "Od, man," she said, "it wasna thunner ava; it was jeest ane of our lads that fired it wi' a match." The traveller told his fool's errand to the old dame, who sagely remarked—"It's jeest like the three craws;" and then he trudged on through rain and mire to the nearest furnace, which happened to be an old haunt in Lanarkshire. There he found what he set out to seek—sparks. There are two ways of viewing this story. Here is a great thing beside a little thing—a meteor and a match—and they may be contrasted or compared. Here is a big fallacy turned into a little fact, and a man mocking himself. But there is a moral in the tale for those who can see it. There was light at the end of this train, if it were but a feeble spark, and beyond the match was the will of the man who lit it. Between them is a great gulf which no man can leap; for no philosopher pretends to explain how a man's will moves his hand, or how that lad thought about lighting the coal-gas. Beyond them lies that "great ocean of truth" which the greatest of men have seen stretching out before them at the end of their earnest lives. Sparks of truth were worth all the trouble of the trip: "the play was worth the candle," though it was a burlesque.

Though this hunt failed, plenty of meteorites may be seen at the British Museum. A printed catalogue gives a list of 134 specimens of "aereolites," "meteorites consisting for the most part of various silicates interspersed with isolated particles of nickeliferous native iron, meteoric pyrites (troilite), &c.," which are exhibited in one case. Of "siderolites," "meteorites consisting of nickeliferous native iron in a more or less continuous or *sponge-like* state (with schreibersite, &c.), cavities in which are charged with silicates, &c.," nine specimens are exhibited. Of "aerosiderites," "masses of native iron generally nickeliferous and containing phosphides of nickel and iron (schreibersite), carbon, troilite, &c," 73 specimens are shown. These represent 216 meteoric falls, previous to August 1, 1863, when the list was printed by Professor Maskelyne of the mineral department, where all these may be seen. The heaviest specimen weighs 2800 lbs.

On the 14th of May 1864 a meteorite fell in France. Mathieu (de la Drôme) in his almanac for 1865 gives an account of the fall, and a paper on meteoric stones by Louis Figuier which gives a great deal of information in a small space. Chladni, Arago, Humboldt, Herschel, and many other eminent men, have described these visitants from the outer world, and in spite of learned slow coaches, who long refused to accept evidence, it is now admitted that from early historic times small planets and fragments of planets—bodies which moved in space in obedience to the laws which govern the movements of the earth, and other members of the solar system—have passed within reach of the earth's attraction, and have fallen as stones fell in 1864. The received opinion is that cold masses, attracted by the earth, are heated by friction while passing rapidly through the earth's atmosphere, and shine as fire-balls and shooting stars, which explode and fall as hot meteorites at last. The structure of many specimens implies that the whole of each mass was fused before it cooled, and froze, and crystallised, and oxydised, and broke.

Besides the collection at the British Museum, about 1100 specimens are preserved in museums in Europe, and the number is constantly increasing, because attention is directed to this curious subject. The "Bolide" of 1864 was seen at nearly the same hour from Paris to the Pyrenees, and M. Adolphe Brongniart, who happened to be near Gisors, saw the meteor pass from west to east at 15 to 20 degrees above the horizon, and disappear without noise. At Paris and at Gisors it was seen to the south. In the south it was seen, at eight in the evening, a globe of fire as big as the moon, followed by a train of luminous sparks ; it seemed larger as it approached the ground ; it was seen to burst and scatter a shower of sparks, leaving a small white cloud, which lasted for some minutes. At last, the inhabitants of a region between Nerác and Nohic d'Orgueil saw a fire-ball, which seemed larger than the moon, pass over their heads, revolving on its axis : it cast off sparks and jets of white vapour in every direction, and it burst like a shell at last, scattering shining fragments, which disappeared behind a cloud. An observer maintained that after the explosion of brilliant sparks he saw a dark red globe continue its course. After an interval of from five to two minutes, a loud noise was heard by those who saw the explosion. A shower of stones followed, and fell between the villages of Nohic, Orgueil, and Mont Béqui. They were hot : a peasant burned his fingers with one, the grass was singed by others. About twenty fragments were picked up, and they were covered with a black varnish : to produce a like glaze on a freshly-broken surface the stone had to be heated to a white heat. This meteorite contains about 5 per cent of carbon in the state of graphite, and many soluble salts. It was seen by so many observers that a map of its course was made, and its trajectory calculated by M. Lausedat, Professor of the

Ecole Polytechnique. Some of the crumbs which fell from this, the latest of meteorites, are shaped like bits chipped from the crusts of volcanic bombs. They are chambered and pierced with holes, and the solid breaks in two directions, like the upper layer of the lava-crust shown in the cut p. 429, vol. i. It is therefore possible to compare the structure of furnace sparks, volcanic bombs, and small planetary bodies, and upon these three degrees to plant a theory as to the structure of the earth's interior.

The great majority of meteorites are mere angular fragments.

One specimen at the British Museum is composed of three fragments, picked up separately, and at considerable distances from each other, but they fit and form a portion of a shell. In this they resemble fragments chipped off volcanic bombs. These broken bits of a crust are covered on all sides by a vitreous glaze, so in all probability they travelled far after the larger mass burst.

A great many have marks of fusion on the surface. Many are spongy.

One described by Pallas in 1778, at St. Petersburg, weighed about 700 kilogrammes; it had the form of a large bomb, a little flattened, and partly covered with a rude ochrous crust. The interior was made of soft iron full of holes, like a coarse sponge. These holes contain grains of olivine as large as peas. This seems to have fallen entire, and to have the structure of a volcanic bomb. It is like a furnace spark which has cooled without bursting.

In the Smithsonian Institution at Washington, the so-called "Ainsa" meteorite is preserved. It weighs 1400 pounds, and is meteoric iron, with specks of a grayish silicious mineral enclosed. It is now in the form of a great rude signet-ring, but it seems to be a portion of a hollow sphere. The hollow is irregular, and bulges out into concave recesses like those which commonly occur in iron sparks ; like those which are shown in sections of volcanic bombs. The outer surface is spoiled, and if ever a crust surrounded this iron core all traces of it have disappeared. This remarkable meteorite was found at Señora, in the Sierra Madre, in California, and it was used for many years as a public anvil. The greatest diameter is 41 inches. The woodcut in the title-page of this volume is from a rough pencil-sketch made at Washington in October 1864. In some respects the Ainsa meteorite is like the woodcut in the paper by Mr. Barkas above quoted.

A comparison of forms in hollow spheres of hot water; in sparks thrown off by hot silver, iron, slag, and other substances ; in "bombs" projected from terrestrial volcanoes, and in meteorites attracted from space; makes it probable that a flattened spheroid with a frozen crust, through which luminous fluids and hot vapours now escape in all directions, may now have a solid chambered spongy core, packed about bent rays, and about a centre of motion; made of materials which do not easily melt, and which freeze at high temperatures. According to astronomical calculations founded on the earth's movements, the average density of the whole mass is 5.67, water being 1. The specific gravity of iron is 7.7, but hollow iron ships float in water, like pumice-stones, and a spongy mass of any material might have any apparent density according to its structure and state of expansion. Chambers may be filled with the hot fluids and gases which radiate through holes in the frozen crust, and shine with terrestrial light when they follow the paths of rays and strive to escape. Jets of vapour and fountains of sparks so escaped from the fire-ball of 1864, and they so escape from shining furnace sparks.

CHAPTER LV.

TUBES AND SPRINGS.

MAN has been classed as the cooking animal, so most men have boiled something; and whoever has boiled anything must know something of the mechanical force of heat.

Hot solids melt, fluids become vapours, and all increase in bulk when they have room to expand. Softening and expansion begin near a source of heat, and spread; the heat spreads and radiates as light does from a luminous point; and matter moved by heat also spreads and radiates. At a given distance from a source of heat, expansion and outward movement in any material come to an end, and there contraction begins or movement stops. Particles attract each other unless they are kept apart. If sources of light are also sources of heat, they are centres from which a mechanical force radiates, and all light appears to be associated with heat, though the amount may sometimes be too small for measurement.

When water in a kettle is sufficiently heated steam-bubbles form near the fire. While the upper layers of water are cold these collapse suddenly to grow again; the water simmers, and the kettle is shaken. When the upper layers are warmed the steam floats up, the bubble expands as it rises; and at last it lifts up the surface of the water, bursts through it, and expands more freely in air when relieved from pressure. In thus bursting a dome of water, steam drives drops of water before it, and these projectiles describe curved paths while they rise and fall. They are scattered by radiation, and attracted by gravitation. The amount of force applied, and its direction, determine the distances traversed and the curves described by these projectiles. The bursting water-dome starts a whole system of waves, which radiate and spread horizontally. The steam-bubble transfers its charge of heat and force to the air about it, and it starts a movement which spreads horizontally and vertically, as sound spreads in the air. The water particles, which heat separated and drove upwards, attract each other when the heat has passed on ; the steam condenses, and drops, attracted by the earth, fall down.

The particles of air, which repelled each other and rose when heated by steam, attract each other and fall when the heat has passed on to the next shell of air. And so movement spreads, and circulation goes on about a source of heat and light. Who is to limit the movement which begins at a fire under a kettle ?

Whatever the source of mechanical power may be, like radiating and converging movements must result from radiating and converging forces. A spirit-lamp, a fire, a furnace, the earth's heat, and the light of the sun, all cause like radiating movements when used in the same way.

Water in a transparent glass vessel above a lamp circulates like water boiling in a kettle on the fire. Water boiling in a tray full of sand moves on the same principle as water boiling about iron and slag, or about hot lava, or like water in a spring heated by the earth. The sun's rays, collected with a burning-glass and thrown upon metal under water, cause the movements which would result if the metal were heated as much in any other way.

Whatever the substance may be, radiating and converging

forces, of sufficient "energy," produce like movements. Porridge in a pan, glass in a retort, fluid metals and stones at furnaces, mud in boiling springs, lava-floods on wet ground, lava-springs which are volcanoes, all move on one principle; and some retain forms which register the movements which resulted from the forces applied. The heat of a lava-drop spinning in air acts on its surface, and the outside gives a clue to the internal structure of the stone: the heat of the earth acts on its surface, and the forms which result may give a clue to the earth's structure.

If all sources of heat and all materials be alike in these respects, then small experiments help to explain the forms which result from the action of the earth's heat. Materials which melt and freeze at low temperatures, will serve as well for illustration and study as those which only melt at furnace heat.

Oil, water, and mercury, in a glass vessel, make a series of three fluid layers, which are portions of concentric shells, and are at rest at ordinary temperatures. If the lowest layer is heated the whole series is disturbed. If cooled so that one freezes the shapes alter. If water freezes above mercury, in a closed vessel, the fluid metal beneath the solid ice is forced into irregular angular shapes, and globules are squeezed up into the hard crust, where they take the forms of air-bubbles compressed in ice. In like manner water and oil in the same bottle are disturbed by every change of temperature which freezes the one or boils the other. Water and air at 32° react upon each other, as iron and air do at 3000°. In both cases the gas imprisoned within a solid shapes a chamber whose form records the direction in which forces acted. It is easy to tell which side of a plate of ice or cast-iron was uppermost if there be an air-bubble in it. By this rule applied to a bit of lava it is easy to tell which side was uppermost, and in which direction a stream flowed when it froze.

The impression, p. 423, is from a vertical section made through an upper layer of lava, which was flowing from A to B when it set. It was part of a lava surface near Reykjavik. The ridges are sections of great coils which formed about the centre, from which a little spring of lava boiled out, and froze as it spread. The movement was like that of boiling water, but in this case the boiling fluid curdled and froze on the surface, and the horizontal waves remain.

At p. 400 is another impression made from a section cut down through the middle of a set of loops on the surface of a frozen rill of slag. It boiled up through a hole in a freezing crust; and streams spread as boiling water spreads above a centre of ebullition. Each rill flowed fastest in the centre, and froze first at distant points and at the sides, and the flow is marked by curved loops like string. In these two cases materials and dimensions differ, but the forms are alike though produced by terrestrial and furnace heat. Solar heat properly applied produces the same forms on sealing-wax or asphalt. Slag can always be seen flowing and freezing, sealing-wax can be melted at home; and forms on these explain lava-forms, and like forms of any dimensions anywhere.

Solder and sealing-wax, like boiling lava, take a shape and retain it; and these and other materials, which are easily managed, serve their purpose as well as iron. Plaster-of-Paris sinks in cold water, becomes a plastic mud, and then sets hard; it is moved by streams and by currents in water, and when it sets it retains the shape which it took while moving. Water and silt, plaster, sand, or clay, in small quantities, illustrate the action of hot or cold water on larger quantities of like materials ; and so models illustrate natural phenomena.

The Geysers may be compared with a geological toy ; and forms which result from the earth's heat may be explained by forms which result from the heat of a lamp applied as mechanical force.

A working model of a hot spring is very easily made. Some flat broken plates of slag, and a pile of sand and fine dry earth, laid upon an iron tray, may represent the country about the Geysers, which consists of shattered strata of lava, volcanic sands, and loose soil. A pile of broken ice and snow laid on the heap is placed like glaciers, which crown high mountains in the region ; and a gas lamp under the tray acts the part of the earth's heat, which boils water beneath the surface in Iceland. So far this model imitates a natural arrangement of a bit of the earth's crust, situated between regions where the upper temperature is less than 32°, and the temperature under ground is more than 212°, the freezing and boiling points of water. It is a region of Frost and Fire. Soon after the lamp is lighted, the pile of ice begins to melt and slide upon the sand and stones, as glaciers do on sloping hills. A heap of iron tossed into a furnace melts and slides for the same reason at a higher temperature ; and ice and iron flow when they are fluid. The water flows and sinks through loose sand, and through cracks and holes in the plates of slag ; and so it finds a way to the lowest depth of the iron vessel. Iron finds its way through lighter cinders to the bottom of a furnace ; it sinks through slag as water sinks in oil ; and all fluids of different specific gravities which do not mix find their respective levels and take their places in a series, like oil, water, and mercury in a glass. In the model, only one solid is melted, and a wet pile of sand and stones remains in a pool of water, supported by an iron tray, which a lamp heats but cannot melt. So far the heat of fusion enables gravitation to move ice more speedily from a higher to a lower region. The melting snows of Iceland form large rivers which reach the sea; but great part of the water sinks down through sands and shattered lavas. The water which sinks where it falls finally reaches some region where water boils, some lava-crust which stops it, as a hot iron tray keeps water from sinking deeper. A column of water, sand, and lava, with a base near the region whence lava-springs rise, must be intensely heated, so as to exert a powerful mechanical force, which radiates from the earth's centre upwards. At one end of this series "perpetual snows" crown the hills; at the other is steam; and between these two, water circulates as it does in a tray full of sand, or in a kettle. When water is boiled in sand, steam forms below within six inches of unmelted ice upon the surface, and water boils furiously within a few inches of water which is scarcely warmed. Shallow water cannot be much heated so long as ice floats in it; but sand and stones impede the movements of water, and steam, and heat. It follows that the temperature of a hot spring is no measure of its temperature deep under ground.

But though these movements are retarded, they are still the same in kind as the movements of water boiling in a Florence flask. There is circulation; currents sink and rise, though snow and ice are at one end and fluid lava at the other.

Because hot springs are found in most regions of the earth, great underground heat is not peculiar to Iceland or to any district. There is a great store of heat and force within the earth's crust, ready to act wherever a weak point is found. Currents in water move solids. Sand retards circulation in hot water, but is equally urged by the force which it resists.

When the force accumulates, sand is driven by boiling water, and steam builds it up into heaps and scatters it in the air. A heat insufficient to fuse solid sand melts solid ice and turns it to steam, and so it projects the sand like shot from a steam-gun. When water is rapidly heated in a narrow tube, steam forms so as to scatter a column of water like a charge of shot. When water is heated in a kettle with the lid on, steam formed below rises to the top, and there expands till it either drives the water out of the spout or lifts the lid. The mechanics of the Geyser have been explained by these two modes of action. According to one theory, the base of a column of water becomes so hot that it flashes into steam, and blows out the charge above it. The other explanation supposes a steam chamber communicating with the base of the pipe, so as to force water out of the spout of this giant kettle when the steam gets up. Both theories may be correct.

In models the latter action commonly results. The melted ice becomes steam under the slag roof, and forces water out, while cold water is pressed in by weight. The water is repelled by heat and attracted by gravitation, and so an alternating outward and inward sidelong movement results, because the slag roof of the steam chamber prevents the steam from escaping upwards. When a bubble of steam escapes it carries off a charge of heat and force, and water enters the chamber; when the water is heated sufficiently steam drives out the water and forces it through sand and chinks in the slag; and so, after a short time, jets and fountains of hot water, steam, and sand, burst through the cold wet surface where ice remains; and these, after playing for a moment, stop suddenly when the steam has blown off, and the boiler is re-filled. This is a result of heat-force, for the height of the jet is decreased by decreasing the quantity of gas burned, and the action stops entirely soon after the gas is turned off. Another result is the packing and sorting of sand. The boiling water sorts coarse and fine, heavy and light materials, and packs them in stratified beds ; it drives water fountains through beds of sand, makes hollows beneath the surface, and it piles mounds of definite shape upon the top of the heap. In nature, as in this model, water is dragged down by weight and driven up by heat ; cold makes it a solid in one region, heat makes steam of it in another ; it moves from the earth towards the sky, and from the sky back to earth, as it is heated by the earth's radiation, or cools by radiation into space. Vapour in air becomes a cloud, and a snow shower, melts and sinks, turns to steam and rises again ; and so a cloud becomes a glacier and a geyser in Iceland, because the world is hot, and space about it cold ; and the action is the same in a tray full of sand and stones heated by a gas lamp.

The action of a boiling spring may thus be imitated ; but something more is wanting to complete a model. When a jet of water has forced a way through sand, the loose sand falls back, and the passage fills. It is so in the model. Near the foot of Krabla are several large, deep, funnel-shaped hollows in loose volcanic debris. These sandy craters are partly filled with hot sulphurous green water ; but every shower and breeze of wind disturbs the sand, and the holes through which water rises are filling rapidly from above. In sandy bays, where burrowing shells flourish, a certain so-called "spout-fish" thrusts his long neck through sand when the tide flows. His mouth is level with the surface, but his body and shell are far down. When the tide ebbs and danger approaches, the shell-fish retires, and in shrinking, spouts water and sand at the foe. He leaves a small crater, but the next wave fills it, and so all trace of the spout-fish is lost.

Like this creature, a boiling spring would leave no trace if it only spouted through holes which filled as fast as they were made. There may have been springs boiling in ancient sands, of which no trace remains in sedimentary rocks.

Many of the hot springs in Iceland deposit solids when the water cools, and these form permanent tubes and craters, which could be recognised anywhere. Some are deep, still, hot wells; some are always surging about; some are great fountains spouting at short intervals; some explode occasionally;—and all these have craters and tubes of definite forms, which result from movements in the water. These forms are no accidents, for they can be copied in models, and they recur at different places in Iceland. When the tide flows over the sand below Granville in France, thousands of sea-worms emerge from holes, and their long bodies and active feelers stretch and wave in search of food. When the tide ebbs, these creatures shrink back; but loose sand sticks to their slimy bodies, and in shrinking each adds a ring of sand to the tube in which he hides. As multitudes live together, a mound of sand, pierced like a sponge, forms at last. Like these, hot springs add to their tubes by every movement; and the form of the tube results from movements in the boiling water.

Geyser Tubes.—Of all these tubes, the best known and the easiest to get at are the Geysers. They are only seven days' journey from Leith, and situated near the base of a volcanic hill somewhat smaller than Arthur's Seat; a cone of lava is at the top of it; sand and cinders are on the sides. To the east is a wide, flat, wet valley, beyond which, some ten or fifteen miles away, is a low range of hills; and behind these the top of Hecla may be seen in clear weather. At the head of the valley, far away to the north, are dark, bare, high peaks, amongst which are enormous fields of snow and ice. To the west, behind the volcanic hill, at a distance of about a mile from the springs, a range of high ground begins, which extends a day's journey to Thingvalla, and includes a number of high rocky volcanic peaks, and great lava-floods ; and Skjaldbreið, the great centre from which these flowed, is to the northwest (see p. 409).

To the south-west the wide valley opens out into a great boggy plain, which reaches to the sea. It is covered with grass and marsh-plants, traversed by large rivers flowing nearly south-west ; large lakes are in it ; and every here and there rocky hills spring up in the moor like distant blue islands in a firth. The whole country rests upon heated strata ; for in a calm evening the white steam of hot springs may be seen blowing off at intervals in the marshy plain. To the east Hecla is still hot, and beyond it lies Skaptar Jökull ; and hot springs are in that direction. Many are in the plains to the south ; one is half-way to Thingvalla ; a little geyser is near Reykjavik ; a spring is near the town itself ; and further west are many more hot springs. The whole country is volcanic, even to the Westman Islands, far out at sea ; and even under the sea volcanic eruptions occasionally break out. Streams of lava have flowed over beds of loose materials, and now roof in and confine hot water beneath the surface ; and so steam is forced to escape through vents, rifts, holes, and cracks, like those which pervade the upper lava-beds. To the north also is sufficient evidence of extinct volcanic action : the land is high and snow-clad, and cold reigns there now ; but beyond the mountains are many more hot springs.

All these have one thing in common :—they are all in low grounds near the base of volcanic hills, midway between cold and heat, ice and steam ; where the water which flows from the jokulls, through ashes and porous strata, shivered lava and volcanic caverns, stands nearly level with the surface of the flat marshy ground. Heat is below to boil it, a tough lava to keep it from sinking deeper; a region of heat, sufficient to keep the great kettle boiling, is below that; and a great lid of mountains is piled over the steam-boiler.

There is then every reason to expect that steam should escape where the weight is least, and that springs should burst out at the foot of the hills.

The tubes have still to be explained.

Above the great spouting Geyser, distant from it about 100 yards, and on the top of a steep bank of loose sand and ashes, are several still quiet pools of water which are a few yards wide, and which look as if they were puddles of rain collected in hollows at various elevations. An active man might leap over them; and the wonder is how water can rest at all on such porous ground. These are, in fact, springs hot enough to boil food, and their depth is unknown. The water is beautifully clear and green, and the sides of the well are seen through it, darkening as they descend, till they are lost in a black hole fathoms down. In August 1861, an emerald green tongue was anchored by a string in one of these wells, quietly boiling for dinner; while a kettle of soup, with a big stone on the lid, was simmering up to its ears in hot water on a natural bridge of stone which spans the pool. Far away down on a sloping shelf reposed an old copper coffee-kettle, which some former traveller had dropped in, and the boiling water was slowly welling up in the middle, rising every now and then, a smooth greasy mound, like the swirl which a salmon makes when he rises at a fly and wags his broad tail in derision at the cheat. A small steaming rill, the waste of this well, and the measure of its supply, trickled steadily down the bank, depositing stone on the ashes. As the coffee-kettle had been on its shelf long enough to gather a crust, it is clear that this spring, though boiling, boils quietly. It is of great depth, and such a column of water would burst through the loose ashes of which the ground about the spring consists. Two such columns could not exist within a few feet of each other at different elevations, in mere tubes formed in porous soil. But the columns do so exist, side by side, in these natural wells. They are enclosed within rough stone tubes, hardly pervious to water; and the question is, how these rugged irregular stone tubes came to be formed at first.

If the question is answered for one tube, the formation of similar tubes, wherever found, may be referred to the same agency; and similar tubes are to be found in all stages of construction in many parts of the world, and more especially in Iceland.

Rough Stone Tubes.—On the ridge above Thingvalla, to the eastward of that valley and close to the track, at about half a day's journey from the "kitchen," on a hill-side, and below a considerable mountain, in a country whose surface is wholly composed of bare cinders and lava, there stands a rock which rises some eight or ten feet above the loose rubbish. It might be carelessly passed as a clinker which had rolled down the mountain, and a little way up the opposite slope. It is in fact the protruding end of a rough stone tube of great but unknown depth, and it is very like the tube of the kitchen. It contains no water, and apparently never has, for it is too porous to hold it. So far as the chamber can be seen it seems to be a large conical hall of rough black lava, covered by a small conical roof, with a hole in the side through which a man could creep. All round are scattered traces of great heat. It is evident that this tube was made of melted stones, and that the force which modelled it cast stones out of it, for there they lie scattered all about it as fresh as if they had fallen the day before. It is probable that this is a chimney, which is or once was connected with a subterranean chamber.

Within a mile or two of this tube a roof of lava has fallen into a cavern, over which the track leads. It is a large hollow blown in the lava, but no one has explored it. About seven or eight miles away the plain of Thingvalla has sunk down over an area of more than a hundred square miles, leaving broken edges to mark the original level of the roof (vol. i. p. 93). If the lava could be raised up again, and the rifts mended, there would be a chamber in the valley some hundreds of feet high beneath a roof some hundreds of feet thick (vol. i. p. 90); and if such a lava-boiler were filled with the lake and boiled, the steam-power would be sufficient to account for many of the phenomena in the district. In particular, steam might well blow vertical tubes in soft lava, and so shape Tintron, with its roof of clinkers and its spreading lava-waves.

A couple of days' journey to the north is Surtshellr. It is the best known of Icelandic caverns; but every lava-flood in the island seems to be honeycombed with great caves. At p. 426, vol. 1, is a map which shows the position of Surtshellr; and the nearest iron-foundry will show how such horizontal caverns are formed. The large one extends along the lava-stream, and is at the edge of a slight fall in the ground. At page 429 the edge of a broken roof is shown in the foreground, and here the case of Thingvalla is repeated on a small scale. The roof having sunk, small cliffs surround a hollow. The entrance to the cavern is to the right, and there the roof, though much shattered, has not fallen. The cavern has been explored for about a mile; the roof has fallen in several places, and the cave is partially filled with snow and ice. At furnaces, slag commonly runs in a trench scraped in ashes. As it flows it freezes; first at some considerable distance from the outlet. A bridge of stone spans the stream, and then the tough surface gathers behind the bridge, and forms a series of wrinkled loops, which look like coils of string. This upper crust grows up stream, while an under crust forms below; the hot slag flows on between them, and if the supply is stopped, the fluid interior of this tube flows away till it cools and stops. When

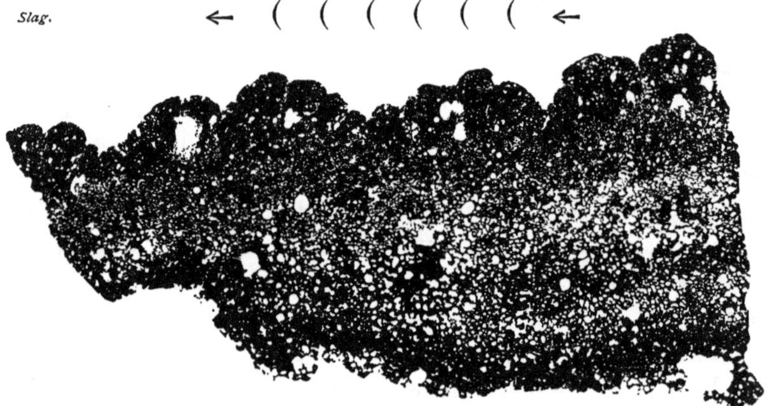

Fig. 105.—Vertical Section through a Frozen Stream of Wrinkled Slag. Printed from the Stone.

Folds on the surface are like heavy drapery. The stream moved to the left, and folds gathered up stream towards the right. The fluid froze at the surface, and crusts and folds which formed under and behind each other can easily be traced from the structure of this crust (see pp. 390 and 423).

this happens, the lower end of the tube is filled with the same material which makes the sides and roof. The workmen break up this slag stream to let the fluid escape more freely, and hundreds of broken pipes about the size of drain-tiles may be found about any ironwork. It often happens that a tube of this kind splits along the roof while cooling, and then a whole series of loops of slag—))))))—are torn through the middle. The roof of Surtshellr is covered with similar loops and coils, which show the directions of the flow. They are thick as cables, but exactly like coils on slag (see p. 423). In many places these wrinkles are torn through, and the whole roof is shattered. In a section the uppermost layer is prismatic; layers below are stratified horizontally; the under surface, where it is preserved entire, is hung with pendants of spongy lava, with a vitreous crust. These froze while dripping from the newly-frozen roof. The growth of this horizontal chamber is fully explained by its structure, and every lava-stream is full of such hollows.

Myvatn.—Vertical chambers also abound at Myvatn: many project through the hill-sides near the lake, and have the most fantastic shapes. They suggest ruined castles, turrets, and such-like edifices, but they were all built by volcanic heat. In this region the lavas are disposed in beds, which have been much broken, and cold water now flows in hollow chambers beneath lava roofs.

Similar tubes may be seen at an earlier stage of growth.

Vesuvius.—In 1842 there was a tube at the bottom of the crater of Vesuvius; smoke and air and sulphurous vapours were then rising from it, as from a chimney, with a loud rushing sound like the noise of a great wind. Far away down in the earth a dim redness was seen glowing through the smoke: it was earth-light seen through the dark crust. Heat was converting some material into vapour, in some underground chamber, and the expanding vapour had burst through the earth, and made a tube by plastering melted stone upon the sides. The same force had cast out some of the spare materials; for half melted and even burning sulphur, scorched cinders, and bits of lava and pumice, were scattered about the great hollow basin which surrounded this hot tube. The mouth of the hole itself was about the size of a coal-pit; and for size, shape, and material, it was extremely like the empty pipes described above, and the hot wells above the Geysers.

It was choking work to get down to the bottom of the crater of Vesuvius then: shoes were charred, sulphurous fumes were swallowed, in passing over beds which were visibly burning; eggs were baked for luncheon, and sticks were burned in red cracks in the lava. No man could have approached the spot where the giant Fire was at work in his tube, like a great sea-worm in a sand-bed.

In 1844 a small cone and crater had grown about this pit, and through it more red-hot stones, and fountains of dust and vapour, were thrown, as fountains of steam and drops are thrown by bubbles of steam from boiling water. The solids either fell within the hollow, and rolled down, to be again blown out, or they fell outside, and rolled to the side of the old crater. A small "cone of eruption" was growing in the crater of the older cone of eruption, which stands in a still older broken cup which, as it is now believed, grew under the sea.

A few years later the work could be safely watched from the upper edge of the crater, and it was thus described:—

The place where the mouth of a stone tube had been in 1842, the bottom of the crater, was filled by a pool of seething lava, and a small lava river was slowly oozing through the side of the cone about the level of the pool in the crater. The stream flowed down outside, and froze as it flowed, as water flows from a spring and freezes in winter. But every now and then the red-hot viscid pool, which was doing its best to freeze in the basin, got a fresh supply of heat from below. It grew white-hot, and then the whole crater seemed to fill with a purple haze, and then the surface burst, and a fountain of hot vapours rushed up into the air through the hole, carrying with it a thin flake-like stony material, which fell in showers within and far beyond the edge of the crater. Lava was then bubbling, and simmering, and boiling over in the ashes; heat was blowing a new tube amongst the cinders, making great stone bubbles and breaking them, and scattering the fragments far and wide; and as the finished tube resembled the Icelandic tubes, it is probable that the tubes at the Geyser were first made like the tube in Vesuvius.

In 1857 lava had risen in the crater of Vesuvius to the level of the edge, and had formed a plain. On this two small cones had risen; they were hollow, and through them hot vapours escaped; they were like Tintron with its extinguisher roof. Later a fresh crop of hollow cones grew up; and then the plain, with its miniature cones and craters, was burst up and destroyed.

Tubes radiating from the earth's centre are commonly formed by the escape of hot vapours through viscous hot lavas, and mounds of definite shapes grow about these open tubes, from overflows of lava and fountains of projectiles which rise through the tubes.

Filled Tubes.—All these are as it were living specimens of a common species; their habits can be studied and their growth watched, though they are dangerous neighbours.

Extinct varieties of the same tribe—fossil pipes and chimneys, springs and chambers—also abound; and they are as easily known as a fossil bone when the others have been seen.

In a quarry near the Drachenfells, on the Rhine, near the top of a conical hill, such a tube was visible in 1846.

It was made of stone of one kind, and filled with stone of a different colour. It was in the condition of the tube in Vesuvius when it had filled with a new overflow of lava, and such strings are very common in igneous rocks of all ages. They exist in granite, as well as in lava, and tell their story of past action by their form, as clearly as fossil bones tell of extinct life.

Small Natural Tubes.—To understand fossils it is necessary to study living animals, and active volcanoes are not always within reach. To understand the formation of tubes by heat the action must be watched; and there is a very lively, harmless young specimen, whose operations can be watched, close to the Geysers. A little mud spring is in a hollow to the north of the Great Geyser; it is almost hidden amongst the ashes, and about as big round as a stew-pan: in it the formation of tubes by hot vapour is going on. The spring was betrayed by a ploutering, poppling sound, which, to a hungry Scot with the brevet rank of cook, was absurdly suggestive of boiling porridge. A vision of a nursery and a rosy maid, a stew-pan and a fire, rose up as if by magic amongst the cinders; but there is no porridge to be had in that benighted land. A deaf French traveller, who was supposed to be dumb, was startled into speech, and exclaimed, "Chocolate!" The spring was full of half liquid boiling tough clay, through which steam and other hot vapours escaped; and as the vapours burst through the surface and rose, the mud flowed back and filled up the holes as fast as they were made. This small tube-making engine was like Vesuvius when the lava was soft in the crater and vapours were escaping through it. If the material gets tougher the soft tubes will be finished, and the poppling will cease, as it had ceased in Vesuvius in 1842, when the lava was hard though hot, and vapours were escaping freely through a rough tube. In course of time the mud may be baked into stone, and the tubes will then resemble larger tubes in the same neighbourhood. They may become vents for hot vapours, or for hot water, or lastly they may be filled up with some other material and become strings like those which abound in all parts of the earth's crust. The little natural engine is making tubes of the same pattern as those which are made by larger engines moved by the same force. By watching it the whole process may be learned, as the action of a large steam-engine is learned from a model.

Experiment.—If a small spring thus tells the story of a big one, the growth may be studied at home. Any material which will melt and take a new form, and retain it, will answer the purpose. About a pound of common red sealing-wax was melted at a slow heat in a tin vessel four or five inches deep, and the mass was allowed to cool. Cold water was poured in till the mound of sealing-wax was covered all but the top. A gas lamp was then placed under the vessel, and a slow heat applied. The cold water in contact with the sealing-wax kept the surface tough, while the lamp melted it below, and in a few moments the wax began to boil on the dry spot. It not only boiled, but overflowed because of the downward pressure of the water, and the upward force of its own expanding vapour. But as it boiled over, each successive overflow cooled and hardened when it met the water; and so a wall of hard wax grew about a pool of boiling wax. To make the wall grow higher more water was slowly added, and the circle rose and kept pace with the rising water. The pressure on the surface of the wax increased as the water deepened, and the lamp kept the wax boiling in the tube as it rose. Downward pressure outside forced up the fluid, and expansion within drove it higher; so the wall grew to be a hard tube containing the same materials in a fluid state. It was like the Vesuvius lava-tube during an overflow of hot lava. If this process had been continued to a certain point, the heat would have ceased to act, and the tube would have cooled into a solid pillar; but the form to be produced in this experiment was an open tube, so the lamp was extinguished when the wall had risen about three inches above the mound of wax.

Gravitation and cold came into play; the tough surface of the wax hardened and became a roof which resisted the pressure of the cold water; the vapours inside condensed, and the hot wax diminished in volume, so as to leave hollows beneath the crust; the atmosphere pressing upon the fluid in the hard tube forced it back into the hollows whence it came, and the hot wax sank in with a rushing sound. Presently some crack opened in the cooling roof of the chamber, and water flowed in and rose up, filled the tube, and replaced the melted wax. The wax tube had become a water spring.

The outer surface of the tube so made was wrinkled, each fold corresponding to an overflow of wax and a rise in the water. The inner surface was smooth where the air plastered it against the hard sides. The opening was wider above than below, and of irregular dimensions; but generally a horizontal section was an oval or some rounded figure, while a vertical section showed chambers and pipes winding about under the surface of the wax. This experiment explains the making of larger tubes, and gives some notion of the invisible mechanism of the great Icelandic fountains. The model tube was joined to a chamber, and so are the geyser tubes.

Experiment 2.—Plaster-of-Paris will take a form while plastic, and retain it when it sets; it is easily moved by water, and serves well to illustrate the working of mud-springs and the formation of tubes and cones in lava. A shallow tin tray was filled with dry plaster and heated over a lamp ; an equal bulk of cold water was then poured in, and it boiled when it reached the tray. The plaster set quickly ; but, before it hardened, steam had blown a large chamber, and pierced two holes in the roof. This contrivance, when set to work, imitated the action of intermitting hot springs : water poured over the plaster sank and filled the chamber ; when it was heated, steam drove water spouting out of the holes which steam first made. The action was like that of a kettle boiling with the lid on, and with water above the level of the entrance to the spout.

By sprinkling dry plaster over the surface while water was boiling out through these two holes, two craters were made which differed materially in form. One was like the Strokr, a deep conical pit ; the other like the basin of the Great Geyser, a shallow bowl. In one, the water was always far hotter than it was in the other. On breaking up the model the reason was found. The roof of the chamber was so formed that steam escaped towards one aperture, when a certain amount of pressure was overcome. It only escaped in the other direction after the water had been forced out, so as to dry a lower arch, and so open a passage into the second tube. As most of the steam went one way, one spring boiled furiously when the other was hardly warmed, though both opened into the same boiler. The shape of the basin formed about the tube resulted from the movements of the water. The hottest radiated most directly from the source of heat, and so made the steepest walls.

It would be tedious to describe all the plans tried and all the models made.

Sealing-wax heated under dry sand boiled up, and made

tubes with cones and craters, from which eruptions of sealing-wax flowed like lava. When water was poured on, the tubes became miniature hot springs. When the model was cooled, the same holes and ducts were cold springs when water poured on higher points had sunk in. When a mound of any material rose high enough it was sealed by cold, and then fresh vents opened near the base of the mound where resistance was least. At the top of the volcanic hill near the Geysers is a sealed tube, and probably the hot fountains play through vents which opened below, when the hill was made, and the power greatly spent.

Similar Forms.—The same thing probably happened wherever there is a hot spring under a hill, and wherever there is an open tube or a circular lake, near the base of a conical hill whose top is of igneous origin.

The same power, though decreasing, would continue to drive mud or water through tubes till the rocks underneath cooled. Duddingston Loch below Arthur's Seat, and the spring in it; two round lakes below Benknock, in Islay; round lakes at the foot of the Jura mountains, and similar forms elsewhere, may all be traces of the same decreasing igneous action which raised up hills. Even cold springs flowing through underground channels may be relics of the same force.

Tubes can be made by pouring wet plaster into a hot tray. Steam drives the plaster away, and it grows up a hollow chambered mound with tubes and basins, each a miniature hot spring. The movements and the forms which result are like those which resulted from the freezing of silver.

The same forms are produced by shaking dry plaster into boiling water, as meal is shaken in to make porridge: the plaster is moved by currents, and takes a cast of the ray-force which moves them. Potters' clay, paste, porridge, asphalt, glass, slag, iron, lava, or any other material through which vapours can force their way, will take these casts; and the form is a record of the force of heat radiating from the earth outwards. The highest mountains in the world contain tubes; they pierce the crust in all regions, and they can be made at will experimentally, by setting radiation and gravitation to work upon fusible solids, and vapours which can be frozen.

In all these examples—in furnace-sparks and refuse, in volcanic bombs and lavas, and in terrestrial volcanoes—radiating tubular forms result from radiating movements caused by force radiating from sources of heat and light.

FIG. 106. THE GEYSERS FROM THE HORSE-TRACK.

The hill in the middle is volcanic. From it the top of Hecla may be seen to the right. To the left, behind the hills, and out of sight, is Skjaldbreid. Glaciers are amongst the hills in the background. (See pp. 396, 413, and 432.)

CHAPTER LVI.

SPRINGS, CHAMBERS, TUBES, CRATERS, AND CONES.

CHAMBERS in a crust often communicate with the outside by tubes; but these are often partially or wholly filled with vapours, fluids, and solids, which escape from the interior of a cooling mass. Sections of volcanic bombs (p. 379) show this structure; the growth of it may be watched in models; and hot springs in Iceland give samples of this work in all stages.

These tubes differ from rough stone tubes near them; they are smaller, less porous, of regular shapes, and lined with materials deposited by water. Some are partially filled, others are choked up.

It has been shown that the Great Geyser and springs about it probably communicate with the interior through tubes blown in lava near the base of a small volcano. The cut, p. 409, shows the position of these springs at the foot of a hill. The Great Geyser now spouts through a smooth vertical shaft, which is chiefly made of silica deposited by the water. The mouth of this steam-gun spreads a little near the top, somewhat like a "bell-mouthed blunderbuss;" and about this muzzle is a shallow saucer. The woodcut, vol. i. p. 12, is from a drawing made in the saucer after an eruption. Beyond the rim of the "crater" a conical mound spreads and slopes every way at a small angle. The woodcut, p. 414, is from a drawing made at the base of the mound during the eruption which emptied the crater, but did not empty the pipe. The dimensions ascertained by measuring with a salmon-line and a fishing-rod are:—breadth of basin, when filled, 57 feet at the widest place; breadth of pipe, about 20 feet, but somewhat less where the walls are vertical; depth from the surface of the water in the centre when the crater is full, 75 feet; ledge upon which a plummet rests on one side, 45 feet. The diagram, p. 415, is drawn to scale from these measurements. The Strokr or Churn is a conical pit, 36 feet deep. At about 22½ feet is a hole in one of the sides; at 19 feet is a hole on the opposite side. Water generally fills the pit to within 6 feet of the top; but after an eruption both side vents are occasionally seen. The mouth of the pit is surrounded by a raised wall of silicious stone (see title-page, vol. i.), in a shallow saucer much broken, because it is usually dry and exposed to frosts and the feet of men and cattle. At the mouth the pipe is 8 feet wide; it is less than a foot wide 30 feet down. A third pipe spouts occasionally; the mouth is about the size of a hat, and the hole seems to expand as it descends. Besides these three, many other smooth pits and pipes, of various shapes, contain boiling water and mud of various colours; and these, within an area of a couple of acres, are near about the same level. Higher up on the hill-side are springs which do not boil and spout now; and still higher, old tubes are covered or filled, and their sites are marked by petrified grasses and twigs and ripple-marked stones, like those which surround the Geyser. All these forms result from movements in the water, and these from the earth's heat.

The Great Geyser is generally full up to the brim, and movements at the surface suggest two forces nearly balanced: these are weight and heat. From time to time the water rises a few inches, overflows a little, and sinks quietly down, to rise again after a pause. It is like mercury in a barometer when gusts pass. Atmospheric and steam pressure may regulate these slow movements, and the eruptions. Every day, sometimes every hour, the kettle simmers. Bubbles of steam either form in the tube or escape into it somewhere near the bottom, and these condense suddenly in colder water. The sound is like that of a blast in a mine—a quick, loud report, which shakes the ground to a great distance. When fires are lighted in a steamboat, the noise of simmering is very like this natural artillery : vibration passes through boiler and ship to water and air about it, and waves spread horizontally from the sides of the ship. The sound is commonly heard in houses warmed with hot-water pipes ; and walls are shaken when bubbles of steam collapse in boilers. Steam may be watched in a hot spring at Reykholt. There the water is very clear, and about three feet deep in the basin ; bubbles, large as cricket-balls, rise at intervals out of a hole ; and above this vertical tube a dome of water rises on the plane surface. From it water spreads in radiating streams. The pool is shaken when bubbles collapse ; when they reach the surface a dome bursts, and a fountain of drops and steam spreads and scatters in the air. In larger springs the bubbles cannot be seen, but they can be heard. They do not always reach the surface, but they start an upward current, which makes a dome and flow in the circular pool which fills the crater. This movement follows the well-known sound of collapsing steam simmering on a large scale. The radiating flow makes beautiful curved patterns of streams, eddies, whirlpools, and waves, which are reflected from the sides of the basin. The brink is wetted by every rise, and dries after every fall ; and after each change vapour leaves the solid which hot water had dissolved. The edge of the crater and the outside of the cone grow continually, while currents shape the tube and basin by rising and falling, by spreading and converging. As in a model, the shape of the tube is a cast of the currents which move in it.

Of all unpunctual exhibitions the Geysers are the most provoking. In 1861 the grand fountains went off as a party of travellers came in sight of the place (p. 409); they saw white clouds of steam three miles away, and that was all they saw. The tent was pitched and a watch kept; but the watchers fell asleep, and it is said that the Great Geyser exploded without rousing the tired sleepers. Every few hours came the warning—thud, thud, thud—which kept expectation on the stretch; but nothing came of it all next day and all next night. One man was packed up in a bag of mackintosh cloth, and laid out with his face to the spring, to make sure of one sentry; but he saw nothing. He looked very picturesque, somewhat like a mummy extracted from its wooden case. All next morning the water rose and fell, and sank and rose again, balancing. Tired of waiting, the party set off at last, and met a fresh party going to the place. They arrived in the nick of time, saw an eruption, and returned next day. In 1862 the disappointed returned. One party, who had very little time to spare, rode in hot haste to Haukadal, and saw many eruptions in a few hours. Those who followed more leisurely waited for three days; but this time they did see the show. It was a grand display, and well worth all the waiting. Instead of ending suddenly or gradually, the steam-salute shot faster and faster; thuds followed each other rapidly, and the whole ground shook; then the sound of dashing water, the music of waves, was added to the turmoil. A great dome rose in the

Fig. 107. The Great Geyser boiling over.

middle of the pool, and frequent waves dashed over the edge of the basin, while streams overflowed and drenched the whole mound. Great clouds of rolling steam burst out of the water domes, and rose in the still air, swelling like white cumulus clouds against a hard blue sky. Up they rose, whirling rings and spheres of vapour driven by the earth's radiation; and down they came, showers of drops dragged back by gravitation. The underground artillery was silenced, for steam had the mastery of pressure, and the kettle boiled over. At last the whole pool, 50 and odd feet wide,

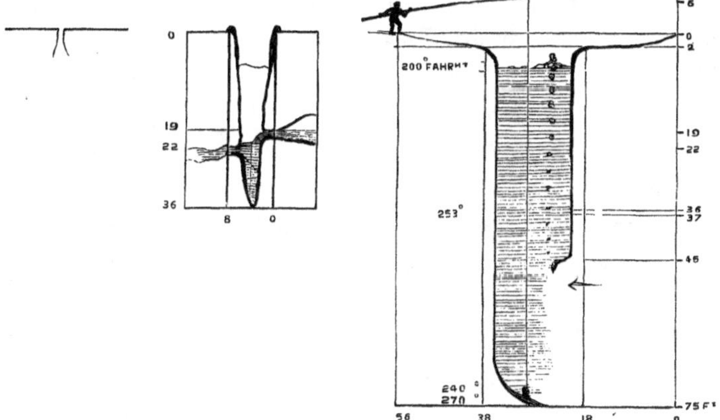

FIG. 108. STROKR AND GEYSER.

rose up a single dome of boiling water and burst, and then the column in the tube, 70 feet deep and 20 wide, was shot out of the bell-mouthed blunderbuss with a great burst of steam. The charge scattered as shown in the woodcut; it rose about 60 feet, and most of it fell back, and sank in with a rush; and so the glittering fountain rose thrice like some mighty growth. After the last effort, the pool was empty, and the pipe also for a depth of 6 feet; the spilt water was steaming down a stone aqueduct of its own building, and it tumbled into a cold burn in the wet muir at last. By this eruption the tube was scoured and smoothed, and something was added to the basin and the mound; for mutton-bones, feathers, and suchlike, were covered with a crust in a year. Each drop, large and small, had its own motion while it flew; it described a curved path, revolved, and threw off part of its mass in steam. If it travelled far enough, it might freeze; if hot enough, silica held in solution by water would be left by steam in the air. Inner surfaces grow inwards, upper and outer surfaces grow upwards and outwards; and so this pipe will choke at last, if the growth continues. The mechanism of the Great Geyser cannot be seen, because the water is too deep. The Churn is sometimes emptied so far that the works are seen.

Strokr is a conical oval pit, less than six inches wide near the bottom. The size of the plummet used makes a difference in the soundings, and possibly there may be some small steam-pipe at the end of the cone. The water is always surging, growling, and frothing about within 6 feet of the top. Steam rises through a hot column 13 feet deep, and never collapses, because there is less pressure to be overcome; this well boils, but does not simmer. By turning a barrowful of turf into the pit, this kettle is made to boil over; steam is stopped, the water is stilled for some minutes, and the mud is greatly heated below. Then a dome grows and bursts, and wad and water and steam from the gun grow up like a giant sheaf of corn. First the water in the well makes a furious swirl, like an eddy from a stricken whale in shoal water; and then the column rises and overflows slowly with increasing swiftness, till the dome rises up and bursts, to make way for a steam-bubble as big as a balloon. Up go the projectiles, and down they come in showers and streams, to rise again with furious bursts; and woe betide the spectator who gets within range of this scalding spray.

After one of these displays the water-level was more than 20 feet from the edge, and then at 19 feet the mouth of one tube was seen. From this hole, which was about half a foot in diameter, boiling water and steam jets squirted into the pit at intervals; and it soon filled to the old level, and hissed, and growled, and frothed, as before. Another hole was seen by an Icelander in the opposite side of the pit at 22 feet from the top. The spouting of Strokr is caused by the shape of a steam chamber, and the mechanism is the same as that of a closed kettle or the models above described (p. 405). The shape of the pit results from the movements of the water, and these result from temperature and hydraulic pressure. Because the movements are violent and very irregular this tube is rough, and layers deposited in it are strangely contorted (see title-page, vol. i.)

In all probability the mechanism of the larger fountain is built on the same principle of steam chamber and tube. The lateral steam-pipe in Strokr has a projecting roof; on the north side of the Geyser-pipe a plummet rests on some ledge;* and when the tube is filling steam-bubbles rise at the place where they would appear if they came from under this roof. By long practice a fisherman is able to tell what goes on at the end of his line. An old comrade, a salmon-rod which has earned many a good meal, was used to get a large thermometer into the middle of the Geyser tube. When the weight was near the ledge, after it had fallen from it and sunk a few feet some force appeared to lift it, and drive it about, for it struggled like a fish in a flurry. When it was hauled up it had

* Mr. Bryson of Edinburgh was the first to discover this ledge, so far as I know. His discovery was tested afterwards, and the ledge is a fact.

burst. The explanation suggested by the shape of Strokr, and by numerous models, was that steam, or currents of very hot water, were spouting sideways into the tube under the ledge. When the plummet sank lower it ceased to struggle, and pulled steadily at the rod. According to experiments made by Mr. Bryson in 1862, the temperatures marked in the diagram were overcome by the pressure.*

A column 37 feet deep prevented the formation of steam at 253° of Fahrenheit. A deeper column of 75 feet made steam bubbles collapse at the high temperature of 270°, but soon after this temperature was got the Geyser exploded. It seems impossible that a layer of lava or of any other material only 75 feet thick can still continue hot while the surface has been cool ever since the Geysers were first discovered. The source from which the heat comes must be far deeper; and probably steam rising from great depths heats all these kettles and makes them boil over.

The Little Geyser spouts occasionally without any warning, and rises, 50 feet at least, like a fountain, from its narrow pipe. The rest of this family bubble and sputter, each on a different plan.

The Oxhver, like the Geyser, is near high ground in a district of recent violent disturbance, but on the north side of the island, about 140 miles away. A number of pipes, with craters and cones formed about them, are near a marsh at the foot of the hill; of these, one is called the Bath-house, because, according to tradition, it burst up through the floor of a house.

* Mr. Bryson's plan of taking the temperature was ingenious. A number of thermometers were filled but not sealed. These were lowered, and part of the mercury was spilt. When it cooled it left an open space. By heating the tube till the space was filled again the temperature was got. A common maximum thermometer made for a high temperature (260°) burst or was smashed at the first trial.

The woodcut, vol. i. p. 16, is from a sketch made in 1861. It is a small copy of the Geyser, and the water balanced in the same way while dinner was cooking in the overflow. Close to this pipe, in the same stone mound, is a copy of Strokr, a rough warty irregular basin, with a wall about a conical pit, in which water seethes furiously within about six feet of the top. The Badstua explodes occasionally when the steam gets up; the other is always expending all the force it borrows from some chink or hole in a steam chamber under ground. A third pit is called the Oxwell, because an ox fell in and was boiled. Bouillon came with the first eruption, bouilli at the second, and a third effort cast out bones. This well is within 100 yards of the other two, has an intermediate shape and depth, and works on a different plan. The shallow conical miniature churn is always boiling furiously. The deeper Oxwell boils over at intervals of ten minutes: the basin is rough, and the tube somewhat conical. The deepest of this set—the Bath-room—simmers and shoots underground, and balances on the steam, but explodes occasionally when the steam gets up. The shape of it is like that of the larger pipe, which plays on the same plan.

The level of the Kitchen, above described (p. 397), is considerably higher than the level of the Geyser, and therefore steam has a greater pressure to overcome. The water balances, but neither seethes nor simmers, nor boils over. The shape of it differs, for it has reached old age. The sides of the tube are never above water, so they gain nothing by evaporation, and grow slowly inwards. The waste is small, so the pipe must be narrow below. The chief growth is at the inner edge of the highest layer, where the stone is alternately wet and dry, and for that reason the large rough tube of the Kitchen is roofing itself with a slab. A bridge spans the pool already, and the edges are growing horizontally. When this flat roof is built, it will either burst or keep down the steam in a closed chamber of large size. Many such caverns are hidden under loose rubbish. About the Kitchen, holes open occasionally, and betray them; and, on a still cold evening, white columns of vapour rise up and hover like ghosts of buried Geysers above their hidden tombs.

So far, one result of terrestrial radiation is to build chambers, tubes, basins, and truncated cones, with materials held in solution by hot water, brought from below to the surface, and deposited there at low temperatures. The same action carried further makes a sealed cone. Near Reykholt, about 50 odd miles to the N.W. of the Geysers, a spring has built a mound in the middle of a cold river. Steam rises through the gravel, and the spring boils furiously, and boils over every few minutes. It rises through tubes with small basins at the top of a steep gray mound some 10 or 12 feet high. Neighbouring hills, which make one side of the strath in which the river flows, are made of bedded trap, the beds dipping towards central high hills to the east of the place. A fault cuts vertically through these beds, and it seems to run towards the place where this hot spring has built a stone mound in cold water. Some few miles away, a whole cluster of springs have been spouting for many years, and at Reykholt is the bath in which Snorro bathed centuries ago. Opposite to the spring is another "fault" in the old beds. In No. 1, p. 379, a whole system of "faults" may be traced from the crust to the centre of a stone, and many of these pass through chambers which were hot. The terrestrial heat which boils all these springs may be at a great depth, and faults may be ducts for superheated steam. The hot region certainly is lower than the sea-level. A large spouting spring is close to the sea at the southern shore of the great bay of Faxefjordr. No near ground is high enough to account for this fountain, and the sea would have cooled this point long ago. The fires which work these engines at so many distant points must be far down, and the power the same which builds mountains. Sixty miles about north from Reykjanes, Snæfells Jökull is built on the end of a point. It is 5808 Danish feet high, and the shape of it is very like that of a mound built by a hot spring. A sketch of Snæfell is at p. 85, vol. i. All these forms, which are seen growing slowly about hot springs—chambers, tubes, craters and cones, domes and streams—abound in lava and in mountains in Iceland.

At Myvatn, in the north of Iceland, is a cluster of extinct volcanoes. These rise 6 feet, or 10 or 12, or 50 or 60; and near them are mountains of like shape, which would cover half the site of London. Fifty or sixty of the small hills are within a square mile, and great streams and lakes of frozen lava cover neighbouring districts as big as small counties. Some of these are bare; others are covered by sandy and marshy plains, by large lakes of water, and by dry deserts of gravel and sand. Through these, large glacier-rivers cut channels, and they build stratified deltas, pack silt, and make sections. A few days spent in this country are worth whole years of geological study elsewhere. It would be easy to cut through many of the small mounds; but their structure is so evident, and so many samples of them in all states of growth and decay abound, that to dig would be loss of labour.

In the first place, many chambers are open.

Close to the small cones—so near as to make it evident that one set of forces shaped the whole—the upper crust of the lava was blown into small domes, like bubbles blown on metals or on boiling water. Many of these domes are broken, so that hollows beneath can be seen. When snow covers this tract in winter, swelling forms remain to show what is beneath ; and if the earth has an igneous crust, upthrows in sedimentary rocks may, in like manner, betray buried chambers of like origin. Silt-beds are now forming in the lake, above molten lava-domes, and the sea and its sedimentary formations may cover larger hills of the same kind. The whole of a large undulating plain near Myvatn is thus chambered. Near a church on the west side, a track leads over a series of vaults, most of which are split at the crown of the arch, and through these rifts water is seen flowing over the next layer of a series. A section of one of these vaults is exactly like a low flat bridge spanning a pool, but it is part of a bubble, formed as bubbles form on the Geyser before it explodes, or on a kettle when it boils. The upper crust is three to four feet thick ; the surface is wrinkled ; the roof of the chamber is smooth ; and a section of it shows a series of bent layers like those which roof in Surtshellr (vol. i. p. 429). The floor is rough and wrinkled like the outer surface. The dome was blown while the floor was fluid, and the floor flowed and froze after the roof was made. If two concentric shells have thus formed, any number of them may exist at any depths, and chambers may be of any size. The crust of the earth may be like the crusts of the stones, p. 379. If such large chambers exist, it must be a question of power and resistance—heat and the strength of the boiler—whether the roof shall bend or burst, leak, yield, or resist.

The same lava-domes, the same vaulted lava-ice, abounds at Reykjalid, on the other side of Myvatn. A stream poured over some rough ground, and froze to a thickness of four or five feet : it poured on below, and left the ice stranded. It is rough and broken, cracked, starred, and uneven, like " blind

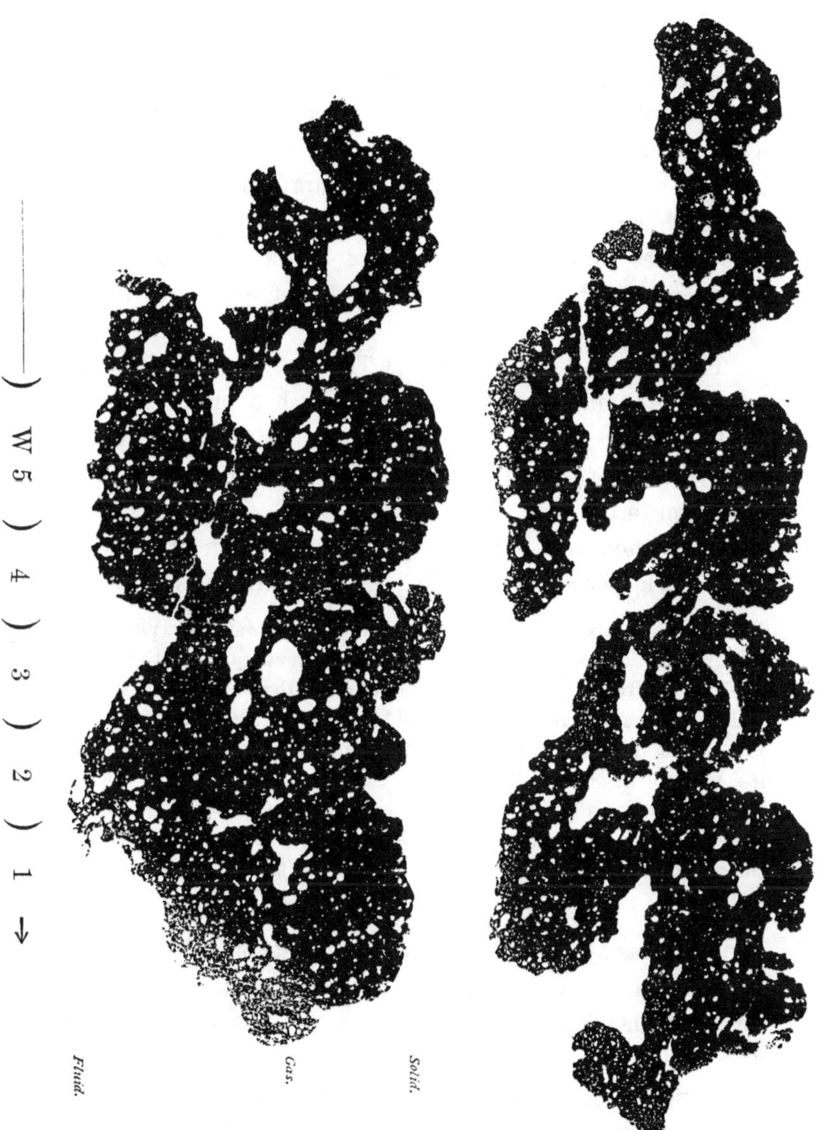

FIG. 109. SECTIONS THROUGH THE SURFACE OF A FROZEN LAVA-STREAM, which flowed downwards (in the impressions on this page). See pp. 390 and 400. The surfaces are to the right, and show the characteristic form of a lava flow of small size.—Printed from the Stones.

ice" on a pond, or ice stranded by the ebb; but here every movement is recorded by wrinkled folds on the surface. A little way from this shattered crust the horse-track leads over a dome-shaped, swelling, wrinkled surface, starred and torn, but not broken up. Under that roof are chambers, and the tramp of horses rings hollow as they pace along. Cracks in these domes show that some upward thrust tore them while they were tough. These are "craters of elevation" in all stages of growth.

The lava at Surtshellr and at Thingvalla has sunk, so as to make a "crater of depression," if such a phrase may be used; and the broken edges at Thingvalla are hundreds of feet thick. Forms which resulted from freezing can be seen in section in the rifts.

In Henderson's *Iceland* is an account of a great eruption which took place in 1783. At page 225 is this passage:—

"The torrents (of lava) that continued to be poured down proceeded slowly over the tract of ancient lava to the south and south-west of Skâl, and, setting fire to the melted substances, they underwent a fresh fusion, and were heaved up to a considerable elevation. It also rushed into the subterraneous caverns; and, during its progress underground, it threw up the crust either to the side or to a great height in the air. In such places, as it proceeded below a thick indurated crust, where there was no vent for the steam, the surface was burst in pieces, and thrown up with the utmost violence and noise, to the height of near 180 feet."

Here was an upheaval of a tough surface, and the bursting of a hard crust, by imprisoned air and steam expanded by heat, and the action was on a large scale. At page 228 it is said:—

"With respect to the dimensions of the lava, its utmost length from the volcano, along the channel of the Skapta down to Hnallsar in Medalland, is about 50 miles; and its greatest breadth, in the low country, about 12 or 15 miles. The Hverfisfliot branch may be about 40 miles in length, and 7 at its utmost breadth. Its height, in the level country, does not exceed 100 feet; but in some parts of the Skapta channel it is not less than 600 feet high."

A tract of about 1500 square miles was covered with fluid lava in a few days to a depth equal to the height of moderate hills, and that amount of matter was pumped out from under the earth's crust, and flowed over it, leaving, it must be assumed, an equal hollow beneath.

It is hard to guess what is the power of an engine whose boiler may have the dimensions of the Firth of Forth or the Firth of Clyde, and whose furnace is hot enough to fuse lava.

If lava-bubbles were blown by steam generated in small cracks and caverns, what would the steam of the larger cavern accomplish under the pressure of such a roof?

In old lava-streams near Skjaldbreið many samples of like work may be seen. One great bubble, as big as a cellar, with a roof two feet thick, has a large open angular gap in the top. It was burst, and the keystone of the arch was blown to a distance of ten or twelve yards, where it now rests upside down. It must weigh some tons.

If domes on a biscuit are reproduced in lava hundreds of feet thick, similar domes of greater dimensions may build volcanoes in proportion to their size. The crusts which are seen in cliffs along the coast of Iceland may roof in caverns from which Hecla grew; for cones of like shape grew from smaller lava-crusts at Myvatn. It is not possible to get at the works of the big engine, but it is very easy to dissect a little one; models can be made and broken; and cones and craters near Myvatn are as easily seen as models.

Chambers abound. Tubes of lava like Tintron (p. 398)

also abound in the district. Near the church are cones and craters of every pattern.

Some are truncated cones, with a conical hollow in the top: these are "cones of eruption"—mere ramparts of black frothy cinders without one solid block or stream of lava outside. They are regular in form, and grass is beginning to sprout on their smooth sloping sides. Rain is beginning to furrow the slope; and in winter the mound is covered with snow. The little volcano is then like Snæfell, or any other high cone of eruption. The shape is enough to betray the extinct volcano in the Andes, or elsewhere. In this case a circular rampart of ashes conceals the tube through which a fountain of vapours and stones played. Vesuvius and Hecla are like this specimen. It would be easy to cut through the little mound, but a walk of a few yards does equally well.

One regular truncated cone of eruption, made of loose cinders, stands with part of the base in the lake, and it has been tilted bodily to one side, but so quietly that this mound of loose ashes still retains its shape. It is now covered by a fine sward. In the centre of the crater, the end of the lava-tube, through which the fountain played, is seen. Six strange weird-looking blocks of dark rough lava, like the roof of the Tintron tube, peep through the turf like a circle of stones about a hero's grave. These mark the source whence the cinders came—the place where a choked tube is buried under a circular barrow, which a miniature volcano piled over its own head before it expired. If the mound were in England it might pass for a work of art. It is no work of human skill, but a sample of a cone of eruption—a tool-mark of a natural engine worked by terrestrial radiation. It would be easy to dig out the buried tube, but a walk of a hundred yards does better.

Close at hand is another specimen of the tribe, which has not grown so far as to hide the lava core of a cone of eruption. In the middle of a circular mound of loose ashes stands a truncated cone of lava, with a plain on the top. In the middle of the plain is a depression, with a set of radiating cracks, and round the edges of the plain is a raised rim. The work stopped at the stage which Vesuvius had reached. When the crater was full to the brim (p. 403), it was like the basin of the Geyser before an eruption (p. 414); and the last movement was downwards, as in the case of the sealing-wax tube described above (p. 406).

In the first of these three mounds the tube is hidden by the stone fountain which rose from it and fell about it; in the second the end of the pipe projects; in the last case the top of a lava-cone frozen about a lava-spring, the frozen lava-pool in the lava-crater, and the choked up lava-tube, stand together in the centre of the ring of projectiles, which scattered as the drops are scattered from the craters of springs, or from boiling water anywhere. If the power had been sufficient to keep this tube open and continue the work, the ring of ashes would have risen till the edge of the tube was at the bottom of a funnel, like that which surrounded the tube of Vesuvius in 1842 (p. 402). But the power was spent before this hill had grown; the fountains ceased to play, the spring froze, and the shape remains to tell its own history of the works of Frost and Fire. This lava-mound is about the size of a small glass-house chimney; but within sight of it is a mountain of the very same pattern, which, though not so high as Vesuvius, covers more ground. It would be easy to quarry a hole in this specimen, and as it sounds hollow, there may be a chamber within the mound. It would be easy to cut a trench through the circular mound of ashes, but sections of similar mounds are close at hand.

At Bonn, on the Rhine, the seven hills are larger specimens of this class. In 1853 the river was crossed from Bonn, and several of the hills were scaled. They are truncated cones, with plains on the top, and one at least has part of a circular rampart about the plain. If these ever were surrounded by rings or mounds of projected ashes, they have been washed away ; but ancient lava-streams which flowed from these old lava-springs can be traced along the slopes opposite Bonn. The Castle of Godesberg is on a mound of the same description ; and all these sound hollow, though made of rock. At Myvatn small lava-cones are in all stages of growth, and some are in fact hollow cones, like Tintron.

Many of these have no mounds of ashes about them ; others have. One stands in a ring about 160 yards across ; the lava-cone is about 30 feet high, and it has a circular plain on the top, with a rim about the edge, and a hollow above the place where the tube ought to be ; it rings hollow. The sides are steep, and it was no easy matter to reach the top. The plain seems to consist of balls of lava as big as grape-shot, set in frozen lava like plums in pudding, or barley in broth. Close at hand is another specimen without the roof. It is about nine feet high, and shaped like a glass-house or a lamp-shade ; it is made of rough clinkery lava, and rises through a plain of cinders. Near it is another about the same size and shape, but one side has broken down, leaving a shell about three feet thick.

It is easy to creep into these and others like them. In some the inner surface is smoothed, and grooved, and plastered by fountains of vapour or fluid, which first blew them and then spouted through them, and so rifled the gun. Close to one of these a lava-bomb was found (p. 379). Near to these are domes which have burst, bubbles which have not burst, and frozen lava-springs, with a dome surrounded by frozen wrinkled streams, which radiate from the source.

The growth of a volcanic mound is thus illustrated by small samples in all stages, and the mechanism of the small engine is well seen.

A lake of lava froze while boiling. Chambers formed under the crust, and hot vapours which made the chambers struggled to escape from them. In some cases a bubble was blown; in some the bubble became a hollow cone; in other cases the chamber leaked. Tubes were blown, and through them springs of lava, or fountains of stony froth and vapours, were driven by the earth's radiation, as fountains of steam and hot water are driven by it through geyser tubes.

Large specimens of like work are in Iceland, and may be seen in a couple of months.

Near Myvatn is Krabla; and one set of rocks on that mountain appears to be parts of a hollow cone of lava, through which hot vapour escaped and fused the inner surface, to make obsidian. The place was seen late in the evening, and this may be an error.

At the foot of this mountain are many old craters and many boiling springs, and from it old lava-streams diverge in many directions.

From the top of any hill in this neighbourhood scores of larger cones of eruption may be counted, and small ones may be reckoned by hundreds.

In crossing the island from Hecla, by way of Sprengisandr, still larger specimens rise up through snow and ice on all sides.

Hecla is a cone of eruption, and round the base of it are enormous tracts of lava, great frozen plains without a blade of grass, in which strange weird solid fountains of frozen lava stand up like black monsters where they froze. The base of Hecla is wide, and the crater is small in proportion ; another effort would finish the cone, and roof the tube like Tintron. But the tube is there, though buried ; and as soon as the power accumulates sufficiently it will burst, as it did a few years ago. Where it will burst is a question of power and resistance. The last eruption broke out near the top, and a considerable lava-stream flowed down a hollow, froze suddenly, and formed clinkers. The only substance to which these can be compared is "pulled bread"—crumb torn to bits and baked hard.*

All down the Snæfell peninsula, on both sides, are cones and craters of many shapes ; but specimens like them all may be found at Myvatn in a morning's walk.

From Helgafell a great yellow mountain is seen. It was a cone and crater of eruption ; but one side of the crater burst out, and the fallen rubbish makes a stream of heaps, sorted apparently by a water-flood. Perhaps a lava-stream did the work, and is buried under the floats.

At the head of this regiment of volcanoes is the great cone of Snæfell, with its plains of basalt.

All round Faxefjord are small lava-craters, surrounded by lava-streams, which rose and flowed every way as from a spring. One of these is Eldborg (fire castle.) It is made of lava, disposed in beds which dip every way from the edge of the crater. The stone is spongy and brittle, and it must have seethed like Strokr when it overflowed. At the bottom of this great cup is a boss of hard lava, the crown of a solid pillar, which froze in the tube. For miles around this frozen lava-spring streams radiate. The newest are clinkers, piled

* For a sketch of Hecla, see title-page, vol. i.

in the wildest confusion. To climb over them is almost impossible. It is exceedingly dangerous ground, for the stones are hidden by mosses and lichens, and feet and hands slip into unseen rifts. The stones move easily, and break ; and the surface cuts like shivered glass. Older and larger streams, which came from this source, are like other lavas in Iceland—compact, firm stone, with a wrinkled surface. At a guess, the crater at Eldborg may be about 400 yards wide, and 200 feet deep. No measurements were taken, but sketches were made.

Most of the valleys which drain into Faxefjord have small cones of eruption and streams of lava, and in many cases the cone stands in the middle of a far larger broken-down crater, of a different colour and make. Each of these would be a study, but mental pictures alone were brought home from this region. To the right is a low marshy plain, reaching to the sea ; to the left, tall cliffs of bedded igneous rock, with faults and fissures, and all the marks of weathering old and new. As the day wears on, glen after glen opens in this great sea-wall; and far away in the distance a bare red mound glows like a heather hill in autumn. On either side of it are yellow hills, fragments of the old crater ; and from these, down the glen, comes a stream, black and gray and green, like a peat-moss in the Highlands. A turn brings in a bright silvery stream of water, the river which the lava-stream has driven to one side. All that will grow in Iceland—birch, fern, moss, and grass—grows best about these lava-streams. Either the black colour gathers more heat from the sun, or the debris of lava makes good soil, or there is a store of earth-heat in the lava which warms the plants like a flue in a hothouse. The only specimen of mountain ash found in the island was found near Eldborg, growing on modern lava. But all these are tiny springs to some of the old giants of their race.

From the Geysers to Brunar is a ride of about forty miles. The way leads up hills, to the left, in the cut, p. 409. It passes over a small lava-stream, far larger than the largest about Vesuvius, and then a goat's track leads out of a glen up a steep slope through a notch in another range. The dry course of a burn, or a natural rift in this hill, gives a section of the country. The hill is made of layers of ashes, plastered over with lava. The rock is cracked, and full of holes; and it rings hollow under foot. To ride over it is like riding over vaults, and great hollows are open where the sand has been washed away. At the top of this strange pass the edge of a lava-flood is reached, and for the rest of the way to Brunar the track crosses the stream. One branch of it flowed to Thingvalla, and it seems as if part of it reached at least as far as Reykjanes, about seventy miles away. The bottom of the sea is made of lava, according to the report of fishermen, so there is no certain limit to the flow. At p. 90, vol. i., is a view from Thingvalla. In the centre is Skjaldbreið, and the way from the Geysers to Brunar crosses the shoulder of that dome from right to left. As it seems the lava radiated from Skjaldbreið; and that mountain is a frozen spring, the top of the pillar which froze in the tube from which all this vast flood of molten stone rose and flowed. But if so, there must be a chamber in proportion left somewhere under ground. There is no cinder-heap about this source; it overflowed and froze without spouting, for lava-surfaces are well preserved in all directions. This hill is from 4000 to 5000 feet high, but no measurements given in the map.

This was a large lava-spring in its day, but the older igneous rocks which make the large mountain tracts and the whole island came out of some larger well and some bigger cistern. It may be that the broken walls of rock which hem in Faxefjord, and dip away from it with the radiating glens which drain into the fjord, are remnants of a crater 60 miles wide. The highest mountains in the world are volcanic, and their shapes are but large copies of mounds at Myvatn. A force now active raises molten stone 28,000 feet above the sea-level, or 28 feet, or the same number of inches, according to the amount of force applied ; but, in all these cases, the force is the earth's radiation, resisted and controlled by gravitation.

Far out at sea, the Westman Islands are cones of eruption like those which abound all round the coast. Some are bare ; grass grows on others ; and some are broken all round by the sea. The cliffs are high, and give beautiful sections of the structure. There is no room for speculation ; the facts are there patent and manifest, drawn in coloured lines like a geological section. The mounds consist of layers of ashes, tuff, and overflows of lava, which rose from many vents. They seem bent in every possible direction, but really they slope away from old craters which were buried by later eruptions, so they form a complicated pattern of waving lines. Sealed tubes, pillars of lava now frozen where lava-springs rose, are seen in the cliffs, with faults, and dykes in the faults. These are harder than the rest of the mound, and they are not bedded. Millions of birds rest in shelves weathered out of the stratified series. No bird can perch on the side of the hard compact lava, which froze in holes and chinks. One of these islands, Erlandsey, is a study in itself. No drawing can give any true notion of its complicated structure as shown in the cliff ; but the form of the truncated cone which rises in the middle is but a repetition of mounds at Myvatn. Like forms have been made repeatedly by boiling sealing-wax, water, and plaster ; and sections made in these models are miniature copies of the structure of Erlandsey. To describe each model of a whole series made, in order to copy each of the forms described in this chapter, would be waste of time and space. Let one sample suffice, and let those who take an interest in the subject cook volcanoes for themselves.

After working at models for many years; after these last chapters, written some years ago, had been rewritten and printed; the following arrangement was made, with the intention of imitating the forms and movements of hot springs and volcanoes:

An iron pan, 17 by 13 inches wide, and 2 deep, was placed 2½ inches above a gas-burner, with 4 rings, of a diameter of 9 inches. A layer of fine sand, about half an inch deep, was spread over the centre of the pan above the burner, and a ring of dry plaster-of-Paris was made about the sand. A pound of coarse sealing-wax was laid on the sand. The gas was lit, and the sealing-wax was slowly melted upon the sand. It boiled, and made a pool of melted wax upon a foundation pervious to water. In this it resembled the natural arrangement of a sheet of lava upon a bed of dust, which recurs so often in volcanic countries, and in particular at the place above described (p. 432). When all the wax was melted it was covered with a layer of dry plaster, through which the sealing-wax rose. It raised domes, and burst them, as lava-domes are burst in Iceland. The crown of the arch was starred, and then from the middle of the star a bubble of wax rose, which burst and overflowed, covering the plaster.

This resembles a possible natural arrangement. A bed of limestone may be covered by hot igneous rocks and burned. If water then gets to quicklime it will set. The craters thus formed were "craters of elevation." Copies of like forms constantly recur in slags and lavas; and according to Von Buch and Piazzi Smyth, Monte Somma and the outer ring of the Peak of Teneriffe were so raised from under the sea.

To get more power, water was now poured in round the edge of the pan, and more plaster was dusted in, to keep the wax in the middle. When this charge had set, there remained a plain of wet plaster, pervious to water, surrounding a lot of springs of boiling wax, which covered a layer of sand. The plaster was at rest, but the fusible wax heaved and swelled, and burst and bubbled, and sank down again, like any other boiling material from metal to water. By adding cold water till the level of the wax was reached, these wax-springs were made to grow and become tubes, as in the experiment (p. 406). While water on the surface was at 60°, water below boiled furiously, and steam burst through the wax, throwing up sand through miniature tubes, which communicated with steam chambers. In order to concentrate the power, dry plaster was poured over all vents but one, and there steam blew off, driving out wax, which froze in the water when it flowed down. The vessel was now filled to the brim. The surface water was at 100°, but steam escaped through several pipes in soft wax, which boiled up and rose more than an inch above the water. A thermometer placed in the steam rose to 212°, but probably the temperature was higher. At this stage, sand, wax, plaster, and water, were thrown to a considerable distance by steam, which hissed and sputtered through this miniature crater. In the neighbourhood of the crater the white plaster cracked, and dykes of red wax rose, while fumes from the wax rose through the porous plaster, and discoloured it. These fumes spread in the air, and travelled far; for the smell of wax pervaded the house. In all volcanic countries fumeroles abound. In particular, near the Geysers, fumes rise and are condensed amongst the ashes. By adding cold water the temperature was kept about 60° to 100°. Plaster does not melt at 212°, so when it had set a hard shell was formed about a fusible mass. Sand neither melts nor sets ; without digging into the model it is plain that a chamber was thus formed equal to the amount of wax and sand which was driven to the surface. Where the roof was weak and fusible it sank in, and cones of plaster and mounds of wax sank into the chief crater and disappeared. So craters of eruption have disappeared after rising above the sea. If there had been enough of sand a sand cone of eruption would have formed about the wax tubes. To make a cup and cone, dry plaster was sprinkled about the crater. Steam and boiling water drove it away from the centre, and the basin and mound of the Great Geyser were copied in plaster. When the first layer had set, more plaster was sprinkled over the mound, and so it grew. But when it had grown to a certain height the boiler burst, and a new crater opened in a starred dome veined with dykes. Water, wax, sand, and steam, burst out and broke up the crust, throwing balls of soft wax to a distance. The boiler could now be filled by pouring water into one of the craters, and so a good head of steam-power was kept going. By shaking dry plaster over both, two truncated cones, with cups and pipes, grew. Boiling water rising through wax tubes moved on a definite plan, and sorted the loose plaster, which set and took a cast of the currents. When these two mounds had grown so high that the pressure of columns of water in them equalled the strength of the boiler, it burst once more, and a third crater opened at a low level amongst the plaster. The operation was so far completed in about two hours, at a cost of about 80 feet of gas, and the materials. When cooled, water stood at the same level in all the pipes, and the lowest of the series flowed as a cold spring, if water was poured into any of the rest.

They all communicated with each other, and met in a common source. But when the model was heated again, water stood at various levels, and rose in the large tubes far above the edge of the pan. Moreover, one spring was always hotter than the rest; it boiled first, and spouted highest of the series. A model once made works for a long time, but this one was doomed to destruction from the first: the toy was broken by overturning the pan, and the works were dissected. The layer of sand had disappeared; part of the wax had taken the shape of lava clinkers; part of it was plastered on the roof and sides of a steam chamber in the plaster, and formed the lining of long steam-pipes, which wound about through the mass; part of it was in the open craters, in choked tubes, and in hollow cones, which rose through the plaster, but did not pierce the surface. These were the vents which were stopped to concentrate the power at one spot. The roof of the chamber was so shaped that most of the steam must have gone towards the pipe in which the water was hottest. It was heated and forced up by the steam, and the steam took the easiest way to escape from the gas fire which worked this engine. So far this model illustrates a theory, formed upon a careful study of natural forms. On the outside of it were upheaved strata, dome, overflow, and fountain; cup, cone, and pipe; and these were miniatures of movements and forms at the Geysers, at Myvatn, and elsewhere. Inside were tubes and chambers, like those which abound in the crusts of volcanic bombs (p. 379). The conclusion arrived at, so far, is that the igneous crust of the earth, and the mechanism of hot springs of water and lava, are like these miniatures, and like them were shaped by radiation and gravitation, directed by laws which govern the universe.

CHAPTER LVII.

RAYS.

A mental quality, which phrenologists term causality, drives men to seek causes. In 1851 and 1862 this turn of mind drove many visitors into the department of machinery in motion ; they were attracted by sights and sounds and smells which repelled others. It may or may not be true that certain bumps on their heads were large ; they certainly had like tastes, and they formed a class. Amongst them were members of all classes in society, drawn together by a common wish to learn how things are made, and to see work done. One who haunted the world's fair got to know where to find faces, with certain trains of thought mirrored upon them. Simple wonder, with round eyes, staring agape, was in faces clustered about the big diamonds ; amongst the engines, even wonder looked somewhat wise, or seemed to try.

The rattling, grinding, clashing, grating, thumping discord of many engines spread from the place. Following sound, a door was reached, and there a beam of electric light struck full into the eyes like a stinging dart. To look was more painful than pleasant. Most men blinked and rubbed their dazzled eyes, looked puzzled, and stepped out of the line of fire as soon as they could. Some who looked too long injured their sight. All around was a noisy maze of wheels and axles, strings and bands, rods and pistons, whirling and turning, rising and falling, advancing and retiring, moving and hard at work. No visitor ever hoped to comprehend all the engines which moved and worked in that one department; but every one who chose to think could find whole trains of causes there. Those who went far enough found out that the commissioners supplied steam-power to the exhibitors gratis.

Without striving to comprehend the maze, it was easy to look through it, and see, beyond it all, a furnace-fire, a light, and a man's thought—three distant links in a vast chain of causes, but links within reach. Leaving the first idea of the exhibition, and the spark which kindled the fire, a more immediate cause of all the movements was in a boiler-furnace, and one result of this Fire was Frost.

One engine was making ice all day long. An air-pump exhausted a vessel so as to lift pressure off ether; the ether boiled and expanded, and became vapour, which the air-pump removed, to be condensed elsewhere. The vessel which held ether thus boiling at a reduced pressure was under salt water, in which tins filled with fresh water were plunged. In these water froze. It froze first next the tin, and the solid crusts grew towards each other, forcing air before them, so as to shape chambers and tubes in a transparent shell of ice. The last drop of fluid was in the middle of each 'shape,' and the shape of each system of air-bubbles showed the directions in which force had acted. The furnace-fire became force, and force was set to draw heat out of water in the vapour of ether; and so this engine froze water because water was boiled. One day a rough-fisted man with big brows and bright eyes watched the proceedings in silence for some time, and then remarked promiscuously to all who cared to hear, "I've seen that mony a time in the pits." "That" might be seen in a coal-pit near Glasgow in 1863. Air was driven down to the "face" by a steam-engine. It was compressed in a pump, and in long pipes; and heat was squeezed out of it, for the pump and the pipes were warm. When the compressed air escaped below, it expanded and took up heat so fast that vapour froze and became hoar-frost in the coal-pit. So fire turned into force causes frost in some cases.

Leaving all the spinning, weaving, grinding, rolling, packing, folding, hammering, squeezing, carving, sawing, modelling contrivances, which shared in the force of one fire, certain engines illustrate parts of this book; for fire and weight, expansion and contraction, were set to move air and water, and other substances, with engines.

Amongst the engines were many for blowing air into furnaces. These howled like a winter storm in a forest, or roared as they only can roar. A hand with relaxed muscles fluttered like a flag in the nozzle of the bellows, and felt that air is a fluid of sufficient density and weight to do the work of a hurricane, balance a column of mercury, and work an engine. Part of the force caused waves in the air, which produced discordant sounds; part of it made harmony, for all the great organs were blown by engines. The force of fire was so directed as to move air in many ways; part of the force produced sound waves in air, part of it moved currents of air.

Another set of engines lifted water. In the middle of the department, a broad cascade fell over a tall screen, with all the dash, and spray, and froth of a burn falling over a rock. But this fall had no burn behind it. A centrifugal pump was whirling in a basin; it lifted water through a flat tube, and water fell over the edge back into the pool. There, from constant friction, the circulating water grew warm and steamed. Fire, turned into force, caused waves and circulating currents to move, and part of the force became sensible heat again.

Part of it became visible light in the electro-magnetic engine, which cast sharp arrows of light and rays of sensible heat through a distant doorway. That light was produced by the passage of a powerful electric current between carbon points (see Introduction). These do not touch, but when they approach each other, they become intensely hot, and very luminous. Bright crackling sparks then fly off at some angle to the course of the current, and these sparks describe paths which depend on the laws which govern the flight of all projectiles. Many were gathered when cool. Under a microscope, they appear as minute black globules with a lustrous glassy surface, with cups and cones and craters, like other sparks. Some of these adhere to carbons which have cooled, and they too are spherical.* After many complicated changes, force caused, or became radiant light, heat, and motion. Force and light radiated from luminous spheres, and from sparks thrown off from a luminous current.

Another variety of the same light was produced by passing the current along a stream of falling mercury.† Thin as a wire, it flowed continuously till the electric current took the same path, and then the stream burst and shone. Globules and jets of vapour dashed outwards, driven by radiation. This light has a strange ghastly colour, and the spectrum is peculiar; the breath of it is poison, so it has to be shown through a glass; the fumes condense on the glass, and obscure the light, as earth-light is hidden by the earth's crust. By these electric lights all the chemical and other results of photography are produced. One furnace-fire was a source of rays: rays took many shapes: light, heat, cold, waves, sound, elec-

* May 27, 1862. Holmes' light. † Way's light.

tricity; galvanic, magnetic, and chemical action; actinism, fusion, sublimation, motion, condensation, freezing, repulsion, attraction, work, and recording forms, were all found at this one focus—this one luminous point in a maze of engines—this source of rays.

The forms resulted from the turning of a wheel; from force, from a spark, and from human will; for the action stopped when the steam was turned off at the end of each day.

From these engines, and their work, it appears that radiation and gravitation are mechanical powers which men can set to move and shape gases, fluids, and solids, including all matters yet found in the earth or in meteorites, and all those which spectrum analysis has found in the sun. In the department of machinery in motion, gravitation and light, force and human will, could be seen through an incomprehensible maze of engines:—without knowing all that sprang from one thought, and all that made it grow, this much could be seen. The source of motion, the origin of force, is out of reach; but through all the tangled mazes of the incomprehensible engines which move in space, gravitation and light, force and Divine will, may be seen even with dazzled eyes.

One remote cause of motion seems to be in rays of light.

A certain clever maker of filters used to attract custom by filling his windows, near Temple-Bar and in Regent Street, with all manner of quaint waterworks. One contrivance was a fountain, on which a striped ball hung suspended under a glass shade. It hung on one side of the water-pillar, it turned horizontally round about it, and while it turned slowly with the sun, or "widershins," as the case might be, it also whirled rapidly about an axis of its own, which changed place continually, but apparently on a definite system. Perhaps the poles changed also. The ball had three distinct movements at least:—rotation about its axis, revolution of axis about the axis of the fountain, and revolution of poles about some unknown point or points. Besides these, the ball and the fountain revolved about the axis of the earth once in twenty-four hours; and the earth and this little satellite have been round the sun many times since the satellite was first observed near Temple-Bar, more than ten years ago. In these regions the ground is shaken by heavy traffic; the engine was disturbed, and the ball fell now and then. When it did the fountain rose higher, struck and spread upon the dome of the shade, flowed down the walls of it into a marble cup, and into a pit, where it disappeared. Like the water, the ball fell into this miniature crater and rolled to the bottom of it ; but there it fell against the fountain, which rose through a tiny brass pipe in the midst of the pit. Struck on one side, the rolling ball rolled the other way ; it turned like a whipped top, and it soon rose again whirling, because one side of it was lifted faster than the other side fell. It whirled as the water circulated from the fountain in the middle towards the wet circumference where streams flowed down ; and it rose slowly to a place where attraction and repulsion were nearly equal, and there it hung balancing. It rose or fell an inch or two when the engine was disturbed, or when it was shaken too much the ball fell into the cup ; but, generally speaking, the ball has kept its place for many years. To watch it was pleasant pastime for a law student who studied sparks, but never could see the beauty of "scintilla juris."

Apparently that engine was worked by a single force, divided and diverted so as to make it act like two opposing forces. It was a "gravitation engine." The fountain rose because water in falling from a higher to a lower level pushed water in a bent pipe out of the way, and drove it up. So the fountain was repelled by the earth's attraction turned back by the engineer who had learned to manage this force. But some other force had lifted the weight ; so this engine worked by two forces, and the sun's rays helped the earth's rays to lift the ball when it fell. The hand which winds it up moves a clock, so light made this fountain play.

The ball whirled for the same reasons, but the man who made it whirl could not comprehend its movements, and no man does.

One of the best mathematicians of the day is wont to encourage and amaze "young men from the country" by showing them, at the first of a series of lectures on physics, a series of mechanical tricks which are explained by known laws of force expressed in numbers, or in symbols which mean numbers. His climax is to spin an egg-shell—a hollow oval with a big end and a little one—upon a fountain, with this comment:—"All the mathematicians that ever were cannot explain that." Nevertheless the youngest members of the class delight to repeat the experiment, chiefly because of the splash. They can reproduce the movements without fail, and they can perceive without much effort that the force which works this engine is the converging force which makes a stone fall, and stretches a plumb-line at every point on the earth's surface ; but behind that force is the other which raised the weight—and it is light.

If so many different movements result from movement towards one point, and from the action of one force, two opposite forces may do complicated work. If experiment precedes the full explanation of it, the most ignorant may try what forces will do with matter; for the wisest can do no more when he gets to unknown ground.

Learned geographers, geologists, and famous navigators,

lately met to settle the best route towards the North Pole. They differed as to the route, but all agreed that the pole might be reached. Their question turned on the movements of ice floating in a revolving circumpolar sea. The best route for a ship is where the sea is most open, the best for a sledge where ice is most compact; and that question turns on the movements of floating ice, on the law of its growth, and on the shape of the cup which holds it. The worst route for a ship would be to start about lat. 36° 10′ N., long. 39° W., where the last iceberg was seen (chap. xliii.), and to sail over the banks of Newfoundland, where ice abounds, up either coast of Greenland, against the Arctic Current, through heavy ice there. The best would be to sail after the warm Equatorial Gulf Stream, past England and Scandinavia, to Spitzbergen, and seek for open water beyond. It has been found in that direction (vol. i. p. 363). If the ice which drifts past to the west of Iceland comes out of the arctic basin, it seems reasonable to expect to find an equal open space somewhere in the basin, and the most probable place for such an opening is near the centre of revolution, which is the North Pole. This was an important subject; but one of the ablest of the able speakers, in addressing a grave assemblage, compared the Arctic Ocean to a whirling mop. A great authority, who thus compared great things with small, encouraged one who compared the Arctic Ocean to a top and a whirling mop in chap. xxvii., to venture further on the same path. The most ignorant may try experiments, even though he must leave their explanation to those who are better informed.

A trundling mop is an old and apt illustration of pure centrifugal force. If turned slowly it makes little splash; if rapidly whirled, water radiates from it, spreading in rings of spray; each drop sets off at a tangent to some circle described about the axis of the whirling mass, by some part of it which holds on to the rest with a firmer grip. But when the mop spins as a carriage-wheel turns, vertically, drops do not follow straight paths. The centre which attracts is not in this centre of rotation and centrifugal force, but in the earth's centre ; so each drop describes a different curve when a mop is trundled vertically. The man who can calculate the paths of these projectiles must be an able mathematician ; but any child can make the projectiles draw part of their own curved paths, and so take a practical lesson in the laws of force.

At page 96, vol. i., is a drawing made by a drop of ink on a block of wood. The engraver cut away the bare surface and left the rest. From the shape it is easy to see how the fluid moved, to see that these drops struck the target on which they splashed, fairly, at right angles to the plane. In fact, they fell upon a block laid horizontally to catch them, which was moved aside a short way to make room for each new drop. If, instead of thus striking a plane at right angles, a drop strikes it sideways, it takes another shape, which gives like information as to movements and directions of force. To make more woodcuts of this kind would be waste of trouble and cash, for anyone may drop ink from a tube and slope white paper at various angles to see the effect.

A drop is spherical, and if it be laid on paper it draws its own section, and dries a round spot. If it falls it takes a new shape; it becomes a star if it hits fair ; an oval like a leaf with prickles round the edge if it hits the surface obliquely. The falling drops threw off little drops, and some of these are shown in the cut.

The faster it moves, and the more it hits sideways, the longer is the oval. The drop is moving both along the surface and towards it; so, when it moves fast, and hits a surface at a very small inclination, a drop becomes a very long oval, with a line and a dot in front. So far a drop recorded one vertical movement and one reflection—a movement caused by the direct force which makes it fall, and a reflection from the paper. A fluid may then be made to draw diagrams of its own movements, and to record the action of forces.

In the case of a mop, turning like a carriage-wheel, fluid projectiles are moved by two forces at least: by centrifugal force, which projects them at a tangent to a circle, described vertically about an axis of rotation; and by the earth's gravitation, which may be taken to act perpendicularly in vertical parallel lines. The curves which result may be learned by trundling a mop near to a wall; by watching mud drops thrown by wheels against carriage windows; by studying mud upon house windows or walls in a street through which carriages pass. Some years ago a French philosopher invented a very clever toy called the gyroscope, from which, amongst other things, a taste for spinning tops grew. One man furnished the public with "patent metal tops," copied from a Japanese pattern, and he made a small fortune. These tops were set to draw as soon as they appeared. To get mop curves a hole was made in a whitewashed wall, and a metal top was spun vertically, so that it whirled near the wall. A saucer of ink was placed under it, and raised till it covered the whirling edge. The result was a diagram more than six feet wide, which showed at a glance how movement along straight lines—tangents drawn from the circumference of a revolving wheel at right angles to a spoke—gradually bent into movements towards the earth's centre. Thousands of drops drew as many diagrams on the wall. It would cost a lifetime to calculate curves which fluid projectiles draw in a moment. There they remain, curves drawn in all angles which two straight lines will make in one plane—curves which vary as the projecting force varied in direction and intensity. Two forces drew these diagrams, but they did not oppose each other directly. Something more was wanted.

Some of these tops will spin for ten minutes. When spun horizontally, projectiles are not so much disturbed by the earth's attraction. Lines drawn by them curve downwards, like the ribs of an umbrella ; but they are not bent sidewise. A top with a disc of paper on it was spun in a concave lens to keep it on one spot, and a sheet of cardboard was placed horizontally, so that the edge of a circular hole in the middle of it was close below the edge of the disc. Ink dropped on whirling paper was thrown off, and fell on the cardboard obliquely. The result was a diagram in which thousands of minute drops had become as many long ovals, with long lines in front. A ruler laid on any one of these touched the edge of the disc of paper, when it was pasted over the hole in which it had revolved. So far the experiment only demonstrated the well-known effects of centrifugal force on projectiles. This diagram was drawn by two forces ; but by forces acting in different planes. Something more was still wanted.

The first point to be illustrated, if possible, was the action of two forces—one pure centrifugal force, the other a force acting from the centre of a revolving wheel, as a volcano at the equator acts on projectiles, along rays. The top, with a disc of paper, was spun as before, and a drop of black ink was allowed to fall on it near the centre. It described branching spirals from E. to W. as it moved to the circumference, and it flew off at tangents from W. to E. when it got to the edge and was scattered there. Drops of red ink were then squirted at the edge of the disc from a point near the centre, with a syringe. In this case the red ink was driven by two forces—by one which drove it away from the centre along a spoke; by another which tended to throw it at right angles to a spoke; and drops of red ink showed the direction in which they were moving when they fell on the plane. A ruler laid on a red drop did not always make a tangent to the disc, as it did when laid on a black drop. Within a parallelogram drawn upon a tangent and a ray, the red lines converged upon the end of the ray along which the red ink was projected.

The aim of this spinning was to get opposing forces to act in one plane;—centrifugal and centripetal, radiating and converging forces:—and gravitation, still acted at right angles to the other two. Some other expedient was still wanted.

The woodcut is a fac-simile of a disc of paper, on which black and red ink drew curves, as described above. The shaded border is red. The drops are fac-similes of drops which were projected by discs, but to bring them within the size of a page they were cut out, and pasted on lines which touch points on the disc, at which drops aimed from considerable distances.

A drop of black ink fell at A, and described the spiral figures in travelling from the centre to the circumference of the revolving disc of paper. One portion of the drop travelled to W, making a turn and a half, and it was projected towards B. There, if the centre of attraction had also been the centre of revolution, the drop would have been attracted towards C. If, instead of falling on the paper at B, it had returned to C, the path described would have been a curve drawn within the angle W B C.

A drop of red ink was projected at R in the direction of the arrow R 2, and part of it travelled to R 3. If it had

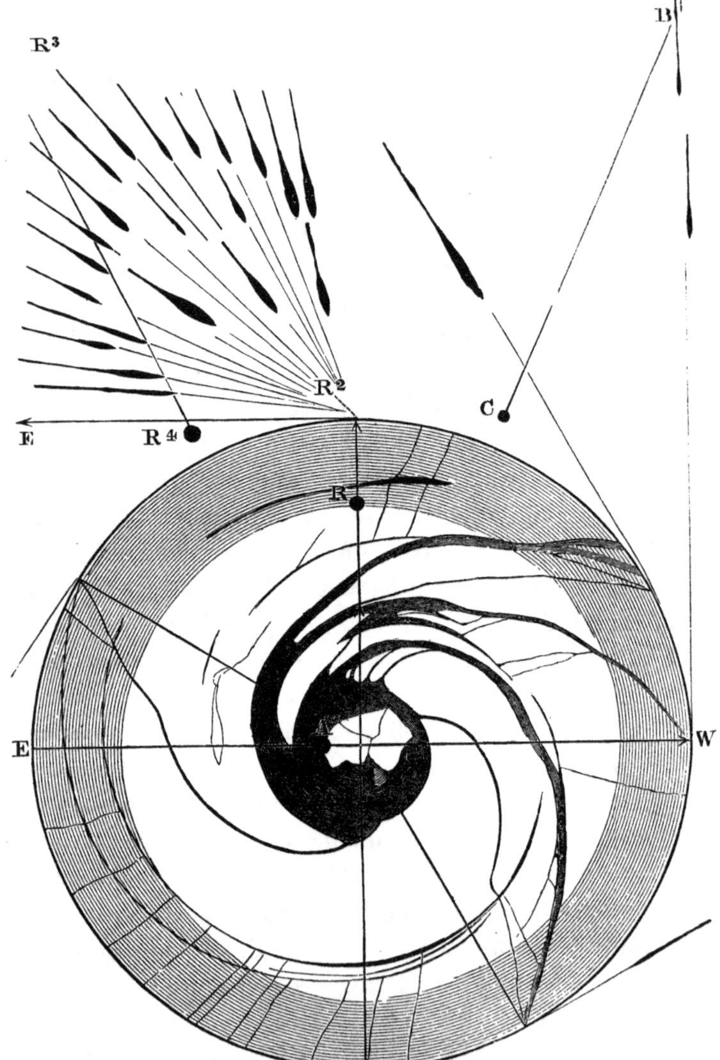

FIG. 110. HORIZONTAL SECTION.

returned to R 4, the curve described would have been contained within the angle R 2, 3, 4.

The first might be called a trundling mop curve ; for it is a result of centrifugal force and gravitation. The second is a result of three forces, and one was on a ray.

From this diagram it seems to follow that a stone projected vertically from a volcano at the equator does not move off at a tangent to the circle described about the earth's axis by the top of the mountain, but moves off on a line which divides the right angle made by a tangent and ray. If the radial and tangential forces produced equal velocities, the line would divide the right angle equally, and the stone would set off at an angle of 45° to the plane of the horizon, eastwards. But at every point in its flight, a stone is pulled sideways by the earth's attraction, as a drop of mud is pulled down when thrown up by a carriage-wheel. In mop-curves drawn on a wall straight lines are bent by gravitation. The straight line is bent into a curve. In the case supposed the curve described is a result of radiation, centrifugal force, and gravitation—a combination of force acting in three different directions : 1, from centre towards circumference ; 2, from circumference at a tangent in the direction of revolution W R E ; and 3, from circumference towards the common centre of attraction and repulsion. In drawing this second diagram, two of these forces acted in the horizontal plane ; the third at right angles to that plane. The object aimed at was to get forces to act, so as to illustrate the action of rays opposed by another force. A volcanic bomb describes a curve like any other projectile cast in the same direction with equal force : the path of every projectile is matter of calculation and of speculation till the experiment is tried ; but without calculation, it seems plain that a bullet aimed at the zenith point from the equator ought to fall to the west of the gun ; from either pole into the gun; from any intermediate latitude to the west, and at some place further from the nearest pole than the starting-point—south or north :—and west.

In the diagram, p. 450, a drop travelled from the centre to the circumference of a disc of paper revolving horizontally in the direction W R E, as the plane of the equator does. Ink travelled from A through W to B, and would have moved towards C in the direction +, if attracted towards the centre. The point A also moved in the same direction about the axis. But in travelling on the revolving disc from A to W, the ink described a backward curve. The paper and every point upon it, and ink adhering to it, moved W R E, but ink rolling along the paper as a bullet flies through the atmosphere moved E R W. It reached a larger circle on which points moved faster, at each stage.

A drop of ink falls perpendicularly. It may be so dropped as to move towards the axis of a disc revolving vertically in the direction W R E. In moving from circumference to centre, it moves forward with the paper, but it describes a curve in the backward direction E R W, because the paper moves faster in the opposite direction W R E. As the first curve was drawn in the direction W throughout, the ink always lagged behind the paper. But if paper moved faster than ink, the point A won the race : the gun beat the bullet ; it could not return to A, but to some point behind it, or to the west.

A drop of ink fell perpendicularly upon the point A, and a drop thrown up through the axis would return into it. Its own centrifugal force does not disturb the path of a rifled shot. Between the equator and poles of a globe, as many discs revolve as there are planes at right angles to the axis. At lat. 45°, the plane of revolution and a plumb-line make an angle of 45°.

A stone aimed at the zenith, driven in the direction R by a ray-force, is subjected throughout its course to the centrifugal force, which acts in the direction T E, or towards the equator.

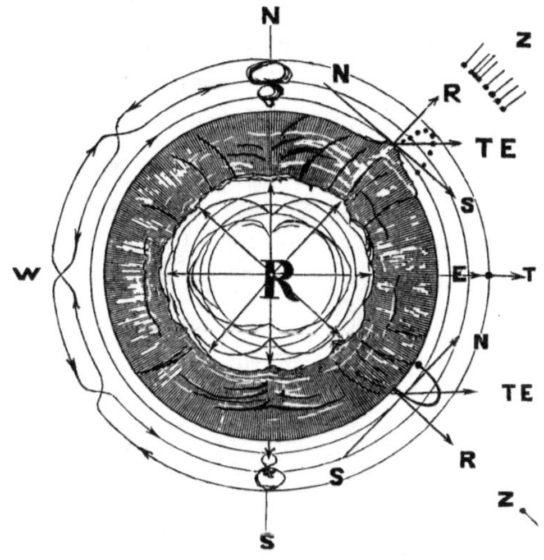

FIG. 111. VERTICAL SECTION.

If this reasoning be right, the longest slope of cones of eruption in the northern hemisphere ought to be south and west of the chief craters, and in the southern hemisphere to the north and west. Polar ice, after passing Spitzbergen, goes not to Archangel, southwards, but towards St. John's, Newfoundland, at something like a tangent, to the circle of lat. 80°. It goes south and also westwards. It describes a backward curve.

The same reasoning applies to volcanic bombs. Ink, in moving from the centre towards the circumference of paper, revolving in the direction W R E, described the curve A W. Lava shaped itself into like curves in the specimens at p. 379. Nos. 1 and 2 revolved in the direction W R E, No. 3 in the direction E R W. The core and crust make backward curves (in the impression), like the curve A W in the diagram, p. 450. The longest axis of chambers, and many systems of chambers in concentric layers, have the same curved arrangement. If the earth has a like structure, a majority of craters ought to be found to the west of the chief cones of eruption at the equator; in the northern hemisphere to the south and west; in the southern hemisphere to the north and west. Oval craters ought to have like bearings for their longest axes; and most of the matter thrown out ought to be found on that side.

In fact, the longest slope of Etna is to the south of the highest point, and most of the matter thrown out is to the south, and to the west of the meridian which cuts the highest crater (15° E.)

The longest slope of the cone of Vesuvius is towards the Bay of Naples, about S. 55° W. of the crater, and the broken ring of Somma is open to the south-west.

The long axis of Hecla, lat. 64° N., bears about S. 60° W. The broken crater is open in that direction, and the last stream of lava escaped on that side, and flowed that way.

At Krabla or Krafla, 66° N., the longest slope is to the south and west. Active water-craters and the latest lava-streams are to the south-west of the highest point, and that is a remnant of a large crater broken down on the south-west.

The longest slope of Snæfell has similar bearings, S. 80 W., near lat. 65° N. (See map, vol. i. p. 85.)

So far as memory and rough notes and sketches serve, this rule holds good for all the large craters noticed in Iceland. Broken craters on shore are open to the evening sun, like the Faxefjord. The eastern edge of the crater, which is the muzzle of the gun, is also highest. These big guns are not aimed at the zenith, but at some point to the south and west of it; and charges of small shot fired from them have fallen chiefly to the south and west of the tubes from which they were thrown.

So far, personal observations and experiments agree, and make a foundation on which to build a theory. Laws of force apply to matter above the earth, and within it; to nebulæ and to worlds, to atmospheres and oceans, and to fluids under crusts; and to mops, maps, and tops when they spin.

Bodies propelled by rays must obey the same laws which govern bullets; and volcanic mountains are chiefly built of projectiles shot from the earth through tubes by rays of force.

Since these pages were first printed, many rude experiments above-mentioned have been repeated with good machinery used in polishing optical instruments, and similar fine work.*

Some eastern artist engraved certain curves on the sun in the astronomical symbol copied vol. i. p. 21. A board was made to turn W. S. E. N., in the same direction as the sun, the earth, and the solar system, that is against the hands of a watch when the face is to the pole-star.

1. A sheet of paper was nailed on the board, drops of ink were placed near the centre, and the board was turned. The ink drew curved rays, bent as they are in the symbol. As a potter's wheel is one of the oldest of human inventions, perhaps this method of drawing diagrams is old.

2. A proof of the map p. 232, vol. i., was placed with the pole in the centre of revolution. Drops of ink were placed within five degrees of the pole, and the engine was started. A drop placed at 90° E. followed the arctic current on the map, touching southern capes in Spitzbergen and the western

* I am indebted to Messrs. C. and F. Darker, of 9 Paradise Row, Lambeth, for permission to use their machinery. April 13, 1865.

point of Iceland. A drop placed at 120° E. described the curve assigned to the Baltic Current. It touched the northern end of Novaya Zemlya, the Waranger Fjord, crossed Scandinavia, skirted the western coast, and passed Scotland from the Moray Firth to Barra. A third drop placed 180° E. passed over the southern end of Novaya Zemlya, and would have crossed the south of England if a fold in the paper had not spoiled the curve.

3. A proof of the map at the end of vol. i. was fixed with a drop of ink on the pole, and spun. It took great speed to start this drop, but when it did move it drew curves which closely resemble, but do not coincide with, curves drawn by hand on the stone from which the map was printed.

4. To test the effect of speed, three drops in succession were placed on the centre, and a sheet of paper whirled thrice at different rates. The curves were not the same, but very like, and it seemed that the ink had started from different circles.

5. To test the effect of distance from the centre, a row of drops were placed on a line on a sheet of section paper. All the curves differed. A second row was placed at equal distances on a line at right angles to the first, and the paper was turned the opposite way. Reverse curves crossed near the bisection of two of the four right angles. The figures produced were heart-shaped, or like the ace of spades. Those which started from the furthest points were least bent.

6. A speed of 1200 revolutions in the minute gave like results.

7. Opposite curves were made with a pencil by ruling a line against a ruler, from circumference to centre, and onwards from centre to circumference (see chap. xxvii.)

As regards a fixed line, the path of the ink was a ray bent

forward by friction against the paper; as regards a moving radius, it was in a ray bent backwards by inert resistance.

As ink moved on whirling maps, so a bit of floating ice revolves with the earth and makes a curve south-eastwards on some imaginary fixed line amongst the stars: it is carried round by friction, and repelled by centrifugal force. As regards the meridian on the earth's crust, the ice describes a backward curve south-westwards, as ink did on the maps. Some eastern astronomer described like curves on the sun's disc: they are repeated on ancient sculptured stones in Scotland; and something like the curves drawn by Maury (vol. i. p. 28) have been found in photographs of the sun (end of vol. ii.)

Centrifugal force acts along a ray, from the centre towards the circumference of a revolving plane, and friction bends the ray forwards.

Sticky gum water dropped on a top spreads along rays, and the path described by each separate drop is but little bent. The paper holds it fast, and carries it round till it gets to the edge. There it flies off at a tangent, at right angles to its path on the paper. Ink, which is more fluid and less sticky, radiates, but the rays are more bent forward when the fluid is sticky. The paper slips past and under fluid; and air, which also radiates from the axis at slower rates, holds ink back. The path of ink on the paper is more bent. In both cases the last course along a tangent is derived from two movements: one along a ray, the other about an axis. The paper which holds together makes most turns, though urged by the same force which also urges gum and ink and air away from the spindle of a top.

This may be shown in many ways. A shallow round cardboard box with upright sides was fixed on a top and spun with it. Burning sealing-wax dropped in took various shapes as it cooled and hardened. Part of it set in bent rays, which started from a ring. The outer crust of the ring was irregular. Projectiles thrown from this circular mound there set off at tangents to the ring, hit the side of the box, and made backward curves where they stopped. The front of each drop was carried forward, and the rest stuck on spots which followed the first spot struck. The target was crossing the line of fire, so fluid bullets made long oval marks on the upright wall of the box. Any projectile must curve back if cast forward from the edge of a revolving disc through still air moving with the disc ; and for that reason volcanic projectiles ought to fall most on the western side of the crater.

By thus watching the effect of rotation on hot wax, some notion may be got of the packing of the earth's viscid and fluid interior within a freezing crust. The round crust formed in a ring, and the fluid was urged towards it by pure centrifugal force. Gravitation acted at right angles to this plane ; the effect of the same force acting towards a point on the axis is matter of calculation.

Flaming drops of wax thrown beyond the box were caught on paper. Some which had cooled were spherical, like iron sparks ; others which had not cooled so much flattened where they fell ; and the long axis of each was in a tangent to the rim of the box in the plane of the horizon. In the vertical plane each projectile described a curve. Other drops fell on water and froze flat. Their structure was chambered like other hot sparks. Each had a core within a crust.

So far these movements and forms were produced by a force which pulled a string at a tangent to the spindle of a top, and by a radiating force which fused and boiled wax ; and the last movement was a tangent to the outer circle, which revolved about the axis.

The next step was to try the effect of rotation, so as to test theories as to the interior of bodies which revolve while changing from a fluid to a solid condition. The top and sealing-wax did well enough ; but better machinery did better work.

1. A circular cardboard box, with upright sides, was spun with a mass of thick wet plaster-of-Paris in it. The forms produced were founded on bent rays.

2. The experiment was repeated in vessels of various shapes, and at varying speed. Forms produced were like those which result from whirling water in a glass bottle, but in this case a plaster cast of the forces employed was taken. While wet, the plaster was a reflector, so accurate as to suggest the making of metal reflectors by the same process somewhat modified.

3. An elliptical cardboard box, with upright sides and a cover, was nailed on and spun half filled with fluid plaster. One end of the ellipse was three inches from the centre, the other six inches. The long diameter was nine inches, the short one six. When the plaster had set most of it was found at the ends. A circle described about the centre of revolution touched the base of a curved wall, which reached the lid and filled the most distant end. At the other end was another wall : it was curved in plan and in vertical section, but not concentric with the other wall. About the centre of revolution was a low mound, from which waves of plaster made backward curves. Enough was done to prove that a hollow shell of fluid may form inside a solid shell, like the water sphere p. 353, and the sparks, bombs, and meteorites mentioned in the last chapter. I will not even attempt to name the curves which were thus produced by whirling plaster-of-Paris while it was setting, but immediate causes were plain enough. Centrifugal and cen-

tripetal forces, an engine, and a man's hand turning a crank, were links in the chain ; but powers which move planets were in that small train of whirling wheels.

The aim of all these devices was to see light through engines. For lack of mental machinery with which to calculate the effects of ray-force, machinery in motion was set to draw diagrams and build models. But some other expedient was still wanting to show the action of rays.

CHAPTER LVIII.

FORCE—MOVEMENT—WORK—FORM.

IN the last chapter various rude expedients, used for learning the effects of rays and rotation, were described. In preceding chapters attempts were made to show that certain forms and movements result from certain forms of force. It has now to be shown that, under certain conditions, radiation causes rotation, and forms which result from that form of force.

Blazing wax dropped on water cools suddenly, and the mass radiates. It throws off streams of vapour, and the recoil moves the parent mass. If the eruption caused by radiation moves off from the centre westwards, the mass moves east. When the eruption is at a tangent to the circumference, or at any angle less than a right angle to the tangent, the parent mass turns on its axis. In this case, the movement only lasts for an instant, but it proves that a cooling mass may be made to rotate by a force which radiates from within.

Camphor set alight and afloat runs about on water, and it radiates while it burns. Gutta-percha, varnish of various sorts, and many substances, move about when heated and free to move. If any substance will float and hold together, and yet part with some portion of its mass at a low heat, this action would be better shown by it.

One substance of this kind whirls. Collodion kept for a

long time in an ill-corked bottle turns into a brittle jelly. It floats in water, a viscous mass in a tough crust, a soft core of ether and collodion in a shell. As soon as the ether begins to escape, each mass begins to move. A temperature of 60°, sufficient to boil ether, sets up radiation, and ray-force causes rotation. As soon as rotation begins, the direction of the escape is determined, and each mass of collodion whirls so long as any ether is left in it. When all the force is spent, the solid remnant sinks, for it is heavier than water.

A mass becomes globular or lenticular at first, and moves by fits and starts. This is an effect of heat. In hot water, the mass becomes a hollow shell, whirls and often bursts. Placed in sunlight, the masses whirl rapidly; small hollow spheres of collodion filled with the vapour of ether form on the outside and burst, and many of these are thrown off. At each effort, the parent mass takes a fresh start. Sometimes it rushes off whirling in one direction, while the small mass whirls off the other way. Generally, each mass rotates, and also revolves about some central point. Small masses are attracted by large ones, move towards them when they get near, and are whirled off again when the pace is sufficient. The pace slackens gradually, and the globular shape often changes to a transparent cup, through which chambers and globular masses of fluid collodion and ether may still be seen. After the force seems to be exhausted in the shade, a ray of sunshine will set a whole fleet of tops spinning faster than ever, and generally in the old direction. After about a couple of hours, the charge of fluid, in a lump as big as a bean, is spent. Then the whirling stops, and the hard collodion, shrunk and shrivelled like a parched pea, sinks. When dried, it is like wrinkled horn. This experiment has been repeatedly tried, in all weathers, and always with like results. The force is a ray-force—the force of heat in the earth's atmosphere, which drives ether away as the earth's internal heat drives water out of Strokr, and lava and ashes out of Etna. In a bright sun the shadow of ether is thrown on the basin through water, and the eruptions can be watched flowing outwards in streams which curve backwards like sealing-wax dropped into a whirling box, or ink on a top (p. 450).

A like result is produced by pouring collodion into a circular tray floating on water. The vessel sails about without apparent reason, and sometimes it whirls. The mechanical force which thus overcomes the friction of water, and keeps a mass whirling for two hours, would suffice to spin the same mass in free space at a greater rate; and motion once begun continues there, if astronomers are right.

The principle of this movement, and the immediate cause of it, are sufficiently plain: to explain and account for the eccentric paths of bodies of irregular shape, thus whirling in water, would be a hard task for any one, and is far too hard for a traveller to attempt. But rules which govern movements caused by spinning a top must also govern those caused by ray-force in whirling collodion, and in rotating worlds.

If the collodion turns sunwise—south, west; north, east—which it generally does when placed in sunlight, it also revolves in the same direction about some point. It rotates sunwise because ether escapes the other way at first, and probably ether takes that direction because the shell is thinnest on the shady side where the heat is least, and evaporation not so fast. There seems to be no fixed rule, for it often turns "widershins," as the world turns.

Other substances illustrate this action of ray-force. Gutta-percha floats on water, but gutta-percha dissolved in chloroform sinks. Heated with a burning-glass under water, a mass boils, leaps up, explodes, and throws off small spheres; some of which hang under the surface, others rise and fall again, others burst and float above the surface of the water. These discs have chambered interiors in a ring, and their structure, though complicated, is regular. Kept in a stoppered bottle this substance is like any other fluid; exposed to sunlight, it grows into all manner of quaint shapes, and throws off projectiles, while part of the mass evaporates, and the rest becomes solid. In these small experiments light acted as force, and caused first radiation, then rotation, and then projection to a distance at angles somewhere between a tangent and a radius in the plane of rotation, and at right angles to the axis, which, in this case, was a ray reaching from the earth's centre to London.

If ray-force will cause rotation, any rotating body will serve for illustration; and for lack of better machinery, a top was used to see the effect of a mass rotating in a fluid at rest.

A metal top was spun in shallow water, so that the disc was in air and the spindle sunk. The vessel was placed on a solid base, where the sun shone on the water, and light reflected on the wall showed that water was as still as water ever is. The whirling spindle set up a system of waves, which refracted and reflected light, and cast shadows. The top "hummed," and while it did waves were small and of strange forms. As the sound changed, so did the shape of the waves. They were like waves which accompany sounds made by rubbing the finger on the edge of a glass. Instantaneous photography would copy these, and that experiment may be tried some day.* Besides these sound-waves, the top started others, which seemed to set off at tangents, and they spread as rings.

* These fluid forms are better defined than curves which are copied in sand by vibrations in metal plates.

Lights reflected from them seemed to revolve about the top W. N. E. S. W.; while the spindle turned the other way, W. S. E. N. W. Lights and shadows thrown on a wall made a complicated pattern of curves, turning opposite ways, while they receded from the shadow of the top rapidly. These were also reflected from the sides of the vessel towards the centre.

The revolving spindle also started a system of slow currents in the water. Burning sealing-wax dropped on the disc was thrown off, and fell on the water. Discs of wax thrown off by the top floated, and showed movements at the surface. These had little in common with the wave systems. The floats moved slowly, in curves, W. S. E. N. W., as the spindle moved. They also approached the spindle with increasing velocity, passed it swiftly, and retired, slackening their pace gradually till they reached a limit, when they returned. They seemed to describe elliptical paths. The spindle was in one focus, and the other moved round it, as the whole system did, W. S. E. N. W.

The simplest and therefore the best plan for showing these movements and curves, is to spin a metal top in a concave lens. This centre, placed in the middle of a round tray, filled to the depth of an inch with water, keeps the top near one spot. "Gold paint" may be got at any artist's shop. Dropped upon the whirling top, this fine dust is thrown off at tangents, and where it falls it floats. It moves round the top in the direction of rotation, but it also approaches the spindle, whirls round it, and sets off again. The nearer a grain of dust is to the spindle, the faster it moves. The pattern produced is like a series of rays bent backwards. The whole system is moving one way, but the outside does not keep pace with the rest, and seems to lag behind. When the top begins to lose speed, the spokes bend the other way, forwards. But every trial gives a different variety of the same pattern; and sometimes eddies near the outside turn the other way.

To unravel that tangled skein of whirling curves would be as hard a task as to explain the movements of an egg-shell whirling on a fountain; but the force which pulled a string and spun the top was a link in the chain of causes which made the puzzle, for the water was a mirror before the top was spun.

Collodion whirls without any force but the force which boils ether, and it whirls fast in sunlight.

In the first contrivance, radiation set up rotation and kept it up for a long time. In this, rotation arranged a fluid and floating solids; two opposing forces acted in one horizontal plane, and the earth's gravitation did not directly interfere with the curves. The top scattered projectiles, as it did on the diagrams above described, but in this case they fell where they could move. Some force, probably friction, attracted them towards the spindle, and dragged water and dust towards one side of the turning cylinder. It raised up a small mound about it. Centrifugal force drove water away at tangents to the other side. The whole moved in one direction about an axis, and separate parts of the system also turned the same way so long as the top continued to spin.

According to works on astronomy,* the sun and the solar system also turn one way. If seen from the fixed axis of

* 1. They (the planets) move in the same invariable direction round the sun; their course, as viewed from the north side of the ecliptic, being contrary to the hands of a watch. . . . .

2. They describe oval or elliptical paths round the sun—not, however, differing greatly from circles. . . . .

3. Their orbits are more or less inclined to the ecliptic. . . . .

5. They revolve upon their axes in the same way as the earth. . . . .

the sun by an observer with his head towards the north, the system would pass towards the left, for it moves as the hands of a watch move when the back of it is towards the Great Bear, or the face of it is turned towards the Southern Cross.

In this contrivance movements were similar and in the same direction. When the top was spun by pulling the string from the left side of the spindle, everything turned W. S. E. N. W. "against the sun," as sailors say, or "widershins." Radiation caused rotation: rotation spread and caused revolution about an axis. Centrifugal force repelled, but some other force attracted the system, and it revolved. Systems of waves also radiated from the central body, and they seemed to move fastest from the left side of it, because they started thence, and were approaching. The waves moved swiftly, and did not interfere with the other movements.

One aim of these and of many other similar contrivances was to set up systems of radiating waves, in order to watch their effect. Light, according to the best authorities, is an effect of waves analogous to sound-waves. But if there be waves there must also be something material in which waves can be propagated. There is no sound when a bell is struck in the exhausted receiver of an air-pump. But if there be some medium in space through which light-waves move, it ought to obey the laws of motion like any other material—like air, or like water. If these waves of light act as waves of force, then force, though directed by a spinning top, may work as force does when it radiates from a whirling star. In this case the waves moved faster than currents, and bodies of different weight revolving about the top moved at different rates in

6. Agreeably to the principles of gravitation, their velocity is greatest at those parts of their orbit which lie nearest the sun. . . . .

Hind quoted, p. 13. *A Handbook of Astronomy*, by George F. Chambers, F.R.G.S. London, 1861.

different curves. In the collodion experiment the whirling resulted from ray-force. It has yet to be proved that rays of force do accompany rays of light; and one way to learn that fact is the old path to a forge.

There sights and sounds prove that force is active. The sky glows; the hiss of steam, the dunt and thud of hammers, the crash and clang of iron bars, the rattle of wheels, fill the air with waves of discord. Thirsty giants in armour, with vizors of steel wire, stand in a spray of iron sparks near the hammers. They are of the class who are now on strike, and they earn their high wages, for their lives are short, if they are merry while they last. With a loud warning shout, an eager boy charges up with a white-hot, hissing, sputtering mass of puddled iron to feed the hammer; and it may be that another urchin charges the other way, trailing a red ingot to feed the rollers. Every one must take care of himself in this den of fire. A giant in steel boots grips the puddler's ball with a pair of tongs, and with a dexterous whirl and swing it flies glowing through the air, and lands on the anvil. There it is crushed and squeezed till slag flows out of it like water from a sponge. The mass is chambered like some meteorites. When the blow comes, sparks radiate like rays from a star; and each in turn radiates light, heat, and force; for the sparks hiss when they touch water, and they burn skin and clothes. Great scissors gape, and nibble off the end of a steel bar, as a horse bites a carrot. Another pair of steel jaws may be found champing the air at your elbow, and when that mouth gets a bar to bite instead of a bone, it snaps it off with a crunch, and gapes for more. Still larger shears shred boiler-plates like silk. At the rollers, a block goes in and a bar comes out, streaming with fluid slag squeezed out. The iron comes charging over iron plates, like a red snake uncoiling; a boy seizes the head, and turns it back, and the bar comes out as thin as an eel or a ribbon. A few more turns and it would be a wire. It is no place to dream in, but there is plenty to see by this furnace-light.

If the engine is worked by steam-power, then all the force came out of the boiler-fire, and went towards the earth : if worked by water-power, rays, which work the atmosphere, lifted water and poured it into the milldam. So in a forge, as elsewhere, part of the force used was in rays of light.

When a large casting is to be made, a furnace is tapped, and tons of metal are run off into great vessels, lined with clay, as men run ale from a vat. It often happens that the metal is too hot for immediate use, and it is allowed to rest for a while in its great caldron. It is a beautiful object. The surface is in constant motion, and it shines and glows. Creamy red islets form on it, and move rapidly, while shining lanes of bright metal curl and twine beautiful patterns of coloured light. The smooth hot fluid is darker than the scum next above it, and the highest points darken before the scum. Every moment some bright spark flies off, whirling and shining like a star ; each describes a luminous curve in the air, and some burst like rockets and scatter a spray of light. There is a force in the fluid, and it radiates like rays of light.

If it were free to move, iron would revolve, because collodion and other substances move and revolve when they cast off projectiles.

To cool the iron, cold scrap-iron is sometimes dropped in, and these masses float deep and melt as ice does in boiling water, or sink if the solid is heavier than the fluid. These are sometimes wet, and when they are, water explodes and drops of iron are cast whirling to great distances by steam. The power still radiates, but it acts more powerfully on this substance. The same amount of ray-force produces different rates of expansion ; but this action, like the first, shapes projectiles, and throws them away from a hot mass of iron. It radiates :—it shines, it is hot, and it throws off sparks.

Before iron is run to be made into shot and shell at Woolwich, the slag which floats in the furnace, like oil on water, is run from the other side. It pours down and freezes like a hollow icicle where it falls, but a large mound of it grows before the day is done. In it is a magazine of ray-force. While the mound is hot, it throws off a spray of shining drops. As the mass cools, these get smaller and do not fly so far. Some about the size of No. 6 shot were thrown more than twenty feet at first, but after ten minutes the range was only two or three feet, and in half an hour the distances traversed could be measured by inches. It was a magazine, but not an inexhaustible magazine of force.

A ton of iron throws shot and shell through tubes in a crust, as the earth does. Hot slag does the same; and when the slag is broken, the guns may be found aiming at the sky, as volcanoes do. In some of these, half-made shot may be found also.

They are generally egg-shaped chambers with the small end uppermost, and the slag is often spongy near the large end. After the slag has ceased to fire these volleys the surface turns dusky red, and darkens. If water is thrown on at this stage the crust blackens and contracts, water boils above and in cracks, and fluid under the crust often wells up as a shining spring of lava wells up from under the dark crust in which hot springs boil in Iceland. The projectiles now are drops of hot water, or fragments urged by steam ; the old guns are changed into steam-guns ; but force which drives the shot is in the slag, and it radiates. When the crust is broken it shines as the earth shines when a lava-spring is driven up by ray-force.

As the charge of force is expended, the action decreases; and when the mass is as cold as the space about it the movement ends. Till that balance is reached the attraction of gravitation is overcome by the opposite force, which radiates where light shines from a furnace, or from the earth.

Where electro-magnetic light, earth-light, and furnace-light shine, there also force radiates. Lava, silver, iron, slag, all radiate force, when they radiate light, and the rays of the sun also are accompanied by mechanical force.

The rays of the sun reflected from the earth's crust and absorbed by it, by the atmosphere and the ocean, at a distance of ninety-five millions of miles, or at some other less enormous distance according to recent discoveries, cause radiating movements. Solar rays furnish most of the power to engines, whose tool-marks are denudation and deposition. The same rays, reflected from a rough convex surface in the moon, and therefore greatly dispersed, still act as force, for they move the index of a thermometer. Piazzi Smyth when on the shoulder of Teneriffe, and above the clouds, got a black bulb thermometer up to 212° in the sun's direct rays: he got about half as much heat from the moon as he got from a candle on a stool at a distance of 15 feet.*

No thermometer yet contrived will measure heat reflected from distant planets; none will measure heat reflected from a window in Calais, and radiated from the electric light on the English coast; but nevertheless heat-rays cross the Channel with beams of light.

The air gets colder the higher we go, and hotter as we descend, but the sun's rays get hotter and brighter as the air

* P. 231. *Teneriffe*, 1858. London. By C. Piazzi Smyth, etc. etc.

clears. At p. 487 is a diagram drawn by the sun, which proves that the atmosphere absorbs the light, the heat, the burning power, and the mechanical force of rays.

If the sun's rays so act at this distance, it seems to follow that they must also act as ray-force at their source in the sun. If they do so act, then visible forms on the sun's disc ought to be a legible index. In order to learn that alphabet, the sun's rays must be set to work.

In order to prove that rays of mechanical force do accompany the sun's rays, they were set to make pictures, to carve wood, to model wax, and to move machinery.

In the first place, the sun was set to make photographic portraits of himself, and these are some of the expedients used instead of an observatory :—

On the flat top of an out-house in a garden a mirror was placed in a flower-pot, and so fixed as to reflect the sun's rays downwards through a hole. The first flower-pot was placed on a second turned upside down, so the sun's reflected rays passed down through a diaphragm. This arrangement stood over a hole in the roof, and over it the lens of a telescope was laid flat. By it the rays were refracted to a focus in a dark room. The image formed was about the size of a BB shot, and it had to be magnified. Below the focus an iron retort-stand was placed, and in it a $\frac{1}{4}$-plate lens by Ross was fixed. The second lens formed a second image. By varying the distance between lens and object, the size and place of an image can be varied. If the lens is near the object, the image is far from it, and larger than the object; when the lens is far from the object, and rays are nearly parallel, the image formed is near the lens, and smaller than the object. The image formed by the first lens was smaller than the sun, which was the object, because the sun's rays are nearly parallel at this distance from the sun. From that image rays diverged, visibly if air was misty or smoky in the room. The second lens, and a sheet of white paper, were so placed as to form and catch an image a great deal larger than the object to be magnified, which was the image of the sun in the focus of the first lens. In short, the photographic lens was an eye-piece. A common telescope fixed in a window-shutter, and aimed at the sun, will give a magnified image, by sliding the draw-tube till the focus is found for any screen, but the vertical arrangement was made with a purpose.

The distances having been found, a sheet of cardboard with a hole in it was fixed upon the iron shaft of the retort-stand, and the light was shut off.

A collodion plate was then substituted for the white paper, and the card was whirled through the beam of light; so that light passed through the hole during some fraction of a second.

A copy of the best result obtained is at the end. It is a negative on glass, so developed as to whiten it. The collodion film was covered with a layer of black oil-paint, and backed with blotting-paper. It tells light on a dark ground, and is a portrait of the sun drawn by himself in black and white. The first mirror tried was silvered glass of the ordinary kind, and it gave a double image; the second was a sheet of plate-glass, backed with black paint, to absorb one of the reflections. It is very easy to describe this contrivance; it was by no means easy to work it. The sun would not stand still, and the reflected rays moved; the image moved; the place for the screen changed at every moment; clouds got in the way at the instant when all was adjusted; and when the cloud had passed, the sun was out of the field till the mirror was set again. Late in the day, the sun got entangled in a tree, and he hid behind smoky chimneys in the morning. A bright morning often changed to a cloudy noon. Besides all these difficulties, the ordinary ills of photography interfered; and lastly, when all was done, a tidy housemaid starred the glass of the picture now engraved.

Fourteen pictures survive, and no two are alike. In those which have double images curves and other forms are repeated with more or less intensity, but the forms are the same. They do not result from photographic manipulation, but from something beyond the mirror which doubled the reflection. Two pictures were taken on one glass, by passing the screen through the beam of light a second time, after waiting long enough for the image to move its own breadth. Even these do not tally, for clouds in the earth's atmosphere and London smoke interfered; but enough remained to show that the forms copied are beyond the clouds, for parts of the forms are repeated though not the whole. In some respects all the pictures resemble each other.

If developed so as to make a "good negative," the sun's image is a black spot. If very slightly developed, so as barely to show an image at first, details come out when the collodion is covered with a thick layer of white oil-paint, and then the picture is safe, though black upon a white ground. Generally, each picture is surrounded by a ring of light, which is dark in the negative. One edge is darker than the other. Edges are often fluted and rough, as if the image were distorted by waves in the earth's atmosphere. These waves are easily seen on a hot day, and they impede telescopic observations; here they are copied on the edge of an object of known angular size, so they can be measured. They show that the air is moving like hot water; rising from the hot ground, which absorbs heat from the sun, and gives it back to space as ray-force.

The sun's disc is streaked and barred, and spotted in patterns, and when a series are placed together the patterns have something in common.

When the strongest side of the ring is to the left, dark bars, which are bars of light, cross the sun's disc, as spots do in zones parallel to the sun's equator.

Shortly after an eclipse, a photograph of the sun was taken with a lens, which gave an image about the size of BB shot. A well-marked band is in this picture. Another observer noticed a similar appearance, of which he published an account, I think, in the *Photographic Journal*, 1858. The band, or one like it, is well shown on another picture, two inches in diameter; and in one about an inch and a half broad more bands are shown. One small picture has a whole series of bands. When placed under a microscope, this picture has several crescent-shaped gibbous spots, which, from their size, may be grains of dust; but they have the illumination which they might have if they were bodies within less than a degree of the sun's disc.

The picture selected for engraving is like Maury's diagram of the winds, copied at p. 28, vol. i.; like ocean-currents in the Atlantic, laid down upon a new terrestrial globe lately published at Berlin. Others are somewhat like portraits of Venus, Mars, Jupiter and his satellites, made by able astronomers, and published in the *Handbook of Astronomy*, by George F. Chambers, in 1861. In some, lines and patterns interlace like lines of light on hot fluids, and some patterns are drawn on the principle of lines drawn from pole to pole on a revolving globe. In one, the sun's disc is barred with straight lines which meet at various angles, and make a pattern like that flashing northern aurora which Scotch peasants call the "Merry-dancers." This very rude photographic eye saw rays which common eyes did not see on the white paper, and it did not see "spots on the sun," which were conspicuous objects on the screen. The conclusion arrived at was that the camera saw through the sun's atmosphere which dazzles eyes, and copied the currents in it against a luminous background of less intensity. Perhaps the black mirror absorbed rays which are reflected by other mirrors.

A heliostat set to reflect the sun's rays, through a telescope aimed at the pole, would cure most of the evils which beset this rude observatory, but there was no heliostat handy. The only telescope owned had a chemical focus, and was sadly battered; and so this troublesome work was abandoned as soon as a result was obtained. Better machinery, constructed on the same principle, may perhaps be tried soon.*

* The plan devised for observing the sun may be explained in a very few words. It was not carried out for lack of a hill and a heliostat, and for other reasons. On some hill-side facing the south—say Arthur's Seat, near Edinburgh; Primrose Hill, Highgate, Hampstead, or Sydenham, near London,—observe the pole-star, and choose a place which brings the true north to the brink of the hill. Mark the place of the eye, and of a sight on the hill-top due north. About this line of sight, which is a straight line parallel to the earth's axis, build a passage, or else dig one below it, so as to make a fixed tube. At either end of the tunnel place a heliostat, with the axis in the axis of the tube, and at the other end place a screen at right angles to the axis. By changing the angle of the reflectors, any ray may be reflected up or down the tube, and any arrangement of lenses may be set in the ray. The only artificial motion required is a clock to turn the heliostat. The earth does the rest. A very little sunlight will make an impression, so one lens of small aperture and long focus would serve for solar photography. Amongst the advantages of this plan are steadiness in the whole contrivance, even temperature in the tube, and cheapness. The chief cost would be that of a passage of equal dimensions, if built, or the cost of driving a shaft through the top of Arthur's Seat, if that were the place chosen.

So rude were these experiments, that no record was taken of the bearings of the plates. The picture selected has been placed on the page with bearings suggested by itself. As the sun is turning from west to east, light-waves ought to travel fastest from the eastern edge, which is approaching, and fastest from the equator. The image ought, according to theory, to be brightest at one spot, namely the place where the equator cuts the advancing limb. That spot has been placed to the left, and all other forms fit. The darkest parts of the disc are to the right, where the surface ought to be receding; and above and below the equator near the poles, where movement is slower, and light less direct, than it is at lower latitudes. The picture may be a fallacy, but it is so like a fact that it is placed here to be compared with others.

Everybody knows that the sun will paint his own picture, but this particular portrait is peculiar.

It joins in with the rest of these whirling diagrams, for it is drawn on the principle of the whole series. It is a form which resulted from the whirling of the sun and from solar radiation; the forms so copied are like those which result from the whirling of the world, maps, and tops.

On the 18th of July 1860, a great many photographic contrivances were tried. An account of the successful operations of Mr. Warren de la Rue is in the *Photographic Journal* for August 1860, p. 297. A scheme tried in London

Two reflections—one towards the pole, another in any other direction—will steady the sun's ray on a point. The ray may be sent up or down—up a tall chimney, or down a coal-pit or an old well, or along a dark passage. The effect of two reflections has not been tried; but two plane mirrors, one small lens, and a clock, might be made true as easily as the numerous lenses of an astronomical telescope, with all its complicated and costly machinery. In one case, the whole structure follows the sky; in the other, the ray is turned into the telescope which the earth turns.

answered tolerably well, though the apparatus used was of the rudest.

A common photographic camera was placed on a stand, aimed at the sun, focussed carefully with the full aperture; and a stop, with a hole about an eighth of an inch in diameter, was placed in contact with the outer side of the object-glass. It was found by experiment that the sun's image alone made an impression on a collodion plate, when the cover was lifted and rapidly replaced by hand, when the sky was clear. By waiting a certain time, the sun and the sun's image moved far enough to separate images on the plate; and the film kept wet for half an hour. Having set this instrument with a plate in position, all the observer had to do was to lift and replace the cover at regular intervals, without shaking the camera. The world turned the instrument more steadily than clockwork. If time is accurately divided, the distance from image to image is a scale divided by the engine which keeps the best astronomical time.

At 1.32½ mean time, according to a neighbouring astronomer's clock, the cover was lifted for the first time, and it was opened and closed seven times, the last at 1.56½. The sky was very cloudy, so the cover was lifted when there was a chance. The first plate was developed by an assistant, a second was placed, and the camera was turned a few degrees by 2.0½, and so on till 2.58½. In all 38 attempts to take pictures of the sun were made on seven plates, and of these 35 trials succeeded. In particular, three out of four trials at

h. m.
2—30½
2—33½
2—36½
2—39½

according to the watch used, and the time corrected from the neighbour's clock, gave three crescents differently placed. They are all within half an inch of each other, but clear and distinct pictures which bear magnifying. The object aimed at was to catch the "red flames" which were caught by Warren de la Rue in Spain. In London the instrument used and the plan tried failed to catch these forms ; but it caught the eclipse, and it cost very little.

In five of these pictures, taken about the time of greatest obscuration, the upper horn of the crescent has a tiny dot beyond it. The relative positions of points and dots vary slightly, at a regular rate. This is the place to find "Baily's beads," and these may perhaps be photographs of that phenomenon. The passage of the top of some tall lunar mountain along the sun's edge would make the horn of the crescent seem blunt or broken. Constellations of collodion "pin-holes" and "dust-spots" on the film interfere sadly with observations on this minute scale.

This method succeeds well under ordinary circumstances, but during the eclipse it produced some curious results. Some of the crescents came out negative or black ; others came out positive or transparent. Of four pictures on one plate, 1 is a faint negative with a bright edge ; 2 is a good negative with a bright edge ; 3 is gray all over, but positive ; 4 is nearly transparent. Of five pictures on another plate, one is black with a transparent edge ; another is equally transparent in all parts ; the rest vary. Diffused light produces this effect, but on other occasions eight pictures of the sun have been taken on the same plate, all of equal intensity.

These photographic expedients are sufficient to prove that the sun's rays will cause movements in photographic chemicals. Everybody now knows that fact, and everybody wants to have a portrait of everybody, except the sun, which seems ungrateful at least.

It is not so well known that the sun will engrave.

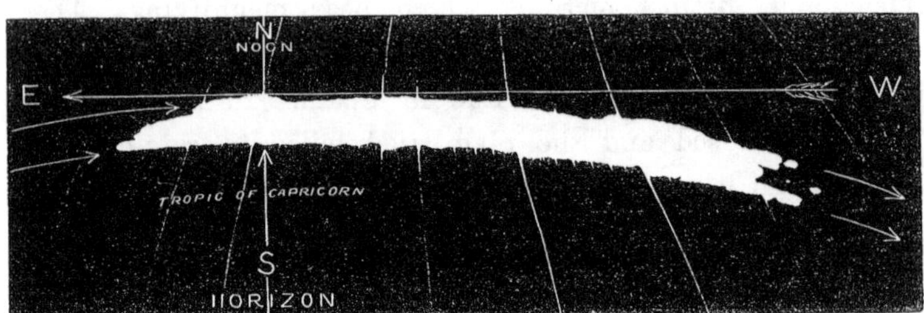

FIG. 112. WOOD-ENGRAVING BY SUNLIGHT. THE SUN'S PATH IN THE SKY.

Lines engraved by the sun on a vertical plane of wood placed in the focus of a spherical lens. Winter solstice, 1863, about six weeks.

The sun was set to carve wood, and here is a specimen from a block engraved by the sun. A glass ball was placed on a stand outside a window, and a wood-engraver's block was placed to the north of it; the printing surface was in a vertical plane, and near the focus of the glass ball. The world turned the block towards the east; the sun's rays turned on the centre of the ball, as a compass-needle turns on a pivot; and the sun's image in the focus travelled eastwards as the sun appeared to travel west. Where it travelled, there it left a deep charred spoor. In the morning the image was at W., in the evening at E., and it made a deep hollow curve. By capsizing and turning it end for end, the impression is righted, the curve is made convex to the plane of the horizon, and the sun's path is from E. to W. on the paper, as it is in the sky to the south. The sun was moving from the Tropic of Capricorn northwards, so the path varied each day. The sky was cloudy, so the spoor was broken. The image moved on a sphere, the surface was a plane; so the sun's round image drilled oval holes.

The diagram proves that the sun's rays set up chemical action, and burn boxwood as a hot iron might; and that they also work as mechanical force, for they tore the wood. It tore along rays which radiate from a centre of growth, but the strongest man living could not so tear boxwood with his hands.

Here is another specimen of the same art: Two dotted lines were drawn by the sun 10th March 1862 and 23d November 1863, when the sky was dotted with flying clouds.

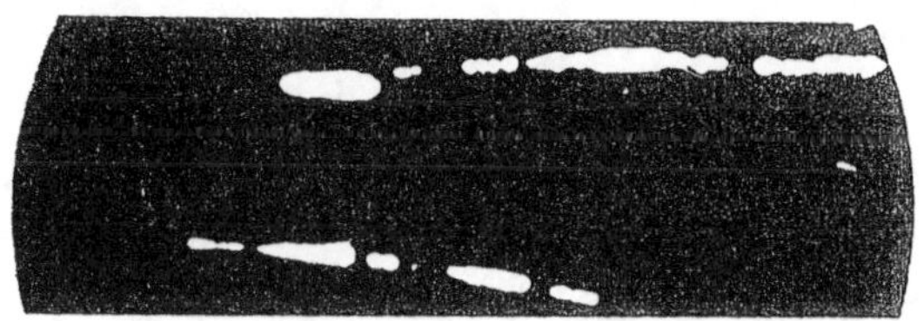

Fig. 113. The Sun's Path on Two Cloudy Days.

The place where each passed the sun is marked by a dark space. The place where the sun was, when the cloud had passed, is marked by a white spot, or by the beginning of a white line. In the wood, the white spaces are at the edges of deep holes and grooves burned away by hot rays. The curves do not coincide, because the block was in different positions.

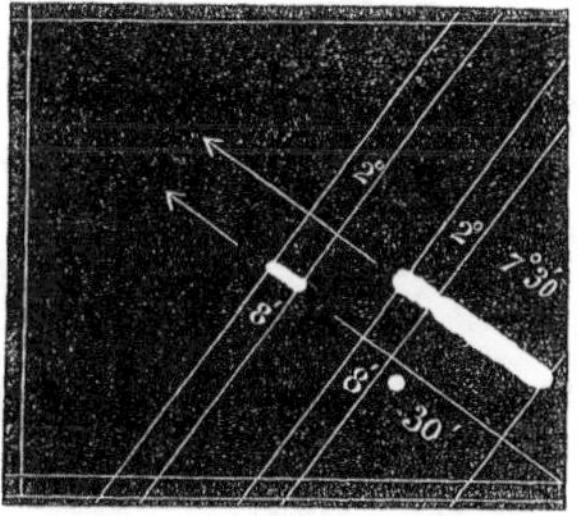

Fig. 114. Solar Scale.

If passing clouds and the sun thus divide a line, space may be divided by making light and darkness recur at regular intervals of time.

Here is a scale made on this principle, April 1865.

A block was placed opposite to a ball of glass, with the surface within the burning focus. At 4.25 the sun burned a dot. The block was then moved by turning the stand on which it was placed, in azimuth. At 4.27 the lens was uncovered, and it began to burn. At 4.35 it was covered, at 4.37 open ; 4.45 shut, 4.47 open ; 4.55 shut, 4.57 open ; 5.5 shut. It was found that the image was too large to show divisions, so the table was turned a few degrees, and the lens uncovered for eight minutes. At 5.13 it was closed, and after that time the sun was hid by clouds. This scale is correctly drawn by the movement of the earth ; on a vertical plane, which is a section of a sphere, with the radius of the burning focus ; and it proves that photographs taken on flat surfaces must be distorted. Thirty minutes of time are equal to 7½ degrees of the circle on which the sun appears to move. Eight minutes are equal to two degrees; and it is evident that spaces and dimensions are unequal on the block.

The same block was first tried in the focus of a ¼-plate lens by Ross. The sun marked the block, but did not burn the wood so as to make a groove. Many scales have been made with the same instrument on collodion plates, and there is no practical limit to the minuteness of a scale thus divided. An image of the sun in the focus of the smallest lens ever made will move a certain angular distance in a given time, and a collodion film will take impressions of it. These blocks are only meant to show how the thing may be done.*

The point to be made good is, that the sun's rays will do the work of hot iron at a distance of a certain number of millions of miles from the source, and these diagrams prove the fact, which anyone can prove with a burning-glass, by

* The principle was applied to drawings made for the Lighthouse Commission, some of which were published in the report 1861 ; and it is a contrivance which may be useful, so it is here described.

writing his name on a walking-stick, if he chooses to take that trouble, on a summer's day.

The pattern which results from the whirling of a spindle in still water is founded on opposite curves : one set is drawn away from the circumference, the other set towards it on the opposite side. Such curves are drawn on watch-cases by engine-turning. As the sun's rays engrave, and the world is whirling, rays may do the work of a steel point on a surface moved by the world, instead of a lathe.

In 1857 the *Meteorological Journal* printed a paper "on a new self-registering sun-dial." It is worked on this principle, and it can be applied to various uses.*

The instrument is of the simplest description. A ball of glass is placed upon a truncated cone of lead, in a hemispherical bowl made of wood or stone, or metal or glass, or any other substance. The centre of the solid sphere coincides with that of the hollow hemisphere, and the dimensions are so arranged as to make the image formed by the glass coincide with the hollow surface. The common centre—the apex of the truncated cone—is the "fixed rest" of the lathe, the sun's image is the cutting point, and the other end of the chisel is about ninety-five millions of miles away, fixed in the sun, for it is a double cone of rays of light. The edge of the bowl must be level, and the instrument placed where the horizon is visible. To use a photographer's phrase, this is a camera with an angular aperture of 180°. The image is formed upon a hemispherical screen, and the high light alone is copied in the picture. The sun's image in the bowl copies the sun's

* At the end of the paper is this passage :—"If it were in general use, the sunny and cloudy regions of the world might be laid down with greater accuracy, and deductions might perhaps be drawn from direct observations bearing on questions of general science foreign to this description of an instrument."

apparent path in the blue vault of the sky, and the shadow of the glass ball moves in the dial, with a burning centre of brilliant light. If the blue vault were a screen, the world's shadow would move round the sun in a year, on curves like those which the sun's image draws on the bowl. When the moon gets in the way, there is an eclipse of the moon.

If the instrument is placed in position when the sun is on the tropic of Capricorn, the image begins to burn on the western side as soon as the rising sun has risen high enough in the eastern sky to clear vapours which absorb light near

FIG. 115. ENGINE-TURNING BY SUNLIGHT.

Here is a section from a block, sawn out parallel to the plane of the horizon from the meridian westwards. It represents the sun's burning power during the morning for about a quarter of a year, at an altitude of about twelve degrees. The depth of the groove may be measured by completing the circle, of which an arc remains. The blank near the middle corresponds to a similar blank on the meridian, and marks foggy weather. (See p. 487.)

the horizon. At this position the image makes a shallow mark. As the day wears on, the image draws a line eastwards; it passes the meridian, and rises in the east. At every step on this path the powers of light vary. The forces which do work in the atmosphere cannot do it over again below; so visible light, heat, and "actinic" power, all vary in something like the same proportion. The shell of air is thinnest over head, and a vertical sun is the most powerful of all. The shell of air is thickest and most charged with vapours and dust towards the horizon, and this sun-dial proves that the sun's burning power is subject to the same law.* Marks burned at about the same distance from the horizon are about the same depth, and the deepest are the nearest to the plumb-line and the bottom of the bowl—namely, marks made about noon and the longest day. By their chemical actinometer Bunsen and Roscoe got the following numbers :—

Total chemical action effected by the sun's rays from sunrise to sunset at the vernal equinox—

Melville Island	1306
Reykjavik	2324
Petersburg	2806
Manchester	3625
Heidelberg	4136
Naples	5226
Cairo	6437

At Cairo the sun's rays at the vernal equinox are nearer to the plumb-line than they are at Reykjavik, and so they do more work on the ground, and less work in the air.

In like manner, rays do most work on the dial when they have least work to do in the air through which they pass. They do less work under a yellow haze of London smoke than they do in the country near London, and they do nothing under a thick cloud. But when the layer of clouds is passed, forms and movements there prove that light is accompanied

* Fuller information on this subject will be found in works on light, especially in papers published by Professor Roscoe. In the *Photographic Journal* (June 15, 1860, p. 256) is an able paper, read by Mr. T. R. Wheeler before the Photographic Society of Blackheath, in which the researches of Bunsen, Roscoe, and others, are referred to. See also *Teneriffe*, illustrated with photo-stereographs, by C. Piazzi Smyth, a book which is very amusing as well as instructive, and expressed in few and simple words. See also papers on Light, by Sir J. Herschel, in *Good Words*.

by mechanical force, which radiates from clouds, and makes them boil. (See chap. v.)

The line drawn on a clear day is part of a spiral on a sphere. Next morning the point of the graver begins again on the west; each noon finds the sun higher in the sky, and the spoor of the image lower in the bowl; each evening finds the sun further north on the western horizon, and the image further south on the eastern edge of the hollow surface on which the burning point revolves about the fixed rest; and so this engine-turning goes on for six months till the longest day. Then the sun's image turns and burns the other half of the spiral design, crossing its former path. Such lines could be drawn by moving a rest horizontally while a ball is turned about a horizontal axis; but the best of turners and rose-engines and tools could not equal the accuracy of this work. One end of the lever is ninety-five millions of miles long, and the other may be an inch, or a thousandth part of one; it is at the focus of the lens.

The object aimed at was gained when the sun had made a spoor; but here is the spoor of the sun on the meridian of Campden Hill for the best part of three winter months in 1859.

The instrument was set on the top of the engine-house at the waterworks at Kensington, 200 feet above the sea, with a clear horizon, where the sun could shine; and London smoke was to the east. The image of the sun was at the Tropic of Capricorn, T, below the edge, H, at noon, and made a shallow mark on the meridian. It drew a groove eastwards, and passed over the edge of this particular plane. As the year wore on, the equator of the bowl rose; and the image cut grooves daily, each of increasing depth when the sky was clear. At a certain time, it fell in with a cloudy atmosphere, and then the work done at noon was less. Just before the equator got to the hot point of the graving-tool, the glass ball was knocked over. It was found resting on the side of the bowl, with deep grooves scored from a different centre at

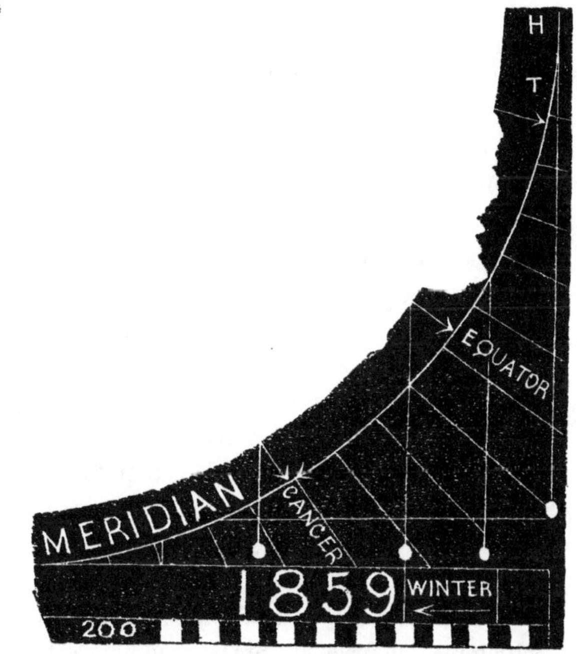

Fig. 116. The Sun's Burning Power at Noon for about Three Months.

wrong places. As this particular register was spoiled, a bit of it thick enough to make a printing block—"a slot"—was sawn out of it, so as to give a section in the plane of the meridian. The deepest groove is a quarter of an inch, the rest can be measured from the outer circle described about the centre of the original hemispherical surface. Of many bowls, this is the only one spoiled by such an accident; the rest are kept in case they may be wanted.

Registers have been kept at No. 5 Richmond Terrace, by Mr. John C. Haile, since the Board of Health was abolished. A shelf was built beside a chimney, and there a new bowl is placed twice a year. The sun and the world do the rest of this engine-turning. Some of the results were published* as part of a sanitary inquiry. Three diagrams made from rubbings show that from 21st December 1855 to 21st June 1856, the sun had little burning power, *though radiant heat registered by a black bulb thermometer was considerable.* During the next half year the sun had more burning power, and marked the bowl at more places. During the next half year the marks burned were the deepest of this series. In all these the smoke of London to the east is clearly shown. Twenty bowls, registers for ten years, have thus been made.

It has been proved in many ways that light has an influence on vegetable and animal life. Anything that impedes light is hurtful to plants and animals; therefore London smoke, which impedes light, does harm; and these observations were placed at the end of a report which aimed at curing the smoke nuisance, amongst other evils.

A small town has sprung up to the west of a garden near London, in which roses flourished. Smoke and houses have cut off 30 degrees of the torrid zone of sunlight from the clearest part of the sky, and many of the plants which flourished ten years ago are withered sticks. A green turf has suffered most where the evening shadows fall first. Only fungi grow in dark mines; and miners are a bleached, short-lived race. Sick persons kept in the Mammoth Cave in Kentucky suffered in

* In the report to the General Board of Health by commissioners appointed to inquire into the warming and ventilation of dwellings (ordered to be printed, 25th August 1857).

the dark and died. Cave-crickets and eyeless fish, which live in that strange region underground, and other cave creatures elsewhere, are sluggish. Plants turn towards light. Many kinds open or close when the sun passes a certain meridian, and of these a botanical clock has been made. A stick or a tree split along the grain splits along a spiral. Systems of branches do not sprout above each other, but are ranged in spirals. Fir-cones, pine-apples, and many flowers, are built on this same pattern. Many creeping plants turn about trees. Honeysuckle turns with the sun in the northern hemisphere.

In short, the pattern which results from the whirling of a spindle in still water—a pattern of bent rays—is the foundation of many patterns, which seem to result from whirling movements and the force of sunlight, which made collodion whirl.

The sun's radiation will cause rotation, and so produce certain forms on the earth; and in the photographic picture of the sun forms are like those which result from the whirling of a spindle in water.

The sun's rays will also model wax.

One plan devised to prove a fact which scarcely needs proof, was an application of the principle of the sun-dial, which engraved blocks in these pages. A sketch of the arrangement is below the picture of the sun at the end.

The glass ball* in the centre has a radius of 50 millemetres; the focus in air is 22 millemetres beyond the glass; and the curve of a picture of the sky formed by the lens in air has a radius of 72 millemetres. Half sunk in water, the focal distance is lengthened to 15 millemetres. So the curve

* Made at Birmingham, February 1861, under the superintendence of Mr. James Chance.

of a picture formed under water by the upper half of this spherical lens has a radius of 87 millemetres. At a distance of 3 inches and 4-10ths from the centre of the ball an image of the sun melted black sealing-wax under water. The wax took a new shape, water circulated about it, and air-bubbles formed about the wax. At the shorter focal distance of 72 millemetres the sun's image sank into black wax like a hot wire.

These movements and changes resulted from the action of rays which had travelled ninety-five millions of miles, and had passed through the coldest regions in the earth's atmosphere.

Do these rays shine out of the sun as the earth's light shines out through the earth's crust; or like furnace light welling up through freezing metals and stones? Or do they shine in the sun's atmosphere as the "Merry-dancers" shoot and shine in the northern sky?

These are questions,—answers can only be reached by expedients.

To see what the sun's rays will do when they act from within outwards, two glass basins were got, one with a radius of m. 0.072, the other with a larger radius of m. 0.087.*

All the circles which made these spherical surfaces were drawn on cardboard and cut out. The outer ring rolled up made a truncated cone for the smaller basin to stand on in the large one; the inner ring made a similar stand for the

* To Mr. Green, the manager of the glassworks of Messrs. Powell, in London, I am indebted for these and other glass contrivances, and for permission to use furnaces in making experiments. A paper published in the Liverpool and Manchester *Photographic Journal* in 1858, contains an account of some of the schemes tried to learn the effects of light and heat on photographic chemicals. One result is, that sunlight will first blacken, and then whiten a negative.

glass ball, and some plaster-of-Paris made a stand for the whole contrivance and fixed it.

It was placed in a window with a southern exposure, and the outer space was filled with water.

By this arrangement an image of the sun was formed upon the inner surface of a shell of glass, the outside of which was in contact with a shell of water. Whenever the sun shone the water circulated about the sun's image, and bubbles of gas formed all over the outside of the glass.

The outside of the inner glass was then coated with a layer of black sealing-wax about a tenth of an inch thick, and covered with a second layer of green sealing-wax varnish, and with a coat of gold paint. When this triple fusible crust had hardened the glasses were placed. On March 19, 1862, the sun only shone occasionally, and while the sun was behind a cloud there was no change; but whenever the sun did appear there was a violent commotion at the inner surface of the crust of wax. There were miniature earthquakes, concussions, detonations, vibrations, waves, sudden movements which radiated from the sun's image at all angles, from the end of the ray which reached from the sun to the sealing-wax—

|
*

On the outside, bubbles of some gas (probably air absorbed by the water) formed all over the surface, to which they were attracted. And here a whole subject for inquiry opens, for the sun's rays affect magnets and electrometers. In the meantime, rays within drove up a dome, and so produced, first a crater of elevation, then a tube. On March 21, the sun was hidden, and the sealing-wax mountains were at rest. The 24th was a bright day with passing clouds. Miniature earthquakes were frequent, and the surface was raised up and pushed outwards by the rays. Blisters became bubbles and burst; and when they did, water entered, and increased the power, by expanding between wax and glass. The outer crust was chambered, and chambers are now seen through the glass. The arrangement was left till the 10th of May 1862, and then moved, after trying the effect of dry sand instead of water. The rays drove wax into sand, but because the nearest centre of attraction was in the earth, not in the sun, and because sand did not cool the wax so fast as water does, the weight of the soft wax dragged it away, and the glass was laid bare. Rays then split the glass along the path of the sun's image in this moving panorama of the sky.

The sun is out of reach, and so bright that human eyes cannot see it; but in this expedient a ray acted as mechanical force. It broke glass, it pushed sealing-wax before it, and so pushed sand; it moulded forms, like those which are modelled by the earth's rays in volcanoes; by furnace rays at foundries; by gas lamps used to make models. The sun's rays modelled forms like those which a traveller's telescope enables him to see on the crust of the moon; like those which a photographic eye saw in the sun. The ray modelled the forms which characterise atmospheric, aqueous, and volcanic action; upheaval; dome and flow; tube, crater, and cone; fault and dyke. It set up circulation in sand, in wax, in water, and in air; in solid, fluid, and gas; and yet the source of the ray of force was in the sun.

Rays from some of the fixed stars act on photographic chemicals.

While engaged on drawings which were published in the report of the Lighthouse Commission in 1861, it was found necessary to construct a scale for the field of the camera used to take pictures.

A solar scale was made and used, but the sun's image covered too much space for accurate measurement. It occurred to the writer that stars near the pole might draw a scale, and the experiment was tried.* A small camera with a "quarter-plate lens" by Ross, was aimed at the north star, having been carefully focussed during the day for the sun's rays. A collodion plate was prepared with extra precautions against dust, and after a long exposure it was developed and fixed. The lines drawn, if any, were too fine for the purpose, so the plate was stowed away in a box for the time. After four years it was backed with black oil-paint, and carefully examined with a lens. A certain number of collodion comets and stars were found; a certain region of hazy light where clouds had reflected rays from the sun or moon; and amongst these imperfections were two arcs of concentric circles, which must have been drawn by stars. According to a rudely-made paper scale, one circle is about 12½, the other 10 degrees from the centre. All photographs taken on flat plates are distorted, and in this case the centre of revolution was not in the centre of the field. The scale was not a success; but the experiment proved that rays from fixed stars act as mechanical force, and move atoms of silver here on earth, after travelling through distances which human minds cannot realise.

Amongst nebulæ, the most distant of all visible objects, are many forms which closely resemble curves drawn by whirling engines: for example, the "spiral nebula, 51 m., Canum Venaticorum; and the spiral nebula, 99 m., Virginis," of which pictures are given by Mr. Chambers in his "*Handbook*," and by an American author in "*The Orbs of Heaven*,"

* "It has been clearly proved that the light of the stars does produce photographic effects." . . . (*On Light*, by Sir J. Herschel, *Good Words*, April 1865. P. 322.)

London, 1858. Without a large telescope it is impossible to try the effect of light from these distant systems; but their forms seem to reveal the action of gravitation, rotation, and radiation, at the limit now reached by human vision.

If a ray will do so much at this distance, it seems probable that it shines, as earth-light does, from hot fluids and solids through heated gases; and if so, the photograph of the sun has the shape which fits this answer to the problem set.* Centrifugal movements, which result from the whirling of a fluid within a solid shell, were illustrated by the expedient described above (p. 459). Shapes caused by them may be seen wherever a fluid whirls; and water whirls in every stream. "Vortices" may be watched from any bridge.

Whirlpools are deep pits surrounded by curved spokes, and the bend shows the direction in which the system revolves. That point is illustrated by expedients described in this chapter. Whirlpools in streams of air moving on a whirling globe are circular storms, and part of the solar system of motion, for they turn as the hands of a watch turn when the back of it is towards the north star, or the face of it † towards the Southern Cross: they turn against the shadow on a dial, against the bright image of the sun, which travels in the centre of the shadow of a glass ball set in a bowl. They turn

* "It has been held that as our trade-winds originate in a greater *influx* of heat from without on and near the equator than at the poles, combined with the earth's rotation on its axis, so the maculiferous belts of the sun may owe their origin to a greater equatorial *efflux* of heat, combined with the axial rotation of that luminary."—Sir J. Herschel, *Good Words*, April 1864. P. 280.

† "At the south pole the winds come from the north-west, and consequently there they revolve about it with the hands of a watch." (Quoted from Maury's Sailing Directions, on p. 23, *Abstracts of Meteorological Observations*, etc., edited by Lieut.-Col. H. James, R.E. London 1855. Blue Book.)

"The wind approaches the North Pole by a series of spirals from the south-west . . . and consequently a whirl ought to be created thereby, in which "widershins," and the old engraver who drew the symbol of the sun (Fig. 4, vol. i.) gave the right curve.

A watch is a northern contrivance, and probably it was made in imitation of a dial, for it was meant to measure time and to be looked at from above. The hands move as the shadow moves on the dial-plate. In the southern hemisphere the hands of a watch move with the storm, because the watch face is turned the other way, and the poles of it are reversed at the antipodes. By reversing the poles of a watch in the northern hemisphere so as to make the face of it aim at the south pole of the sky, apparent movement is converted into real movement: watch-hands and whirlwinds then turn one way. The hands turn about the spindle as the earth turns about its axis and about the axis of the sun, as satellites revolve about their central bodies, as the storms whirl on their axes and move upon the whirling surface of the world. The large engine and the little one, hour hands and seconds hands, all turn one way.

The whirling sun has an atmosphere, and shapes in this photograph are like diagrams laid down by philosophers on maps, after gathering thousands of facts about great whirling storms. In this planet a ball with a solid crust is spinning, and water and air about the crust spin with it, and swing in streams from and towards the axis, crossing the edges of revolving discs diagonally in both hemispheres. The principle

the ascending column of air revolves from right to left, or *against* the hands of a watch." (P. 22.)

"It is a singular coincidence between these two facts thus deduced and other facts which have been observed, and which have been set forth by Redfield, Reid, Piddington, and others—viz., that all rotatory storms in the northern hemisphere revolve as do the whirlwinds about the North Pole, viz., from right to left; and that all circular gales in the southern hemisphere revolve in the opposite direction, as does the whirl about the South Pole." (P. 23.)

of the movement in ocean and atmosphere is the same as in water set in motion by a whirling spindle. The patterns drawn ought to be alike, and they are. Forms laid down on globes ; mountains and coasts, and glens and fjords ; and ice-grooves on hill-tops—tool-marks of denuding engines—agree in direction.

On any sphere revolving, as the earth revolves, in an atmosphere of its own, the pattern outside ought to be founded on spirals, crossing each other like the pattern on the rind of a pineapple, or on the heart of a sunflower, or on a daisy. It ought to be a system of curved cross-hatching, like engine-turning on the case of a watch. That is the pattern which Maury drew in his diagram of the winds after comparing and collating thousands of meteorological observations. It is the pattern which the photographic eye saw on the sun.

Commonly the sun's atmosphere seems to be wrapped about the ball in broad circular bands. On one occasion the bands were broken up and scattered, as by a storm. The bands are seen at the eastern limb about the equator ; and thence they spread towards the poles, in long curved streams, like cirrus clouds and mackerel sky overhead. The light formed long ovals and rings, like whirlpools and systems of bent waves upon water eddying under a bridge, or made to whirl in a tray by spinning a top. The actual dimensions of the shapes figured are of no account ; their proportion to the rest of the disc is the main point. They are reduced by the lens, and drawn to scale ; and they cover space in proportion to spaces traversed by whirling hurricanes and typhoons, and laid down on a chart in the blue-book quoted above. Rotating storms travel over the whole world.

Electric storms, disturbances in currents which affect magnets, are common, and it has been suspected that their occurrence and the appearance of solar spots have some relation to each other. A series of photographs, kept with a register of magnetic and other observations, may settle whether certain forms on the sun's disc indicate storms in the sun's atmosphere, which are felt on the earth as electric storms. Mr. Chambers says—

"We may here take occasion to advert to a very remarkable phenomenon seen on September 1, 1859, by two English observers, whilst engaged in scrutinising the sun. A very fine group of spots was visible at the time, and suddenly, at 11 h. 18 m., two patches of intensely bright white light were seen to break out in front of the spots. It was at first thought to be due to a fracture of the screen attached to the object-glass of the telescope; but such was not the case. The patches of light were evidently connected with the sun itself; they remained visible for about five minutes, during which time they traversed a space of about 35,000 miles. The brilliancy of the light was dazzling in the extreme; but the most noteworthy circumstance was the marked disturbance which (as was afterwards found) took place in the magnetic instruments at the Kew Observatory simultaneously with the appearance in question, followed about sixteen hours afterwards by a great magnetic storm."—(G. F. Chambers, *Handbook of Astronomy*. London 1861. P. 6.)

Amongst eminent men who have turned their attention to telescopic drawing and photography, Mr. Nasmyth's name is conspicuous. He holds that the present condition of planets may throw light upon the former condition of the world.

Mr. Chambers only states facts ; he says, p. 9—

"It has been thought that the prevalence of large masses of spots might give rise to a depression in the temperature for the time being, and thus affect the fertility of the soil. Modern observation, however, would lead us to infer that the contrary was rather the case, an elevation of temperature being contemporaneous with the prevalence of spots."

These shapes may indicate changes in a crust now forming about a fluid, and this observation supports the notion that the sun's rays are like those which shine through the crust of the earth.

Bright streaks and spots of light often break out where dark spots have disappeared. Sir W. Herschel, on December 27, 1799, saw a streak of light which was 2·46″, or 77,000 miles in length (Chambers, p. 9).

The shapes of dark spots projected on paper with a good astronomical telescope are suggestive of forms which result from ebullition in metals, and may indicate the position of solid projections rising through heated fluids and gases. The darkest spots are still so brilliant as to affect photographic plates rapidly.

When a powerful current of electricity passes through certain materials, the form is changed, and the current is changed into light and force ; a wire is broken up, fused, and the drops are scattered as by an explosion. They move off and radiate from the current.

A bell-wire fused by lightning spreads on the wall in radiating lines ; a tree is split by lightning ; when lightning falls in a bed of sand, it sometimes fuses the sand into long, tapering, branching, radiating tubes. Of these, specimens are preserved at the British Museum under the name of Fulgurites. If the light of the sun be electric light, that form of light is accompanied by mechanical force, and it radiates in the same direction as visible light and sensible heat, and actinic rays, which affect chemicals.

In these last chapters force has been hunted through engines of many kinds. If the spoor has been truly followed, light is a power in every engine of human construction, which turns out work, for the power which winds a clock moves the hands. The sun's rays help to move air and evaporate water, so they help to turn all mills ; light of some kind is at the source of power in steam-engines ; plants will not grow without light ; animals cannot work without food ; and the most carnivorous of creatures only extracts power out of fuel gathered by his prey. A horse in a mill is but a link in a chain, and rays also are links in it.

The sun's rays may be set to work directly; they may be set to wind up a clock.

Iron floats in mercury, mercury expands when the sun shines upon the vessel which holds it, and shrinks when the sun is hid. A column of mercury in an open iron tube with a bulb will lift an iron weight when the sun shines, and drop it when the shadow comes; a very small amount of ingenuity will apply the power to a piston, a lever, an axle, or a train of wheels; an index and a needle would register the force applied, and might express it in "foot pounds," for the force lifts a weight.*

The sun's rays evaporate fluids; vapour of ether may be passed through a gas meter, and the index will express the power in cubic feet.† The sun's rays decompose certain fluids, and make certain gases combine. Bunsen and Roscoe applied that power to measure chemical force in light.

The hand which winds a clock moves the train of wheels; the force which causes motion, directly or indirectly, is mechanical force; and the sun's rays have been set to move engines.

The works of philosophers contain a precious essence; they contain truths extracted from fruit and flowers, grain and chaff, gathered by thousands of labourers in a boundless field of inquiry. This book only contains the gatherings of one wandering craftsman; but he has sought for truth, and haply he may have found some grains to add to the common stock.

* For an explanation of modern views on the subject of heat as a mode of motion, see writings by Professor Joule of Manchester, and articles in the *North British Review*, February and May 1864.

† Neither of these contrivances has yet been made; one or other may be set to work before this book is published.

One attempt has been to interpret the meaning of form, to watch work in progress, so as to learn to distinguish the tool-marks of natural engines. If the sun's rays work in the sun as they do on earth, then forms in the sun ought to be a legible index. Read by this alphabet of form, rudely made with rough expedients, they seem to mean—

That laws of force, which cause and regulate movements in gases, fluids, and solids, in the whirling earth, which is only one of many satellites in one of many systems, are good law in the atmosphere of the whirling sun, which is only one of many sources of light and of ray-force.

But if so, wherever light shines there force may radiate, though the eye is the only organ which feels the force.

Even the shapes of nebulæ may betray mechanical force in light.

Thus far this book is an attempt to argue through circles: —an attempt to gain a point by following a ray; and the next point by following another. If the attempt be judged and condemned, the writer can only plead that he has done his best; if acquitted of presumption, he will be content. He hopes to be forgiven for thinking for himself. Many spokes have been tried, many a path trodden; but all paths tried have ended at one spot. By searching backwards from work done, men reach power through engines; by travelling far enough they seem always to reach a source of light. But that is only one centre in an endless train of wheels. The way to see further is forwards: to use light, and try to see if there be more wheels, engines, and powers between work done and the will of Him who made them all and created Light.

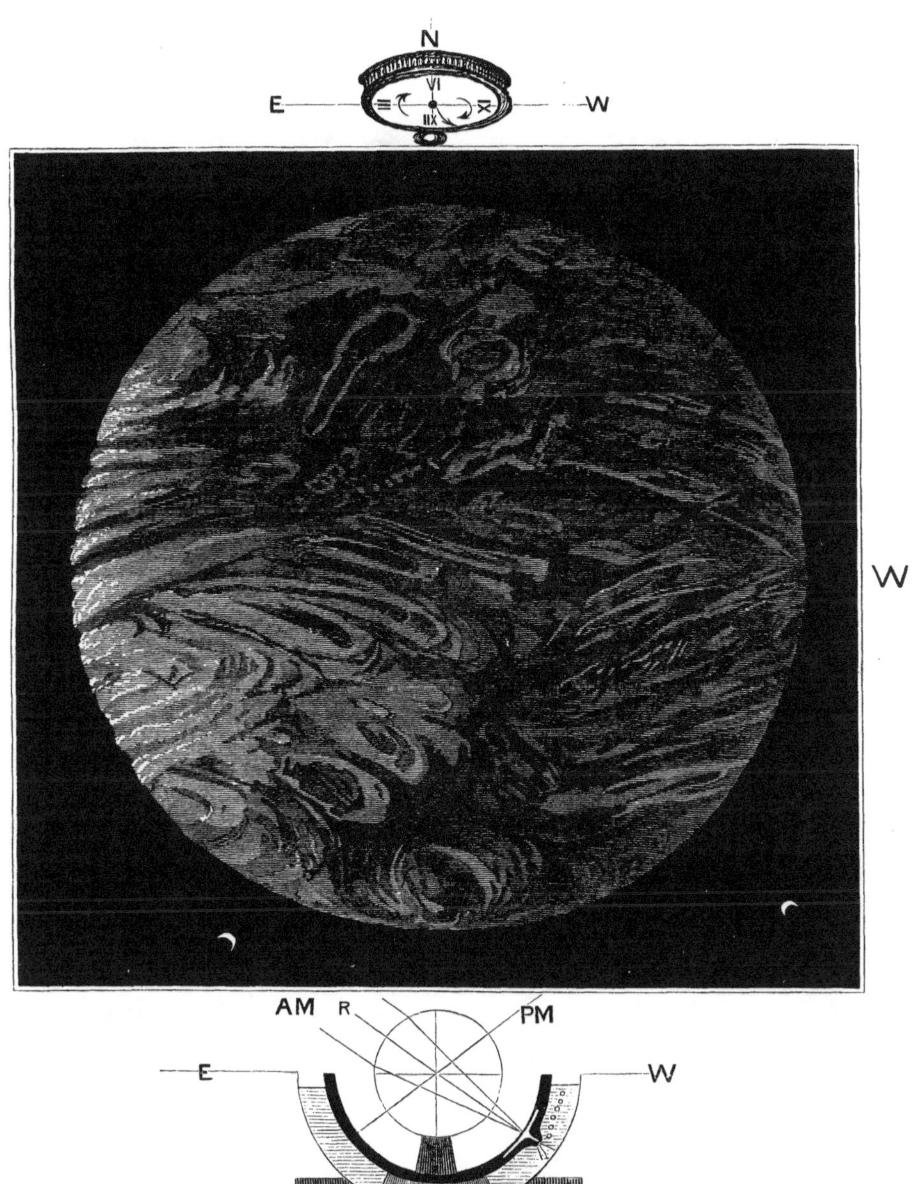

Fig. 117. From a Photograph of the Sun.

Taken on or about March 22, 1859, when a number of spots were visible. On the 11th March the south pole of the sun is best seen, according to Sir J. Herschel.

INDEX.

THE END.

Printed by R. Clark, *Edinburgh.*

www.ingramcontent.com/pod-product-compliance
Lightning Source LLC
LaVergne TN
LVHW021307110826
845150LV00003B/505

* 9 7 8 1 4 2 5 5 5 9 3 4 2 *